SATELLITE INTEGRATED COMMUNICATIONS NETWORKS

SATELLITE INTEGRATED COMMUNICATIONS NETWORKS

Proceedings of the Third Tirrenia International
Workshop on Digital Communications
Tirrenia, Italy, 14–16 September 1987

edited by

Enrico DEL RE
University of Florence

Pierre BARTHOLOMÉ
European Space Agency

Peter P. NUSPL
INTELSAT

1988

NORTH-HOLLAND
AMSTERDAM • NEW YORK • OXFORD • TOKYO

© ELSEVIER SCIENCE PUBLISHERS B.V., 1988

All rights reserved. No part of this publication may be reproduced, stored in a retrieval system, or transmitted, in any form or by any means, electronic, mechanical, photocopying, recording or otherwise, without the prior permission of the copyright owner.

ISBN: 0 444 70372 1

Publishers:
ELSEVIER SCIENCE PUBLISHERS B.V.
P.O. Box 1991
1000 BZ Amsterdam
The Netherlands

Sole distributors for the U.S.A. and Canada:
ELSEVIER SCIENCE PUBLISHING COMPANY, INC.
52 Vanderbilt Avenue
New York, N.Y. 10017
U.S.A.

PRINTED IN THE NETHERLANDS

FOREWORD

Integration of satellites and terrestrial networks began at the start of satellite communications, when terrestrial telephone networks of different countries were interconnected via satellite links. International connections were performed through a limited number of national earth stations of a very large size and cost. In recent years, new emerging technologies for on board and ground applications have suggested new technical and system solutions for a closer integration of satellites and terrestrial networks. This topic has received an increasing attention in industrial, operational, administrative and research institutions. Hence, a forum to assess the status and the prospects of integration of satellite and terrestrial networks was deemed timely and useful.

For this reason the third Tirrenia International Workshop on Digital Communications has been devoted to the topic of "Satellite Integrated Communications Networks". It was held in Tirrenia, Pisa (Italy), from September 14 to 16, 1987 (the previous two Workshops were held in 1983 and 1985, each devoted to other aspects of Digital Communications).

The technical program was divided into six sessions:

1. Integration of satellites and terrestrial networks (ISDN, IBC, LAN, MAN)
2. Intersatellite links: GEO-LEO
3. Mobile satellite communications
4. On board processing
5. User-oriented and specialized satellite networks
6. Transmission techniques and interfaces

In addition a final Round Table was held on the subject "Comparison of conventional systems (presently available or under development) and future integrated systems".

Forty-six papers were presented. There were 127 participants from 15 countries (Austria, Belgium, Canada, Denmark, France, FR Germany, Italy, Japan, Netherlands, Norway, Spain, Sweden, Switzerland, United Kingdom, USA) and three international organizations.

We are particularly grateful to the international organizations ESA, INTELSAT and EUTELSAT that enthusiastically provided their support and sponsorship for the Workshop. We are also grateful to the Session Chairmen for their active participation and to the Authors for their high-quality papers. We are confident that this Workshop contributed to the growth of the satellite communications field and provided an opportunity for friendship and interaction among colleagues from all over the world.

One very sad note in the otherwise excellent atmosphere of the Workshop was the news of the death of Dr. Lou Cuccia. L. Cuccia contributed immensely to satellite communications and will be remembered as an engineer of great talent and extraordinary vitality. Many of the participants were looking forward to the pleasure of meeting him again on this occasion and deeply felt his absence.

Finally, it may be worth adding that it was during this Workshop that the Ariane launcher resumed duties after an interruption of almost a year and a half. On Tuesday, September 15, it launched successfully two satellites: the European ECS-F4 and the Australian AUSSAT K3.

Pierre Bartholomé   *ESA, Netherlands*
Enrico Del Re   *University of Florence, Italy*
Peter P. Nuspl   *INTELSAT, USA*

SPONSORING ORGANIZATIONS

ESA
INTELSAT
EUTELSAT

ACKNOWLEDGEMENTS

The support of the following Organizations and Institutions is gratefully acknowledged:

GTE Telecomunicazioni
IBM
IEEE North and Middle & South Italy Sections
Italian National Research Council
Telespazio
Telettra
University of Florence
University of Pisa

TABLE OF CONTENTS

Foreword   v

KEYNOTE ADDRESS

Satellites: Tomorrow's Solution to Tomorrow's Needs
P. Bartholomé   3

Session 1 — INTEGRATION OF SATELLITES AND TERRESTRIAL NETWORKS (ISDN, IBC, LAN, MAN)
Chairman: G. Pennoni

Electrical Communications Systems Approaching the Year 2000: Cables and Satellites
P.L. Bargellini and G. Hyde   11

Some Aspects of the Implementation of an Integrated Space/Terrestrial Network for Europe
J.M. Casas and P. Bartholomé   23

Economic Comparison of Cables and Satellites
S.J. Campanella and B.A. Pontano   35

Integration of a Payload Enhanced Networking Satellite (PENSAT) and Terrestrial Broadband ISDN
G. Pennoni and L. Bella   47

Satellite Switches
S.J. Campanella   57

Satellite Switching for Mobile Communications: New Issues and Perspectives
L. Bella   65

E.C.S.: Integration in the European Network and Interfacing with the Italian Public Telephone Network
D. De Rosa and B. Vendittelli   73

Internetworking of VSAT and MSAT Systems for End-to-End Connectivity
   K.M.S. Murthy and D.J. Sward   81

Wide Area Networking by Interconnecting LANs and MANs via Satellite
   K.M.S. Murthy and D.J. Sward   91

Session 2 — INTERSATELLITE LINKS: GEO-LEO
   Chairman: G. Mica

ESA's Undertakings in the Field of Optical Intersatellite/ Interorbit Links: Programme, Capacity and Link Characteristics
   M. Wittig and G. Oppenhaeuser   99

GEO — GEO ISL Project of ESA/EUTELSAT/INTELSAT
   D. Dharmadasa, O. Millies-Lacroix, and R.A. Peters   107

Data Transfer over an Integrated Communication Network for an Orbital Infrastructure
   L. Zucconi and C. Smythe   115

Optical Communications Laser Diode Intersatellite Links
   A. Arcidiacono and S. De Vita   123

Communication and Tracking Design Aspects in DRS-LEO User Link
   L. Bardelli and A. Florio   133

Link Analysis of Data Relay Satellite System under 'PFD' Constraint
   T. Tanaka and Y. Tsujino   143

Session 3 — MOBILE SATELLITE COMMUNICATIONS
   Chairman: S. Shindo

A Multi-Beam Mobile Satellite Communications System in Japan
   S. Shindo, K. Satoh, E. Hagiwara, and K. Morita   153

Land Mobile Satellites Using Highly Elliptic Orbits
   B.G. Evans and L.N. Chung   163

An On-Board Processing Satellite Payload for a European
Land Mobile Satellite System
  I.E. Casewell, B.G. Evans, and A.D. Craig   171

A Highly Efficient Multistage Approach to Digital FDM
Demultiplexing for Mobile SCPC Satellite Communications
  H. Göckler   179

Multicarrier Demodulator (MCD) Using Analog and Digital
Signal Processing
  P.M. Bakken, V. Ringset, A. Rønnekleiv, and E. Olsen   187

A Saw-Based Integrated QPSK Coherent Demodulator
  *P. Tortoli, F. Andreuccetti, G. Manes, R. Giubilei,
  and D. Gerli*   197

Satellite Terminals in the Italian Navy and User Experience
  R. Palandri, R. Azzarone, and G.B. Durando   203

Session 4 — ON BOARD PROCESSING
  Chairman: P. De Santis

Performance Evaluation of Regenerative Digital Satellite Links
with FEC Codecs
  N.A. Mathews   213

ACTS: The First Step toward a Switchboard in the Sky
  F.M. Naderi   225

Recent Developments of On-Board Processing Technologies
in Japan
  F. Takahata, H. Shinonaga, and M. Ohkawa   235

Techniques and Technologies for Multicarrier Demodulation
in FDMA/TDM Satellite Systems
  F. Ananasso and E. Del Re   243

The ITALSAT QPSK Burst Mode Coherent Demodulator
  A. D'Ambrosio and G. Alletto   253

T-S-T SS/TDMA System for Services of Different Bandwidths
  G.B. Alaria, G. Colombo, and G. Pennoni   261

An On-Board Processor for ISDN and ISDN-Compatible
Applications
    H.P. Kuhlen   271

Comparison of Digital Transmultiplexer Architectures for Use
in On-Board Processing Satellites
    W.H. Yim, C.C.D. Kwan, F.P. Coakley, and B.G. Evans   279

Session 5 — USER-ORIENTED AND SPECIALIZED SATELLITE NETWORKS
    Chairman: S. Tirro˙

Present Status and Future Developments of Satellite Business
Services Networks in Japan
    Y. Morihiro and S. Kato   289

ERCOFTACS Project: A Satellite Network for Interactive
2 MBPS Communications
    S. Benedetto and S. Tirro˙   297

The European Satellite Business Service: EUTELSAT SMS
    M. Papo   305

Business Services via Satellite: Italian Programs
    S. De Padova and A. Puccio   313

Users and Economical Viability for the Tele-X Business Services
    L. Backlund   319

Session 6 — TRANSMISSION TECHNIQUES AND INTERFACES
    Chairman: G. Tartara

An Acquisition and Synchronisation Unit for a SS-TDMA
Network
    K.R.G. Fowler and M.R.W. Manning   337

A Model for a Frame Synchronisation Unit in SS-TDMA
    K. Olsen   345

FODA—TDMA Satellite Access Scheme:
Description, Implementation and Environment Simulation
    N. Celandroni and E. Ferro   353

Comparison of Digital Modulation Techniques for FDMA/SCPC
Satellite Communication with Small Earth Stations
   P. Sanders and M. Moeneclaey 363

Adaptive Channel Coding as a Fade Countermeasure in
Millimeter Wave Satellite Communications
   G. Tartara and R. Carena 371

Analytical Performance Comparison of Trellis Coded 8-PSK
and QPSK over Hard-Limited Satellite Channels
in the Presence of Interference
   N. Jayamanne, I. Oka, and S. Mori 379

Synchronization Algorithm for Continuous Phase
Modulated Signals
   J. Habermann 387

Pulse Shape Optimization in Partial Response CPM Systems
   M. Campanella, U. Lo Faso, and G. Mamola 395

Land Mobile Satellite Channel — Model and Error Control
   E. Lutz, F. Dolainsky, and W. Papke 403

A Tool for Evaluating On-Board Processing Satellite
Architectures
   S.V. Vaddiparty, W. Doong, T.V. Nguyen, and T.Q. Nguyen 411

DRS Transmission System: Architectural Trade-Offs and Coding
   A. Arcidiacono and R. Giubilei 419

**Round Table — COMPARISON OF CONVENTIONAL SYSTEMS
(PRESENTLY AVAILABLE OR UNDER DEVELOPMENT) AND
FUTURE INTEGRATED SYSTEMS**
   Chairman: B.G. Evans

Round Table Discussion 431

KEYNOTE ADDRESS

SATELLITES : TOMORROW'S SOLUTION TO TOMORROW'S NEEDS
Keynote Address

Pierre Bartholomé

European Space Agency

As the end of the twentieth century approaches, there is a great temptation to try to predict what our society will be like beyond the year 2000. Those whose business is telecommunications are striving now to forecast how their particular field will evolve during the intervening years. The number of technical papers addressing this kind of topic certainly seems to have increased in recent years and a quick look at the content of our programme this week shows that this Workshop will be no exception to the general trend.

Forecasting the future in a technology-intensive industry such as ours is known to be a risky enterprise. Many have tried in the past and have ultimately been proved wrong more often than not. I would like to recall the case of the British Post Office Director General who, when told that an American named A.G. Bell had invented the telephone, commented that he saw no future for such a device in the United Kingdom since that country had plenty of excellent messengers. His US counterpart, on the other hand, reacted very positively to the news; he forecast that, one day, there would be one telephone in every city of the United States!

We all remember Arthur C. Clarke's paper on "Extra Terrestrial Relays" which appeared in 1945 and was regarded at the time as a nice piece of science fiction. I doubt very much whether the author or his readers of the time really expected to see this extraordinary idea materialise in fewer than twenty years. Nevertheless in 1964 SYNCOM 3 inaugurated the era of geostationary satellites.

Although the title of this address might be taken to suggest that I myself will be indulging in a little fortune telling, I have no such intention! I find it much safer and more instructive to take a few steps back in time to re-examine what the future looked like twenty years ago.

P. Bartholomé, Head of Communications Systems Division
European Space Agency, ESTEC, Postbox 299, 2200 AG Noordwijk, The Netherlands
Telephone +31 1719 83141, Telefax +31 1719 17400, Telex 39098

As soon as geostationary communications satellites became a reality, people started philosophising about their potential uses and applications. But, as we shall see, the pace of progress was, in reality, considerably slowed. Apparently, the time that elapses between the birth of an idea and its application is much longer when it comes to using a satellite than that needed to build it and place it in orbit.

In 1965, ATT submitted a proposal to the FCC entitled: "An Integrated Space/Earth Communication System to Serve the US". This title bears a striking resemblance with that of the second paper of Session 1 of this Workshop, which is "An Integrated Space/Terrestrial Network for Europe". Satellite and terrestrial networks now operate side by side, and in some cases complement each other, but nowhere in the world can they be said to be really integrated. I can only conclude that we have been talking about integration for more than twenty years, without actually getting round to doing something about it.

In contrast, another idea, also launched in 1965, materialised very quickly. It came from the American Broadcasting Company, which proposed the distribution of television programmes to local broadcasting stations via satellite. Ten years later, television distribution was already a fast-growing business in North America, and the service had found a much wider audience than was ever anticipated, with cable networks and even private households receiving signals with individual dish antennas. This application has also become popular in Europe, with most EUTELSAT and TELECOM 1 capacity being used for this purpose.

In the field of classical communications, however, it is obvious that the "gestation period" is abnormally long if one uses the time taken to make and launch the first geostationary satellite as a yardstick. It may come as a surprise to some of you that many of the concepts that are central to our discussions today were already described in papers presented at the First International Conference on Digital Satellite Communications, in 1969, in London. The first paper presented at that Conference was by S. Hanell, then with COMSAT Laboratories and now with ESA. Its subject was "Record Message Networks Employing Satellites and Small Earth Terminals", something that in today's jargon would be called VSAT networks. Figure 1 is extracted from this paper. VSAT networks are now developing steadily in the USA and Canada, but it was not until 1982 that they became a commercial proposition. Despite the wide publicity they are now getting, there are still some concerns about their ultimate viability. At any rate, in Europe, apart from some very timid attempts made just recently, the field is still essentially empty, seventeen years after the publication of Hanell's paper.

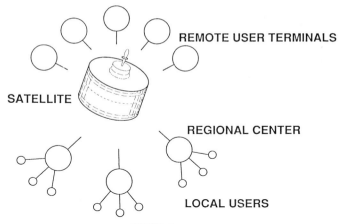

FIGURE 1
Record Message Networks Employing Satellites and Small Earth Stations -
S. Hanell, 1969

Many of the contributions to the 1969 Conference dealt with one or other aspect of TDMA, a new multiple-access technique that seemed most promising and poised to replace the old-fashioned FDMA technique sooner or later in satellite communications. Most authors were reporting results of work that had been going on for several years. They would have found it difficult to accept that what seemed to be such an attractive concept would not begin to be implemented in the INTELSAT network until 1985. Again, about twenty years was to pass between the birth of the idea and its implementation.

Some authors, however, were even more daring and ventured to propose still more advanced concepts, which sounded far less familiar to the listeners than they do today. W.G. Schmidt of COMSAT Laboratories described "An On-Board Switched Multiple-Access System for Millimetre-Wave Satellites", which would now be referred to as SS/TDMA (Fig. 2). Although this concept was implemented in an Advanced WESTAR a few years ago, it has never been used operationally. The next opportunity to put it into practice will be with INTELSAT VI, albeit on a modest scale, and with ESA's OLYMPUS, but again on a purely experimental basis. The SS/TDMA idea still won't have made a real breakthrough even after twenty years.

Finally, I should mention a paper by S.G. Lutz, of Hughes Research Laboratories. It was entitled "Future Satellite Relayed Digital Multiple Access Systems" and made the case for "onboard circuit switching" and "intersatellite relaying". The author envisaged a network of multiple satellites, carrying switchboards interconnected by space links and operated in a tandem configuration, much like terrestrial switching practices (Fig. 3). It is disappointing to realise that ideas like the "switchboard in the sky" and "inter-satellite links", which sound so revolutionary and exciting to us, have already been around for more than twenty years and yet are still so far from realisation.

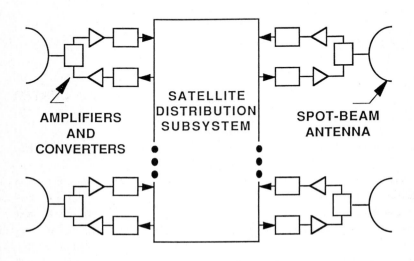

FIGURE 2
An On-board Switched Multiple-Access System for Millimetre-Wave Satellites - W.G. Schmidt, 1969

This brief look back at our progress, or rather the lack of it, leads me to one clear conclusion. Of all the new concepts put forward twenty years ago, only one, namely television distribution, has come to fruition over a reasonable time span and has been very successful. All of the others that I have cited - TDMA, SS/TDMA, VSAT, network integration, onboard switching and inter-satellite links - either took much longer to emerge from the study phase, or have still not yet done so. The reason for the success of TV distribution is I believe that, unlike the others, it was not a technique trying to supplant another, nor was it a technology looking for a market. Moreover, it had little in common with classical communications, and hence its progress was not hampered by the need to conform to standards, to satisfy interfaces and/or to be compatible with existing systems.

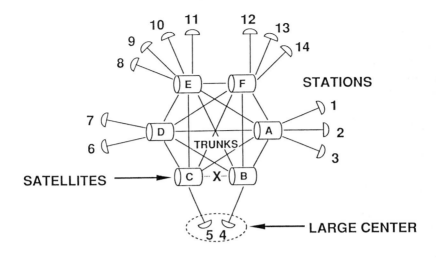

FIGURE 3
Future Satellite-Relayed Digital Multiple-Access Systems - S.G. Lutz, 1969

The history of the last twenty years suggests that, somehow, satellite communications engineers have often allowed themselves to be carried away by their own creativity and imagination. Past experience would certainly seem to indicate that we engineers should keep our own feet firmly on the ground and take a long hard look at our approach to the future. The prospects for communications satellites are no longer as rosy as they once were. Competing technologies are staging a comeback and are beginning to threaten satellites in the very market where they made their first breakthrough, namely that of trans-oceanic links. The new satellite systems that we propose must indeed be tomorrow's solutions to tomorrow's needs. Our solutions must match the market's projected needs exactly, and be available at exactly the right moment.

My conclusion is that time is not on our side and that any miscalculations in timing could be fatal to our future projects. If we underestimate the importance of the non-technical obstacles that lie on the road ahead of us, we may discover that the ever more sophisticated satellite systems of tomorrow turn out to be solutions to yesterday's problems.

SESSION 1

INTEGRATION OF SATELLITES AND TERRESTRIAL NETWORKS (ISDN, IBC, LAN, MAN)

Chairman: G. Pennoni *(ESA/ESTEC, The Netherlands)*

ELECTRICAL COMMUNICATIONS SYSTEMS APPROACHING THE YEAR 2000:
CABLES AND SATELLITES

Pier L. Bargellini and Geoffrey Hyde

COMSAT Laboratories
22300 Comsat Drive
Clarksburg, Maryland 20871-9475

High communications capacity has been achieved by harnessing ever higher frequencies. Bandwidth has increased from the early telegraph cables (10 to 100 Hz) to coaxial cables (10^6 to 10^8 Hz); radio systems, inclusive of satellites, (10^4 to 10^9 Hz) and fiber optical systems (potentially up to 10^{12} Hz).

Noise, attenuation and dispersion in cables, propagation effects in radio systems, message integrity and security, flexibility in assigning capacity, network reconfigurability, capability of providing new services, ease (or difficulty) of repairing failures, and, last but not least, system economics need consideration.

The paper deals with cable and radio systems. The convenience derivable from the integration of satellite and fiber optical systems is emphasized, evidencing the complementarity of the two approaches.

1. INTRODUCTION

Higher carrier frequencies imply wider signaling bandwidth, B, and consequently higher rates of information transmission, R, as

$$R = 2B \log_2 s \quad \text{bit/s}$$

where s is the number of signal states. Figure 1 illustrates these trends for wired and wireless systems.

Substantial communications capacity can be achieved with coaxial cables, microwave, and optical systems. In order to acquire a proper perspective of the problems encountered, these topics will be presented along their historical evolutionary trend.

2. ELECTRICAL COMMUNICATIONS BEFORE SATELLITES

In coaxial cables operating in the TEM mode, attenuation is proportional to the square root of the frequency because of the skin effect. Phase distortion can be compensated for by means of equalizers, and attenuation can be counteracted by placing repeaters at intervals along the cable in order to restore the signal level. Wider bandwidths require larger diameter conductors and shorter distance between adjacent repeaters. Table 1 illustrates the case of overland coaxial cable systems, and Table 2 that of submarine cables (transatlantic cables laid from 1956 to 1983).

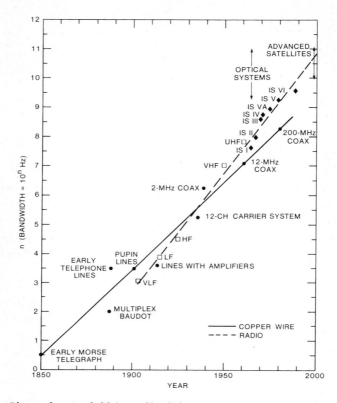

Figure 1. Bandwidth Availability: 150 Years of Progress

Table 1. Characteristics of Coaxial Cable Systems

Cable Type	Diameter (mm)		Bandwidth (MHz)	Repeater Distance (km)	Capacity	Range (km)
	Outer	Inner				
Normal	9.5	2.6	200	1.6–1.9	Large	1,000–3,000
Small	4.4	1.2	100	1.6–2.1	Medium	100–1,000
Micro	2.9	0.7	20	2.0–4.0	Medium	10–100

Before the 1920s, VLF and LF radiotelegraph systems provided bandwidths of a few kilohertz and signaling speeds up to 30 to 50 words/min, i.e., only about 15 bit/s. By the mid 1920s, HF systems achieved 10 times higher telegraph signaling speeds with transmit power down by one or two orders of magnitude with respect to that needed at VLF or LF. Antenna size came down also and radiated energy could be beamed towards wanted directions.

Table 2. Transatlantic Coaxial Submarine Cables

Year	Designation	Diameter (in.)	f_{max} (MHz)	Repeater Spacing (nmi)	Number of Telephone Channels (with TASI)
1956	TAT 1	5/8	0.164	38	48
1959	TAT 2				
1963	TAT 3	1.0	1.05	20	138
1965	TAT 4				
1968	TAT 5	1.5	6.0	10	845
1976	TAT 6	1.7	30.0	5	4,000
1982	TAT 7	2.25	125.0	2.5	16,000

After World War II, HF continued to be used with improvements made possible by better understanding of ionospheric radiowave propagation, by the use of diversity reception, of frequency shift keying (FSK) in lieu of ON/OFF in telegraphy and single sideband (SSB) techniques in telephony. Unfortunately, during periods of adverse propagation (sudden ionospheric disturbances induced by high-intensity solar flares), HF communications are interrupted for hours and even for days; thus continuity of service does not exceed 80 percent. Bandwidth is limited to a few tens of kilohertz due to dispersion in the ionosphere; hence the bit rate does not usually exceed a few hundred kilobit/s.

During World War II VHF and UHF techniques advanced greatly due to radar systems development. The extension of these techniques to communications clearly was only a matter of time. However, link availability above HF is limited to line-of-sight conditions (LOS). An exception is tropospheric scatter systems in which reliable voice communications (up to 240 channels) over distances up to 650 km can be established. In LOS systems with antenna towers of economically acceptable height (30 to 70 m), the range over flat terrain is limited to about 30 to 50 km for each link. For greater distances, it is necessary to cascade links.

Overland wideband LOS systems in the range 2 to 10 GHz or higher were developed after World War II and their use became widespread. Bandwidth increased to several hundred megahertz, and transmission capacity up to hundreds and even thousands of megabits/s (thousands of voice channels).

Thus, until the advent of communications satellites in the mid 1960s and of fiber optical cables in the early 1980s, microwave LOS systems and copper coaxial cables were the two major competitors in high capacity overland transmission facilities. When comparing these two techniques, pros and cons are encountered for each. Both went through various advances and have continued to coexist in a competitive scenario.

A digression should be made at this point by distinguishing between analog and digital transmission. As noise is cumulative in cascaded analog links, each link must be overdesigned with respect to the desired signal-to-noise ratio at the end of the cascade. On the other hand, digital techniques circumvent this difficulty by means of regenerative repeaters, and the trend toward the digital approach has been accelerated. Because of the interaction of transmission and switching techniques, digital systems are cost effective in medium- and short-haul applications. In long distance (overland and submarine) coaxial cables, analog techniques have remained predominant for a long time, and a similar situation has occurred in LOS microwave systems.

3. SATELLITE COMMUNICATIONS

Satellite communications systems were made possible by combining rocketry and microwave engineering. Following experiments from 1959 to 1963 with spacecraft in low and medium altitude orbits, the unique characteristics of the geosynchronous equatorial orbit, proposed as early as 1945 by Arthur C. Clarke, were actively focused upon. As soon as adequate rockets became available and the know-how of injecting payloads in geosynchronous orbit had been acquired, geostationary satellites predominated; their advantages are well known and need not be repeated here.

For fixed services at the international level, the INTELSAT system has been an outstanding success; Table 3 illustrates the advances of its space segment. Communications capacity in a single spacecraft has gone up by two orders of magnitude, while mass and primary power increased by only one order of magnitude. Satellite systems have become fully integrated with terrestrial facilities, and, until the advent of fiber optical systems, their capacity exceeded that of any other communications system. Finally, their unique flexibility remains unchallenged because, as soon as a satellite is positioned in orbit and becomes "visible" from n earth terminals, $n(n-1)/2$ links can be established. The evolution of spacecraft design has permitted the reduction in size and cost of earth station antennas.

At the national/regional level since 1976 many countries have developed national/regional satellite communications systems. High capacity communications links were established quickly and economically across formidable geographical barriers.

With regard to mobile services, the MARISAT system in 1976 led to the international INMARSAT system in 1979 for shore-to-ship and ship-to-shore communications. Aeronautical and terrestrial mobile services are expected to be developed in the near future.

In present fixed satellite service (FSS) the frequency bands of 4/6 and 11/14 GHz are used with 500 MHz of nominal bandwidth available for the up-links and another 500 MHz for the down-links. Frequency reuse techniques (orthogonal polarizations and space separation by sharp antenna beams) make the actually usable bandwidth greater than that nominally assigned. Thus communications capacity up to several gigabits/s (about 60,000 voice channels) will be available from a single INTELSAT VI spacecraft. Other improvements are foreseeable in the future: on-board switching of up- and down-links via a dynamically controlled RF matrix will yield increased communications capacity with only modest increases of spacecraft mass and power. By performing the switching function at baseband rather than at RF and by demodulating and then remodulating signals, additional advantages can be obtained. Advanced encoding, modulation and multiple access techniques will play increasingly important roles. Transponders will evolve toward more extensive use of solid-state devices and MMICs. Antenna subsystems will provide precisely shaped fixed or movable beams.

Table 3. INTELSAT Satellites 1965-1989

Characteristics	I	II	III	IV	IV-A	V	VI
Year of First Launch	1965	1967	1968	1971	1975	1981	1989
Frequency Bands (GHz)	4/6	4/6	4/6	4/6	4/6	4/6 + 11/14	4/6 + 11/14
Usable Bandwidth (MHz)	50	130	460	480	800	2,280	3,360
Number of Beams	0	0	1	3	3	7	10
Frequency Reuse x	1	1	1	1	2	4	6
Number of Transponders	2	1	2	12	20	29	46
Primary Power (W)	40	75	120	400	500	1,200	2,100
In-Orbit Mass (kg)	39	86	152	732	863	950	1,700
Communications Payload Mass (kg)	13	36	56	250	293	356	560
Number of Telephone Channels	480	480	2,400	9,000	12,000	25,000	60,000
Specific Communications Capacity (ch/kg)	12.3	5.6	15.8	12.3	13.9	26.3	41.0

Intersatellite links will provide improved connectivity and better orbit space and frequency spectrum utilization. When the 20/30-GHz bands are commercially exploited, bandwidth will at least triple. Antenna dimensions, in space and on earth, will be reduced; orbital congestion, which is already noticeable at the lower frequencies, will be relieved. It is well known, of course, that systems working at these higher frequencies are affected by hydrometeors. Much experimental work has already been done in the United States, Japan and Europe on this subject, and results have been compiled on various approaches needed to counteract these effects. In the next few years, new higher frequency systems will be developed such as the

ACTS 20/30-GHz program of NASA in the United States, the ITALSAT 20/30-GHz program of Italy, the OLYMPUS program of ESA and continued experimentation in Japan at 20/30 and 40/50 GHz.

With regard to non-communications subsystems, additional benefits will be derived by solar energy conversion by means of GaAs solar cells, which are more efficient and more radiation resistant than the currently used Si cells. Further advances in battery technology are also foreseeable, but, more importantly, mass-efficient electrical propulsion subsystems for positioning and orientation maneuvers will eventually be used, leading to higher communications capacity for a given total spacecraft mass and longer mission lifetime (up to 20 years from the current 10 to 12 years). Finally, notwithstanding the present deficiency of Western World launch vehicles following the Challenger disaster and the Ariane rocket failures, launch cost reductions can be expected in the future from the interaction of competitive enterprises.

4. OPTICAL COMMUNICATIONS

After attempts in the 1970s to employ oversized metal circular waveguides operating in the TE_{01} mode (losses down to a few dBs/km in the frequency range from 30 to 110 GHz) for high capacity systems (up to 200,000 voice channels), and the appearance of low-loss glass fibers operating at hundreds of terahertz, work in this technology was suspended. The theory of dielectric waveguides had been known for quite some time, but applications had not occurred because of the high losses encountered in the dielectric materials available until around 1970. After a classical paper of 1966 which identified impurities in glass as the major source of attenuation [3], research and development work were directed at fabricating high-purity silica glass fibers. By 1970, losses were down to 20 dB/km; highly significant also was the acquisition of the capability of controlling the index of refraction along the radius of the fibers during the fabrication process, whence the distinction between multi- and single-mode fibers.

Contributing factors to the development of optical fiber communications systems were the inventions of coherent light sources (lasers), photodiode detectors and optical modulators [4],[5]. Commercial systems made their appearance within less than a decade. As a result of the divestiture of the Bell System, in addition to AT&T, other companies have aggressively entered this field in the U.S. Bit rate has gone up from 45 Mbit/s in 1980 to 1,668 Mbit/s at present.

The bit rate x repeater distance product has increased from 10-20 Mbits/s x km of the early systems, using multi-mode fibers and light-emitting diode transmitters, to several tens and even one or two hundred gigabits/s x km in systems with single-mode fibers and semiconductor laser sources. Current digital techniques are characterized by extremely high values of message integrity (bit error rates from 10^{-7} to 10^{-9}). Message integrity is also enhanced by the immunity of optical systems to electromagnetic interference.

Submarine optical cable developments have followed the expansion of the overland systems. Although a tenfold increase of the bit rate x repeater distance product has occurred about every three years during the past decade, it should be observed that the bit rate is limited by the speed of the electronic equipment rather than by the potentially available bandwidth. Advances are foreseeable through the introduction of space- and time-coherent laser sources which will permit the realization of heterodyne and homodyne systems in the future. Two other promising approaches should also be mentioned: integrated optics (all photonic vs the current mixed photonic/ electronic systems) and systems operating in the medium infrared region of the spectrum (λ from 2 to 7 µm). Using heavy metal halogenides, in lieu of silica glass, attenuation values from 10^{-2} to 10^{-3} dB/km have been theoretically predicted. The possible consequences here are mind-boggling; for instance, repeaterless transoceanic submarine cables might be feasible if suitable manufacturing techniques could be developed for these kinds of fibers. Aside from such highly speculative considerations, it is evident that state-of-the-art optical fiber cables have already produced a profound impact on communications systems. While it is obvious that optical systems will continue to expand, the authors wish to take exception to the prediction that optical cables will replace most if not all other communications systems, and, in particular, satellites. Such a prediction, which has been advanced by interested parties, appears incorrect for the reasons presented in the following section.

5. GENERAL CHARACTERISTICS OF CABLE SYSTEMS

Whatever the nature of. a cable, i.e., independently whether it is made of copper or glass, the following holds:

    a. A cable is by definition a rigid two-port network; as such, once it has been laid between two points, A and B, it can carry traffic between the two points.

    b. When the need arises, as it always does, to serve other points beyond A and B, two solutions are at hand: laying additional cables between all point-pairs or using one or both points as gateways to distribute traffic to other points of the network via additional communications links. Both solutions involve additional costs and the latter brings in significant institutional constraints.

c. A cable is characterized by well defined and usually stable propagation characteristics until it breaks. Unfortunately, cable interruptions are, to a large extent, beyond the control of both the designer and the operator of the system. After breakage, communications are lost until repairs are made, and the inconvenience produced by the interruption is proportional to the system's communications capacity.

d. Cable systems design involves not only signal transmission but also problems related to laying cables on land (i.e., acquisition of right-of-way) and costly operations at sea, as well as the need to energize repeaters and to attain high operational reliability and long system life.

While most of the above-mentioned problems may be considered manageable in overland systems, submarine cables remain vulnerable to interruptions. Repairs sometimes require a few days (but, not infrequently, weeks) due to the nature of the marine environment. It takes time for a cable ship to reach the spot where an interruption has occurred; the operation of grappling is time-consuming and possible only under good weather conditions. Table 4 stresses this point by illustrating the situation encountered in the Atlantic Ocean for the period indicated. Fortunately, whenever satellite channels are available as alternate routes, continuity of service is assured. Quick restoration of communications services via satellites during periods of submarine cable outages has become commonplace during the past two decades with tremendous, albeit unadvertised, benefits to the users.

Table 4. First Quarter 1987 Cable Restorations

Cable	Month	Number of Days	Number of Circuits	Comments
TAT-7	January	5	3,192	Repeater failure off coast of Landsend, U.K.
BERMUDA	February	8	79	Cable break 3 miles off U.S. coast, caused by fishing trawler.
BERMUDA	February/March	16	79	Planned down time to rebury the cable to prevent additional outages caused by fishing trawlers.
TAT-6	February/March	22	3,378	Cable break 500 miles off coast of France.
BERMUDA	March	4	79	Planned down time to allow completion of Bermuda cable reburial work noted above.
TAT-6	March/April	17	3,378	Repeater failure 500 miles off coast of France.
Bermuda	April	<u>10</u> (est)	<u>79</u>	Cable break 40 miles off coast of New Jersey.
Totals		82	10,264	

6. ADVANTAGES FROM THE COEXISTENCE OF TERRESTRIAL AND SPACE SYSTEMS

The above-mentioned cases are supportive of the convenience (as a matter of fact, of the need) to have satellite circuits at readiness in order to provide prompt restoration of interrupted cables. The situation will not change with the advent of optical fiber systems because the mechanics of cable breakages are independent of the nature of the cable core, be it copper or glass. It would be incorrect, however, to limit the role of satellites to that of restoring broken cables. It is rather the complementary nature of the two approaches (the rigidity of cable and the flexibility of satellite systems), which calls for their integration. In addition, from a topological viewpoint, a cable is inherently a link while a satellite is inherently a node. This basic complementarity implies that the combination of cables and satellites offers clear systems advantages. Cables will provide fixed trunking capabilities where needed, and satellites will supply nodal connectivity to serve a multitude of points. Undoubtedly the integration of cable and satellite systems, which has proved beneficial already, will become essential in tomorrow's Integrated Services Digital Network (ISDN) [6]-[13].

Consultative bodies such as the CCITT and the CCIR have been instrumental in defining the ISDN and its characteristics. Satellites need to be taken into consideration, not just as another transmission medium, as they have been, but also because of their unique capabilities. Among these are higher connectivity, capability for multipoint services, flexibility, redistribution of capacity to meet rapidly changing traffic needs, capability of providing new and/or immediate services, and readiness to meet emergency situations. For instance, while CCIR Recommendation 614 deals with satellite link performance within the ISDN framework, other international standards definitions recognize the fact that satellite links (notwithstanding their intrinsic time delay) are compatible with the ISDN architecture. It is worth recalling that today INTELSAT satellite circuits offer the only means through which transoceanic ISDN services are available. These means include the INTELSAT TDMA Service, the Intermediate Data Rate Service, and the INTELSAT Business Service. Wideband services with rates exceeding 140 Mbits/s are also presently available through TDMA networks or dedicated carriers. Rates of 250 Mbits/s or higher are in the design stage for future networks, hence compatibility and complementarity with advancing fiber optics technology is assured, from the lower ISDN rates (2B + D = 144 kbit/s) through the future high-rate ISDN hierarchy.

It is in terms of network growth anticipation, planning, design and evolution that communications satellites will maintain a very important role within the overall ISDN scenario and possibly widen it in the future. As soon as communications spacecraft evolve from their early basic function of transparent repeaters in the sky to the more complex role of signal amplification, call-processing and message-switching in orbit with fixed and scanning antenna beams, their integration with terrestrial networks will acquire new dimensions and have far-reaching consequences [13]-[17]. Even before such a stage is reached, communications satellites, albeit of the straightforward transparent repeater kind, have become essential elements in today's communications world, as in the case of VSAT's systems, Business Satellite Services and Packet Satellite Networks.

Twenty-two years of operational experience with geostationary commercial satellites have demonstrated:

a. the acceptability of their inherent transmission delay for all kinds of traffic, once efficient echo control techniques are used for voice and proper protocols for data communications;
b. their trouble-free connection with terrestrial networks over the global scale;
c. an exceedingly high degree of service reliability;
d. their catalytic action in network development; and
e. continuously decreasing costs.

Cost considerations are essential; with advancing technology, costs have come down in all communications systems, yet in no other sector as impressively as in communications satellite systems. Because of the general nature of this paper, the authors will take the position to let the specialists argue about direction in which one should look for a "best buy." It has been, however, a bit amusing to observe papers and reports proving that satellites are more cost effective than cables or vice versa. This debate, which has been going on for decades with coaxial cables on one side and satellites on the other, will undoubtedly continue in the future in regard to optical fiber cables and advanced satellites. Possibly a sensible attitude in regard to this matter is that the conclusion reached in each case depends largely upon chosen "initial conditions."

7. CONCLUSIONS

The considerations which have been presented, although not exhaustive due to the complex problems encountered, should help in sustaining the thesis that the integrated use of cables and satellite systems is a worthwhile goal. From a more pragmatic point of view and considering not only technical but also economical and geo-political factors, the final decision should rest, as we hope it will, in the hands of the users of communications services.

8. REFERENCES

[1] B. G. Evans, "Toward an Intelligent Bird," *International Journal of Satellite Communications*, Vol. 3, Issue 3, July-September 1985, pp. 203-215.

[2] C. E. Mahle, G. Hyde, and T. Inukai, "Satellite Scenarios and Technology for the 1990s," *IEEE Journal of Selected Areas in Communications*, Vol. SAC-5, No. 4, May 1987, pp. 556-570.

[3] K. C. Kao and S. A. Hockham, "Dielectric Fibre Surface Waveguides for Optical Frequencies," *Proc. IEE*, Vol. 113, No. 7, July 1966, pp. 1151-1158.

[4] Technical Staff of CSELT, *Fibre Communications*, New York: McGraw-Hill, 1980.

[5] D. A. Thomas et al., *Lightwave Communication Innovations in Telecommunications*, Part A, J. T. Manassah, ed., New York: Academic Press, 1982, pp. 437-562.

[6] R. Preti, S. DePadova, and A. Puccio, "Integration of a Satellite Switched System With the Terrestrial Network," *CSELT Rapporti Tecnici*, Vol. 9, Supplement to No. 5, October 1981, pp. 551-557.

[7] S. Tirrò, "Satellite and Switching," *Space Communication and Broadcasting*, Vol. 1, No. 1, April 1983, pp. 97-133.

[8] G. B. Alaria and G. Pennoni, "SS/TDMA System With Onboard TST Switching Stage," *CSELT Rapporti Tecnici*, Vol. 12, No. 3, June 1984, pp. 242-255.

[9] P. Bartholomé, "Digital Satellite Networks in Europe," *Proc. IEEE*, Vol. 72, No. 11, November 1984, pp. 1469-1482.

[10] A. Burattin, S. DePadova, and U. Mazzei, "Economical Considerations on a Satellite System in a Small-Size Developed Country," IEEE International Conference on Communications, Chicago, Illinois, June 1985, Paper 40.4.1-40.4.4, *Conf. Record*, Vol. 3, pp. 1289-1292.

[11] A. Casoria, "Tecniche Avanzate nella Integrazione delle Reti di Telecomunicazioni Spaziali e Terrestri," *Elettronica e Telecomunicazioni*, Vol. 35, No. 6, November-December 1986, pp. 265-275.

[12] P. Amadesi, P. Haines, and A. Patacchini, "Satellite Networks in the ISDN Era," 7th International Conference on Digital Satellite Communications, Munich, May 1986, *Proc.*, pp. 247-254.

[13] R. W. Slabon, O. Schmeller, and W. W. Knoben, "Advanced Data and ISDN Services in the DFS Satellite Communications System," 7th International Conference on Digital Satellite Communications, Munich, May 1986, *Proc.*, pp. 107-112.

SOME ASPECTS OF THE IMPLEMENTATION OF

AN INTEGRATED SPACE/TERRESTRIAL NETWORK FOR EUROPE

J.M. Casas and P. Bartholomé
European Space Agency
ESTEC, Noordwijk
The Netherlands

ABSTRACT

This paper examines the role of Advanced Satellite Systems as providers of backbone trunking services in the Integrated Digital Network. The forces shaping the architecture of the new networks are contrasted with the advanced functionality offered by new satellite concepts. This leads to a reappraisal of the way satellites and terrestrial networks should integrate. The advantages and problems of this integration are discussed.

Keywords:
Advanced Satellite Systems, On-Board Processing, Multispot Beam Systems, ISDN, IDN, Network Architecture.

1.   INTRODUCTION

The role of satellites in the ISDN era has been discussed in a number of papers [1-4]. The general conclusion of these studies is that satellites are expected to play a dual role in the networks resulting from the catharsis caused by the process of digitalisation, the introduction of new technologies and the emergence of new services:

(i)   Satellites may be used to implement backbone transmission facilities in the public Integrated Digital Network (IDN). This is the major subject of this paper.

(ii) Satellites provide an excellent medium to implement other forms of Integrated Services Digital Networks. These can be either Specialised Services Public Network e.g. SMS services on EUTELSAT system, or user-oriented networks e.g. that are those that are designed and satisfy specific users requirements e.g. VSAT networks or private networks.

The emergence of these alternative networks in Europe is conditioned to a great extent by the regulatory environment. In this respect, it is important to emphasise the tremendous impact that on-going CEC legislation efforts may have in the development of industry [5]. These applications will not be discussed further in this paper.

Satellite systems have been used to provide trunking services for a number of years. They have been fairly successful as is proven by the fact that a substantial part of the international traffic and two-thirds of the intercontinental traffic worldwide are routed by satellite [6]. However, the emergence of optical fibres has caused some forecasters to predict the eventual substitution of all satellite trunking systems by fibres [7].

The main argument against the use of satellites in trunking applications is the continuous growth of the break-even distance. This is the distance at which the cost of providing a circuit between two specific points by satellite, which is distance independent, is the same as providing the same circuit by terrestrial (e.g. optical fibre) means, which are distance-dependent. As the capacity of OF systems grows, the cost per circuit diminishes and the break-even distance increases more and more. This reduces the applicability of satellites to remote applications.

The break-even distance argument disregards however the fact that the satellite offers a functionality that is not directly comparable with the point-to-point transmission that the optical fibre allows. The satellite allows communication between any points of the earth's surface illuminated by its beams. This results in satellites being flexible in their application, reconfigurable, able to match the requirement of the network operator (or the end user), to provide service over extended areas and to support some special services. How, in the light of all this, can we discard satellites basing our argument on the growth of the break-even distance?

In this paper, we try to project the full functionality of state-of-the-art and future satellite systems to the evolving European network. Section 2 reviews the present and future of the network. Section 3 analyses the major features of the satellite systems being considered and the advantages brought about by the integration. Finally, Section 4 reviews the system problems associated with the integration of the advanced satellite system in the terrestrial network.

2. THE EVOLUTION OF THE TERRESTRIAL NETWORK

The Public Switched Telephone Network (PSTN) from which the Integrated Digital Network (IDN) is emerging by a process of technological substitution, providing the backbone for the ISDN, is a dynamic "organism" subject to constant evolution. These changes may alter not only the way information is treated, but also the economic balance between the different trade-offs available to the network designer, and as a consequence its internal architecture.

Figure 1 shows what can be considered a model of an European national network. Table 1 shows some of its main macro-economic figures. To provide a reference, these values are compared with some available figures made public by several Telecommunication Administrations [8-11].

The introduction of cheaper transmission systems and the extended traffic handling ability of Digital Switching Exchanges favours the recombination of a relatively large quantity of transit switching nodes into a lesser number of larger exchanges which are fully meshed in their interconnections [10, 12, 13].

Figure 2 shows the architecture that may result from the redesign of the network incorporating the new trade-offs. There, a relatively large number of sector exchanges (i.e. local transit exchanges) will be eliminated, their functions being taken up by a few larger exchanges at regional (district) level. For instance, modernisation plans of British Telecom call for the substitution of about 450 transit exchanges by about 55. In Italy, the change is more dramatic: nearly 1500 sectorial exchanges and 250 district exchanges will be substituted by 109 new exchanges.

Other important feature of the technological substitution is that the interconnection between the new transit exchanges is meant to be fully meshed. Further, the transit exchanges in the Italian case are subdivided in two sublevels both serving directly local areas but the higher allowing the overflow of the lower. This will be so since the routes of the lower mesh will be relatively thin and therefore very inefficient in terms of Erlangs/circuit if no overflow was permitted.

The implementation of the network does not require a physically fully meshed physical network. To illustrate that, Figure 3 shows the expected network architecture in the UK.

The evolution of the network will occur progressively until well into the 21st century. The cost of this evolution is staggering. In Europe 500 Billion ECU will be spent in the next 15 years by Telecommunication Administrations. In this scenario, the availability of a technology that may offer new flexibility of future telecommunication networks deserves full attention by the network planners.

	FR	G	IT	SP	UK	MODEL
INHABITANTS	56 M	60 M	57 M	37 M	58 M	50 M
- no. of lines	22 M	25 M	17 M	10 M	20 M	16 M
SWITCHING (no. of units)						
- local exchange	6000	6000	13000		6300	5000
- sector exchanges	1800	525	1400		400	500
- P+S transit exchanges	75	8168	21230		40	50
TRANSMISSION (no. of trunks)						
- local trunks	500 K	1000 K	480 K	250 K	420 K	500 K
- local distance (>100 km) (R+N) trunks	125 K	250 K	120 K	62.5 K	105 K	100 K
- international	12.5 K	14.2 K	6.5 K	5 K	12.6 K	10 K
TRAFFIC (Erlangs busy hour)						
- local	692	1 M	536	280	464	500 K
- inter-urban (trunk)	173 K	250 K	134 K	70 K	116 K	125 K
- international	9 K	13 K	3.8 K	2.6 K	6.6 K	8 K

TABLE 1 - SOME FIGURES OF THE EUROPEAN NETWORKS

3. ADVANCED SATELLITE SYSTEM CONCEPTS

The emergence of optical fibre systems and the full awareness of the functionality of satellite systems coincide in pushing the application of satellite links in trunking roles towards lower layers of the transmission hierarchical structure. There its dynamic bandwidth assignment properties and its flexibility of use can be shown to full advantage.

The implications of this deeper integration in the terrestrial network are manyfold. Perhaps its most immediate consequence is the proliferation of earth stations. It becomes therefore paramount to reduce radically the cost per earth

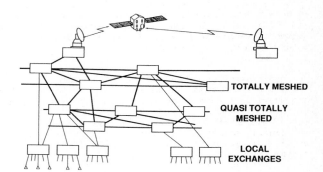

FIGURE 1 – PRESENT STRUCTURE OF TELECOM NETWORKS IN EUROPE

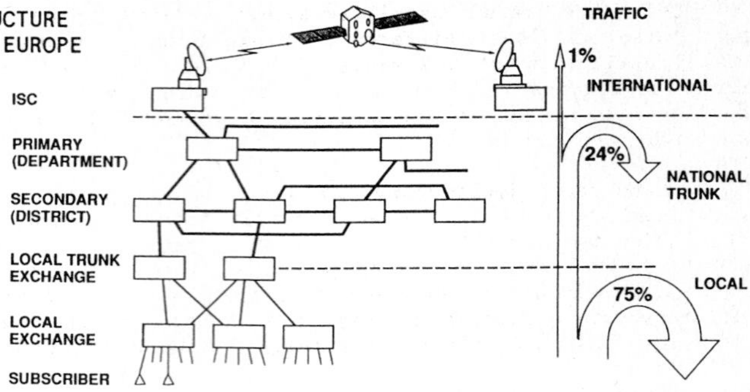

FIGURE 2 – ARCHITECTURE OF FUTURE NETWORKS

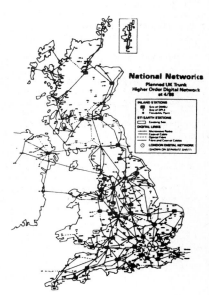

FIGURE 3 – PLANNED UK HIGH ORDER TRUNK NETWORK

station if the overall satellite system is meant to be competitive. Further, the penetration to lower hierarchical levels call for an increase of the traffic handling capacity of the system. It is not the same to share part of the few thousand Erlangs of international European traffic than getting involved in dealing with a part of the hundreds of thousands of Erlangs of the long distance national trunks.

These requirements converge in defining the satellite system with multiple spot beam antennas for lower earth station cost and frequency reuse for increased system capacity.

To maintain the interconnectivity between any pair of beams, switching on the satellite becomes mandatory. And to maintain the burst efficiency, a single burst per transmit station is required, leading to TST/SS-TDMA system.

Since the traffic distribution throughout Europe is far from uniform (the relation between the traffic served by two spots of the same size may be as large as 200:1), the optimisation of the bandwidth assignment calls for the use of beam hopping techniques and eventually dual frequency band utilisation (14/12 and 30/20 GHz). Further system expansion without duplication of the earth segment cost may be provided by means of Inter-Satellite Links, which can also cater for the interconnection of the system with other equivalent systems in other regions of the world.

To provide an element of reference, some system concepts have been combined from several studies and papers [14-17]. The resulting system characteristics are shown in Figure 4 and Table 2. It is interesting to contrast the potential capacity of such a system 64-128 thousand, 32 Kbit/s circuits, with the current total amount of international European circuits $\approx$ 75 thousand (of which a substantial amount corresponds to high usage routes). This shows the extent to which a single satellite system may satisfy both international and national traffic demand.

In the evolutionary network scenario described previously, the advanced satellite system can complement the functionality of terrestrial systems to achieve a harmonious development. The areas in which satellite system functionality is specially relevant are:

(i)   Optimal digitalisation strategy

   The deployment of an advanced satellite system may be the most expeditious mechanism to ensure the adequate provision of digital connectivity (or even access!) to the whole European continent. This may ensure that the less developed regions are not left to the pressure of a very limited demand.

(ii)   Optimal economic design of the network

   The flexibility incorporated by the advanced satellite system concepts allow the dynamic reconfiguration of the network. This leads to optimal design of thin routes and the adaptation of the network to variable patterns of utilisation e.g. seasonal variations of population.

Further, the availability of reassignable transmission plant may be used to realise investments on the network in the measure they are required. This avoids for instance the necessity to equip large amounts of circuits well in advance of their utilisation. The resulting network may be much closer to the traffic demand with the subsequent savings.

(iii) <u>Improved operational features</u>

The availability of circuits on demand allows the use of the satellite system as back-up to the terrestrial system. The ability to pool from a common resource implies that the redundancy arrangements which the satellite offers is one for n, while with terrestrial means this would in general one for one, or two.

Further, the resilence of the network in the face of unexpected events which cause either disruption of services, or instantaneous surge of traffic in specific points is substantially improved.

(iv) <u>Optimal provision of new network services</u>

The deployment of a satellite system would allow the provision of services like point-to-multipoint broadcast, broadband without having to redesign a substantial part of the backbone network on this account.

TABLE 2 - REFERENCE SYSTEM CHARACTERISTICS

SATELLITE SYSTEM CHARACTERISTICS

Access and Process	TST/SS-TDMA
Capacity (DSI dependent)	64000 - 128000 circuits
No. of Transponders	32 x 8 MHz
Frequencies	14/12, (30/20) GHz
Frequency Re-Use	x5, x7
Rates of Operation	131, 33, 8 Mbit/s
Flexibility on Traffix Assignment (spot size ≈ 0.7°)	large traffic, spot beams at 20-30 GHz, small traffic low rate and beam-hopping
Spacecraft Cost Estimate (1 of 5 including launch)	≈ 200 M ECU

EARTH SEGMENT CHARACTERISTICS

System	TDMA
Assignment	on demand
Terrestrial Interface	G732, n x 2 Mbit/s
Rates of Operation	131, 33, 8 Mbit/s
No. of Circuits/System	4000, 1000, 250
Antenna Sizes	8, 5, 3m
ES Costs	400 - 300 - 250 K ECU

MASTER CONTROL CHARACTERISTICS

Signalling	CCITT no. 7
Reconfigurability	dynamic on demand
Clock Synchronisation	Master on ground plesiochronous I/F to ground networks

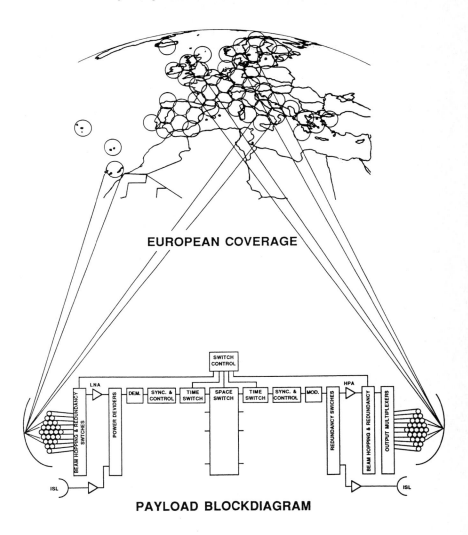

FIGURE 4 - EXAMPLE OF ADVANCED SYSTEM CONFIGURATION

4. THE PROBLEMS OF THE INTEGRATION

Some of the technological aspects of the system proposed are very new and therefore may require in orbit demonstration to validate their feasibility.

However, the really difficult issues of the integration are precisely in the borderline now so diffused between the satellite system and the terrestrial network. Without pretending to be exhaustive, we have identified the following:

- Technical:

(i) Need of optimisation of the process of network design.
- Awareness of the availability of a "flexibility" tool in the implementation of dynamic networking would result in a more economical and robust planning.

(ii) Implementation of network management techniques.
- No network management techniques are currently being used or even in planning stages throughout Europe. Fortunately the satellite system can implement the management of its own resources independently from the fact that the underlying network is static. The satellite system can be configured for a particular application and then would be left to rearrange itself between certain limits dynamically as a consequence of traffic demand.

(iii) Performance.
- CCITT Recommendation G821 sets up, even if very peculiarly, the limits of performance for "standard" 64 kbit/s ISDN channels. For many people, this is the key issue of the integration of satellites in the ISDN [18, 19]. We recognise its importance and specially the impact it may have on the system design if the 20-30 GHz band is used. It is however clear that this is a purely technological issue. The price of this specification has not been assessed yet.

(iv) System reliability.
- The satellite system needs to be designed so as to ensure the grade of service and robustness required from public networks. The analysis of the redundancy required in several subsystems is currently being studied under ESA contract.

(v) Signalling.
- The satellite system benefits from the introduction of the CCITT no. 7 signalling system. It is the nervous system of the network that will allow the reassignment of resources. It further provides the vehicle to convey administrative information which may be paramount in an international system. The best way to interact with the CCITT no. 7 network and the role of the master control terminal as either signalling point, signalling transfer point is the object of yet another ESA contract.

(vi) Provision of new services.
- It is obvious that the satellite may be the vehicle to provide some new services to areas that otherwise may have to wait decades to get them, e.g. multipoint, broadband. However, the way to implement these services over the satellite and specially on the terrestrial tails is yet to be studied.

- Economic:

    A very important problem on the road ahead of the integrated system concept is the demonstration of the economic viability of the proposed solution. While the advantages listed are easy to perceive, they appear in many cases very difficult to evaluate.

    In a companion paper to this conference [20], a simplified economic analysis is presented. It compares the cost of implementing the meshed connections of the second sublevel of the transit networks using satellites and optical fibre systems.

    Other organisations have reported some analysis applicable to specific national applications [21]. Further, ESA has funded a study by ESCO which is currently running and which some preliminary information has been used when writing this paper. Its final results are expected later this year.

- Political:

    The deployment of an integrated system may be performed in relatively trouble-free waters if it is done as a national system. On the other hand, given the relatively short European distances, it may be economically more attractive to deploy it at European scale. This implies that future generations of the European regional system (EUTELSAT III, EUTELSAT IV) could be allowed to have direct access to the national transit networks. While this path may be full of difficulties, since it may collide with the national systems, it presents at the same time the political opportunity to increase the cohesion of Europe in telecommunications networks. Further, its planned introduction might just oblige national Telecommunication Administrations to harmonize and standardise the different national systems. That in itself would be an achievement.

5. CONCLUSION

To design a satellite system for trunking applications was in the past no more difficult than to design the transmission system whose role it was meant to fulfill.

This is now changing rapidly. The evolution of the network, the emergence of very competitive alternatives, and the foreseen evolution of the satellite systems themselves call for a reappraisal of their role in the implementation of the backbone digital network.

This reappraisal requires the investigation of many system problems which did not use to show up on the tables of satellite communications engineers. The problems are there, but the benefits that this integration may provide are very attractive. This is why ESA is actively studying every implication of these concepts.

REFERENCES

[1] A.M. Rutowski: The Role of Satellite Radiocommunication in ISDN. Telecommunications, June 1983.

[2] P. Amadesi, P. Haines, A. Patacchini: Satellite Networks in the ISDN Era. Proceedings ICDSC-7, Munich, 1986.

[3] NASA Technical Memorandum 88867: An Assessment of the Status and Trends in Satellite Communications 1986-2000.
November 1986.

[4] J.M. Casas: ISDN Status and Opportunities for Satellite Systems.
ESA STR-220, May 1987.

[5] Commission of the European Communities: Green Paper on the Development of a Common Market for Telecommunications Services and Equipment.
COM(87)290 final, June 1987.

[6] Handbook of Satellite Communications (Fixed Satellite Service).
CCIR, Geneva 1985.

[7] F. Guterl, G. Zorpette: Fiber Optics: Poised to Displace Satellites.
IEEE Spectrum, Vol. 22, no. 8, August 1985.

[8] CEPT: Survey of Plans for the Introduction of ISDN in Europe.
Report of the Special Group on ISDN (GSI), 1984.

[9] Logica: European Communication Services Towards Integration.
1986, ISBN 0907071 03 1.

[10] ESCO: Study on the Integration of Advanced Communication Satellite System and Terrestrial Networks.
Preliminary Papers, ESA Contract 6781/86/NL/DG.

[11] Spanish Ministry of Transport, Tourism and Communications: Informe Anual sobre Transporte Turismo y Comunicaciones 1984.
Madrid 1985.

[12] B. Armbruster et al: Network Structural Challenges.
IEEE Proceeding of the 3rd International Network Planning Symposium, Innisbrook, 1986.

[13] U. Mazei, E. Miranda, P. Pallotta: "On a Full Digital Long Distance Network".
9th ITC, Torremolinos, 1979.

[14] S.E. Dinwiddy: Advanced On-Board Processing Satellite System Concepts.
Proceedings International Symposium on Satellite Communications, Graz, September 1985.

[15] S.T. Companella, B.A. Pontano, K. Kao: Economics of Multiple Beam Satellites.
Proceedings ICDSC-7, Munich, 1986.

[16] G. Pennoni: A TST/SS-TDMA Telecommunication System: From Cable to Switchboard in the Sky.
ESA Journal 8 (2), 151 (1984).

[17] R.R. Lovell, C.L. Cuccia: Switched Lasers and Electronically Hopped Beam Antennas in Space.
NASA's Advanced Communications Technology Satellite, Space Tech 86, Geneva, 1986.

[18] J. Lewis: Satellites and the ISDN.
    IEEE Conference on Digital Satellite Communications, Phoenix, September 1983.

[19] T. Knight: The Role of Satellite Communication in ISDN.
    Telecommunications, June 1987

[20] G. Pennoni, L. Bella: Integration of a Payload Enhanced Network Satellite (PENSAT) and the Terrestrial Broadband ISDN.
    International Workshop on Digital Communications, Tirrenia, 1987.

[21] L. Zanneti et al.: An Economic Analysis of the Introduction of a Domestic Satellite System in a Developed National Telecommunication Network.
    Proceedings ICDSC-7, Munich, 1986.

ECONOMIC COMPARISON OF CABLES AND SATELLITES

S. J. CAMPANELLA*
B. A. PONTANO*

In response to competition created by the advent of fiber optic digital communications, satellite system operators are seeking more cost-effective methods of operation in order to remain competitive. This paper describes and analyzes the design philosophy of a cost-effective, efficient satellite system which achieves a more economical total earth-space system.

1. INTRODUCTION

Until recently, satellite communications systems have relied principally on the use of satellite communications payloads with low-power transmitters and small aperture antennas located in the geostationary orbit 22,300 miles above the earth in their first two decades of use. To achieve reliable communications with such a capability, earth stations with large antennas, such as INTELSAT's standard A with a diameter of 30 m, are needed to receive the very weak signals from the satellite. Also, the uplinks must contain transmitters powered by 8-kW amplifiers, which transmit signals of a megawatt or more of radiated power [1] when combined with the high spot beam gain of the earth station antenna.

Satellite systems of this configuration result in small, low-e.i.r.p., low- G/T, low-cost satellites operating with a network of large, high- power, expensive earth stations, each costing five to ten million dollars. Such stations can be used economically only when they serve several thousand telephone circuits.

However, the advent of fiber optic digital communications [2] has significantly changed the competitive picture by greatly increasing the number of circuits that can be carried in a single cable installation, thus placing pressure on satellite system operators to seek more cost-effective methods of operation in order to remain competitive.

A significant reduction in the size, power, and cost of the earth stations must be achieved to enable distribution near to traffic origins of many stations carrying traffic loads of only a few hundred circuits. A network of such distributed stations located near traffic origins optimizes satellite multiple accessing capability, or the ability to view almost half the earth's surface from a single vantage point, reaching areas where cable of any type is uneconomical. Also, such a network obviates the need to aggregate the traffic at the cable gateway via terrestrial links. Any competitive future satellite system must fully realize all of these advantages, while maintaining efficient use of the satellite communications payload and the radio frequency spectrum resource assigned to the service.

An efficient satellite system fitting the above requirements can be accomplished by reversing the design roles played by the earth and space segments in current systems [3]. Such satellite systems place the large aperture antenna in space, where its capability and cost can be shared by all users, and the small aperture, low-power station on the ground, where it can be duplicated in large quantities at low cost. This is the design philosophy that has governed the system described and analyzed in the following subsections.

2. TRANSMISSION ANALYSIS

A transmission link analysis for a typical multibeam satellite system with on-board demodulation and remodulation is provided in this section for K_u-band satellites with 1° spot beams. The transmission link analysis is carried out parametrically with the transmission rate and fade margins in the uplink and satellite HPA and fade margins as parameters in the downlink.

---

* Both authors are with COMSAT LABS, Clarksburg, Maryland 20871, USA.

2.1. Interference Consideration

The interference entries due to the effects of adjacent satellite interference, and spatial or cross-polarization isolation between the desired beam and interfering beams, as a result of frequency reuses in the multibeam satellite system, are considered in the transmission analysis.

For these link calculations, a total C/I ratio of 21 dB is assumed, based on a carrier-to-adjacent satellite interference ratio of 27.3 dB for the uplink and 25 dB for the downlink, assuming 2° satellite spacing, carrier-to-frequency-reuse interference from spatially isolated beams of 27 dB, and a carrier-to-cross-polarization interference ratio of 24 dB.

2.2 Performance Requirement

The bit error rate (BER) requirement of a satellite link forming part of a 64 kbit/s ISDN connection is given in new CCIR draft recommendation [4]:
1.0×10^{-7} for more than 10 percent of any month;
1.0×10^{-6} for more than 2 percent of any month;
1.0×10^{-3} for more than 0.03 percent of any month.
For K_u-band applications, the last criterion, 1.0×10^{-3} for more than 0.03 percent of any month, usually limits the performance, except for very dry areas.

2.3. Transmission Parameters

Multiple-carrier TDMA FDMA uplinks and single carrier TDMA downlinks are used in the transmission analysis. Two classes of service are described.

In one, called the Medium/High Intensity Load Service, the uplinks use four 30 Mbit/s burst rate carriers spaced 20 MHz apart in each 80 MHz transponder bandwidth; and the downlinks use a single 120 Mbit/s burst rate carrier. This requires earth terminals equipped with 3.5m diameter antennas and 20w HPAs intended to carry 10 to 100 Erlang traffic loads.

In the other, called the Light Intensity Load Service, the uplinks use four 3 Mbit/s burst rate carriers spaced 20 MHz apart in each 80 MHz transponder bandwidth; and the downlinks use a single 12 Mbit/s burst rate carrier. This requires earth terminals equipped with 1.8m diameter antennas and 4w HPAs intended to carry 0.1 to 10 Erlang traffic loads.

The expected modem performance curve is given in Figure 1. The performance of the coder/decoder is given in Figure 2. Table 1 summarizes all pertinent satellite and earth station parameters used in the link calculations.

Table 1. Summary Of Space Segment Parameters (Ku Band Operation)

Satellite	Beam Size (deg)	
	1	
G/T (dB/K)	1 1	1 1
e.i.r.p. (dBW)	Parameter	Parameter
Down-Link Rate	120 Mbit/s	12 Mbit/s
Modulation	CQPSK	CQPSK
Earth Station	Antenna Diameter (m)	
	E-1 3.5	E-2 1.8
G/T (dB/K)	25.0	22.1
e.i.r.p. (dBW)	64.0	54.0
Up-link Rate	Parameter	Parameter
Modulation	CQPSK	CQPSK

2.4. Link Performance

Link performance results are shown in Figures 3 and 4.

For the uplink cases (Figure 3), the maximum transmission rate is plotted vs the fade margin. The results show the fade margin asymptotes to a limit of about 13.5 dB for the rate 3/4 FEC.

For the downlink cases (Figure 4), the satellite HPA power is plotted vs the fade margin. No limiting effect is observed. However, during deep rain fade, depolarization can limit the performance and cause the downlink fade margin to approach a limit.

3. SPACE SEGMENT CONSIDERATIONS

3.1. Transponder Requirements

The satellite is intended to achieve fourfold frequency reuse of the 500 MHz, K_u band assigned to fixed satellite service. Each frequency reuse is divided into six 80 MHz transponder segments. Each transponder is powered by a 10w HPA. For the medium/heavy load intensity services, each transponder segment will carry four 30 Mbit/s uplink TDMA carriers and one 120 Mbit/s downlink TDMA carrier. Each transponder segment will be connected to a hopping beam as required by traffic demand. For the light load intensity services, the uplink and downlink carriers are reduced by tenfold to operate with VSAT earth terminals. In this case each transponder is powered by a 2.5w HPA.

3.2. Antenna Beamwidth

A 1° beam and 10w satellite HPA is assumed for operation with the 120 Mbit/s downlink carrier. A 2.5w satellite HPA is assumed for operation with the 12 Mbit/s downlink carrier. With a 1.0° beamwidth, 40 beams are provided. This is sufficient to cover a country the size of the United States.

3.3. Spacecraft Definition

A K_u-band multibeam 24-transponder satellite with hopping beams and onboard regeneration, based on the above design considerations, is shown in Figure 5. Six 80 MHz transponders comprise each frequency reuse of a 500 MHz spectrum.

Uplink hopping beams are connected to 24 80 MHz transponder filters and subsequent demodulators and baseband processor by means of an input MSM to distribute the space segment capacity among the earth stations. Bandwidth and dwell are adjusted to meet the traffic need. Similarly, the transponder outputs are distributed to the downlink hopping beams by an output MSM. Any activated uplink beam may provide between 80 and 480 MHz of bandwidth in increments of 80 MHz to a given spot-beam coverage area. Likewise, on any downlink beam, the MSM may connect up to six contiguous 80 MHz transponders and their corresponding 120 Mbit/s and 12 Mbit/s TDMA carriers to a single downlink beam.

The digital FDMA/TDMA carriers are connected to their assigned demodulators. For the medium/high intensity load service, each 80 MHz transponder is divided into four sub-bands of 20 MHz each carrying 30 Mbit/s QPSK modulated carriers. Four 30 Mbit/s demodulators are supplied for each 80 MHz transponder. For the light intensity load service, each 80 MHz transponder is divided into four sub-bands of 2 MHz each carrying 3 Mbit/s QPSK modulated carriers. Four 3 Mbit/s demodulators are supplied for each 80 MHz transponder. The transponder frequency assignment plans for the medium/heavy and light load services are shown in Figure 6.

Demodulated streams are reformatted by the baseband processor into 120 Mbit/s and 12 Mbit/s TDMA streams for downlink transmission. Single carrier, high burst rate TDMA downlink transmission improves satellite power efficiency, because single carrier per transponder operation eliminates the need for HPA backoff.

The baseband processor consists of 24 input and 24 output random access memories, each storing one 16-ms TDMA frame consisting of 2800 32 Kbit/s channels. This corresponds to the traffic of one transponder using a carrier bit rate of 120 Mbit/s and R3/4 FEC channel coding. The 24 input memories are interconnected to 24 output memories via the baseband switch. The input memories demultiplex and aggregate the demodulated channels from the parallel demodulators. A 24 x 24 digital space switch reorders them into the 24 output memories. The output memories are played out onto single 120-Mbit/s or 12 Mbit/s TDMA streams in the appropriate order for transmission on the downlink hopping beams.

3.4. Spacecraft Cost Estimates

The spacecraft cost is estimated by use of the SAMSO cost model [5] based on spacecraft mass and power. The estimates are for all medium/high load intensity service operation. When one set of six transponders is used for light load intensity service operation, cost of the fraction of the spacecraft devoted to the service is assumed to be 10% less than the cost of the same fraction for medium/high intensity service. This is because the light service uses lower power HPAs.

Table 2 estimates mass and power of the baseband processor and hopping switches.

Table 2. Mass And Power for the Baseband Processor and Hopping Beam Switches

No. of Transponders	Baseband Processor		Beam Hopping	
	Mass (lb)	Power (W)	Mass (lb)	Power (W)
24	118	393	100	80

The baseband processor contains 96 digitally implemented demodulators, each requiring an estimated 2w of power and weighing half a pound. For 24 transponders, the input and output memories must each store 17.28 Mbits of data, assuming a ping-pong design. It is expected that by 1990, C-MOS memory will require about 5 µ per bit. Therefore, each memory requires about 85w of power.

Whereas the 24 modulators are estimated to require 24w of power and weigh about 8 pounds, the 24 x 24 baseband switch, implemented with gallium arsenide (GaAs) devices, is estimated to require less than 7w of power and weigh less than 7 pounds.

The mass and power estimates for the hopping MSM switches are based upon the use of ferrite switches, which can provide a switching time of less than 1 µs. They have a higher power handling and lower loss capability than the faster switching pin diode switches.

The mass and power estimates for the complete communications payload, including the baseband processor and overall spacecraft, are given in Table 3. Except for the receivers, these estimates are based on INTELSAT VI technology (e.g., for filters, HPAs, and converters) and, therefore, may be somewhat conservative for a 1990 to 1995 launch date. The receivers assume the use of monolithic microwave integrated circuits (MMIC) to minimize mass.

Based on these mass and power estimates, the cost of each spacecraft, including launch, is given in Table 4.

4. EARTH STATION DEFINITION AND COSTS

To keep both investment and maintenance costs to a minimum, solid-state power amplifiers are used when possible. For the mid-1990's, it is expected that 20w will be the output power for a solid state HPA operating at K_u-band. Based on this assumption and the need to maintain relatively large fade margins at K_u-band (on the order of 8 dB for most locations), the maximum

Table 3. Spacecraft Mass and Power

No. Transponders	Communications Payload Mass (lb)	Total Spacecraft In-Orbit Mass (lb)	Communications Payload Power (w)	Total Spacecraft Power (w)
24	618	2838	1176	1710

Table 4. Estimated Spacecraft Cost

No. of Transponders	Nonrecurring (M$)	Recurring (M$)	Spacecraft Cost, One of Five Including Launch (M$)
24	152	65	126

uplink bit rate is limited to 30 Mbit/s for medium load intensity (10 to 100 Erlangs) when transmitting from a 3.5-m earth terminal. The downlink burst rate for such terminals is 120 Mbit/s. For the light load services, it is necessary to exert extreme measures to reduce the cost of the earth terminals. For this reason, the uplink burst rate has been reduced to 3 Mbit/s and the downlink burst rate to 12 Mbit/s. This will permit 8 dB fade margin operation with earth terminals using 1.8 m diameter antennas and 4w HPAs. All of the previous estimates are based on use of R3/4 FEC with soft decision decoding.

The earth terminal costs used in the economic analysis are summarized below for each major subsystem. These costs are grouped into those which are fixed and those which depend upon the number of channels being carried by the earth terminal. The fixed costs, F, those which are independent of the amount of traffic carried, are summarized in Table 5. The cost of the medium/high intensity service E1 terminal is governed by experience obtained in the design of earth stations of this scale used in customer premesis terminals in private networks in the United States. The low cost of the small E2 station is in the same range as that of the VSAT terminals used today throughout the United States and is partly the consequence of use of lower burst transmission rates. Low burst rates reduce modem and TDMA implementation costs (using microprocessors and integrated circuits) and permit use of low power solid state HPAs. The use of integrated microwave circuit frequency conversion chains further reduces cost. Furthermore, there is an overall cost reduction due to mass production in low production cost regions of the world.

The earth terminal cost per channel, C_T, for N channels is calculated from the expression:

$$C_T = F/N + (C_0/N)([N/60])$$

where F = the fixed cost of the terminal and C_0 is the cost of an interface unit that serves a fixed number of channel ports. For the medium/high intensity service, it is assumed that the number of ports served by an interface unit is 60. The symbol [ ] indicates rounding the quantity between the brackets to the next highest integer value. The value of C_0 for various types of 60 port interfaces is:

C_0 = $33K for analog interfaces with DSI;
   = $42K for analog interfaces with LRE/DSI;
   = $26K for digital interfaces with DSI;
   = $36K for digital interfaces with LRE/DSI.

In the case of the light intensity service terminals that serve between 0.1 and 10 Erlangs, the number of channel interfaces is small; and they are costed on an individual basis of $ 400 each.

Table 5. Fixed Cost of E1 (30 Mbit/s) & E2 (3 Mbit/s) Earth Stations

Cost Item	Cost ($000)	
Antenna Size	E1 (3.5 m)	E2 (1.8m)
Antenna	10	0.5
HPA	20	1.0
LNA	10	0.5
Uninterrupted Power Supply (UPS)	15	0.5
GCE	80	5.5
Shelter	10	0.5
Program Management	20	0.5
Integration and Test	20	1.0
	185	10.0

5. TRAFFIC CONFIGURATIONS

Two different carrier configurations are used: one for medium/high loads, which are between 10 and 200 Erlangs per station; and another for light or thin route loads, which are between 0.1 and 10 Erlangs per station. Each of these is considered below.

5.1. Medium/High Traffic Intensity Configuration

In this configuration, each of the 24 80 MHz uplink transponder bands is occupied by four 30.208 Mbit/s QPSK burst modulated, R3/4 FEC coded, TDMA carriers, each occupying a spectrum space 20 MHz wide. Onboard demodulator/decoders convert these carriers to digital baseband; and the uplink channels are switched and reconfigured onto the downlink channels. The downlink channels are carried in beam hopping TDMA form on 24 120.832 Mbit/s burst rate carriers in the downlink beams. The frequency plan for the up- and downlinks is shown in Figure 6.

The up- and downlink channel capacities are now calculated for the medium/high load intensity configuration using the burst structure shown in Figure 7(a).

Uplink Calculation:

The number of symbols per frame for each uplink 30.208 Mbit/s burst rate carrier (corresponding to a symbol rate of 15.104 Msym/s) TDMA frame (16 ms) is:

UPLINK SYM/FRAME = 16x15104 = 241664 symbols.

This frame must carry the station uplink traffic bursts. A reasonable number of uplink traffic bursts per frame, based on accommodating medium to heavy traffic loads, is 8 per carrier. Since there are 24 carriers per beam and 4 beams, the number of stations that can be served by the system is:

NUMBER OF STATIONS = 8x24x4 = 768.

The number of 32 Kbit/s channels that can be carried in the frame is obtained by subtracting the capacity needed to carry the preambles of the traffic bursts (in terms of symbols) from the frame length. Using the burst structure given in Figure 8, the number of channels per frame is:

UPLINK CHAN = (241664 - 8x160)/344 = 698 chan/carrier

and the total for all 96 uplink carriers is:

TOTAL UPLINK CHAN = 96x698 = 67008.

The number of channels is doubled to 134016 by using digital speech interpolation with a channel multiplication ratio of 2:1. Consequently, the average number of channels per station is:

CHAN/STATION = (67008x2)/768 = 174.

Downlink Calculation:

On the downlink, only one burst will be sent to each dwell; and all stations in the dwell will find their traffic in this burst. Consequently, there will be less capacity lost to preambles; and the downlink capacity will be inherently greater. However, this will be of little or no value from the point of view of the traffic capacity; since the lesser uplink capacity will determine the limit. Nonetheless, it is still interesting to determine the downlink capacity. Following the same procedure used for the uplink, the parameters for the downlink are determined below.

The number of symbols per frame for each downlink 120.832 Mbit/s burst rate (corresponding to a symbol rate of 60.42 Msym/s) TDMA frame (16 ms) is:

DOWNLINK SYM/FRAME = 16x60420 = 966656 symbols.

The number of stations per frame is the same as on the uplink. Thus,

NUMBER OF STATIONS = 8x96 = 768.

Using the burst structure given in Figure 8, and assuming that each downlink beam--and hence each transponder--visits 10 downlink dwells, each downlink frame contains 10 downlink traffic bursts--and hence 10 preambles. (Note that each preamble also serves as a reference burst.) Hence, the number of channels per downlink frame is:

DOWNLINK CHAN = (966656-10x160)/344 = 2805 chan/carrier

and the total for all 96 downlink carriers is:

TOTAL DOWNLINK CHAN = 24x2805 = 67320 chan/system.

This is greater than the uplink capacity by a few hundred channels. However, as noted earlier, the lower capacity of the uplink prevails.

5.2. Low Intensity Traffic Configuration

In this configuration, three of the beams use the medium/high load intensity carrier configuration described above; while the fourth uses a low traffic intensity configuration. Thus, the three medium/high intensity beams serve the 40 beam dwells with a frequency of 13 or 14 dwell visits per beam; while the fourth low intensity beam uses a different arrangement. In the low intensity load configuration, in one uplink beam, each of the six 80 MHz transponder bands will carry four 3.072 Mbit/s burst rate carriers for the low intensity stations, using the frequency plan shown in Figure 6. This yields 24 3.072 Mbit/s uplink carriers. These low burst rate carriers permit the use of small earth stations of the E2 type described earlier. The downlink beam will carry six 12.288 Mbit/s carriers, one in each 80 MHz transponder band. Because of the lower bit rate, the downlink transponder HPAs can be reduced to 2.5w. Since there is only one up- and downbeam carrying the service, it will have to hop in all 40 beam dwells.

The up- and downlink channel capacities for the low traffic intensity configuration are now calculated using the burst structure shown in Figure 7(b).

Uplink Calculation:

The number of symbols per frame for each uplink 3.072 Mbit/s (symbol rate of 1.536 Msym/s) burst rate TDMA frame (16 ms) is:

UPLINK SYM/FRAME = 16x1536 = 24576 symbols.

Because the system is intended for light Erlang loads which involve fractional circuit usage, a single channel per burst concept is used to carry the traffic. In order to make most effective use of the capacity of the assigned space segment, the voice channels use 16 kbit/s source coding. The traffic field is assumed to be R3/4 FEC coded. Consequently, assuming a 16 ms TDMA frame, each burst has the structure shown in Figure 9.

Using the burst structure given in Figure 9, the number of channels per frame is:

UPLINK CHAN = 24576/238= 103 chan/carrier

and the total for all 24 uplink carriers is:

TOTAL UPLINK CHAN = 24x103 = 2472 chan.

Each of the 40 dwells will serve an average of 62 channels.

The above number of channels is available to all users on a demand assigned basis. For small Erlang loads, the low usage per user permits many users to share each channel. This results in channel multiplication. When serving users with loads between 0.1 and 10 Erlangs, the number of users that can be served is significantly greater. The multiplication factor, as a function of the average Erlang load per station, is shown in Figure 10. For 0.1 Erlang load users, the multiplication factor is 19; for 1 Erlang it is 4.5; and for 10 Erlangs it is 1.8. Thus, the number of channels that can be served is 46968 for 0.1 Erlang users, 11124 for 1 Erlang users, and 4449 for 10 Erlang users.

Downlink Calculation:

On the downlink, only one burst will be sent to each dwell; and all stations in the dwell will find their traffic in this burst. As was the case for the medium/heavy load stations, there will be less capacity lost to preambles; and the downlink capacity will be inherently greater. However, this will be of little or no value from the point of view of the traffic capacity, since the lesser uplink capacity will determine the limit.

The number of symbols per frame for each downlink 12.288 Mbit/s burst rate (corresponding to a symbol rate of 6.144 Msym/s) TDMA frame (16 ms) is:

DOWNLINK SYM/FRAME = 16x6144 = 98304 symbols.

Using the downlink burst structure given in Figure 9, and assuming that the transponder visits 40 downlink dwells and each downlink frame contains 40 traffic bursts--hence 40 preambles--the number of channels per downlink frame is:

DOWNLINK CHAN = (98304-10x60)/178= 548 chan/carrier

Note that each preamble also serves as a reference burst. Consequently, the total for all 6 uplink carriers is:

TOTAL DOWNLINK CHAN = 6x548 = 3288 chan/system.

This is greater than the uplink capacity by a several hundred channels. However, as noted earlier, the lower capacity prevails. The demand assignment multiplication advantage previously described applies equally to the downlinks.

6. ECONOMIC ANALYSIS

Annual revenue requirements are calculated in the following, assuming that the annual revenue recovery (designated as ARR) must be such that an annual percentage, equal to a difference between the interest on money and a rate of inflation of 3 percent, is recovered on a debt equal to the initial capital invested, and that an additional 15 percent is needed to cover annual operations and maintenance costs. The result is that an amount equal to the initial capital investment must be recovered every 4 years.

6.1. Medium/High Load Revenue Requirement

Consider first the space segment ARR (SSARR). The satellite delivered to geostationary orbit costs $126M, consequently the SSARR is $31.5M. It is more interesting to determine the SSARR per circuit. The number of channels that can be derived from the space segment was previously determined to be 67008. Use of digital speech interpolation with a channel multiplication ratio of 2 doubles the number of channels. The number of circuits is half the number of channels. It is further assumed that the satellite circuit load averages 50%. Thus:

$$\text{SSARR/CIRCUIT} = 2 \times 2 \times 31.5 \times 10^6 / (2 \times 67008) = \$940.19$$

Next consider the earth segment ARR (ESARR). From the previous discussion, the fixed cost of the earth segment is $F = \$185000$, and the variable cost is $C_o = \$36000$ for each 32 Kbit/s DSI interface unit serving 60 circuit ports. Assuming a 60 circuit earth terminal, its total cost is $221000 resulting in a $55250 ESARR. The annual revenue requirement per circuit is thus:

$$\text{ESARR/CIRCUIT} = 55250/60 = \$920.83.$$

The total ARR for the combination of space and earth segments for the medium/heavy load service is thus:

$$\text{TOTAL MEDIUM/HEAVY LOAD SERVICE ARR} = \$1861.02/\text{YEAR}$$

Assuming that each of the 60 circuits logs 480 call-minutes of service per day every day of the year, the total RR per minute is $0.0106.

6.2 Light Load Revenue Requirement

The space segment for the light load service constitutes one fourth of the cost of that for the heavy/medium load service less 10% because it would use lower power, lighter weight HPAs. Consequently, the light load service space segment cost is $28.35M. Since the system is operated as a centrally controlled DAMA, an additional $5M is added to the cost for a central control facility, thus bringing the total cost to $33.35M. This results in a SSARR of $8.35M. The number of 16 kbit/s channels that can be derived from the space segment was previously determined to be 2472. The space segment constitutes a pool of channels shared on a demand assigned basis among all of the earth stations. Conservatively, it is assumed that the pool is utilized by the earth stations with 50% traffic load efficiency. Also the number of circuits is half the number of channels. Thus:

$$\text{SSARR/CIRCUIT} = 2 \times 2 \times 8.35 \times 10^6 / (2472) = \$13491.10/\text{YEAR}$$

Assuming that the space segment circuits log 480 minutes of service per circuit for every day of the year, the SSARR per call-minute is:

$$\text{SSARR/CALL-MINUTE} = 13491/(365 \times 480) = \$0.077/\text{minute}$$

Next consider the light load service earth segment ARR (ESARR). From the previous discussion, the fixed cost of the earth segment is $F = \$10000$ and the variable cost is $C_o = \$400$ for each 16 Kbit/s circuit port. The number of circuit ports, number of stations served, cost and ESARR are functions of the Erlang load as given below:

TABLE 6
LIGHT LOAD INTENSITY STATION CHARACTERICTICS
AS A FUNCTION OF LOAD INTENSITY

STATION LOAD ERLANGS	NUMBER OF STATIONS	NUMBER OF PORTS PER STATION	STATION COST $	ESARR $
0.1	11700	2	10800.00	2700.00
1.0	1112	5	12000.00	3000.00
10.0	123	18	17200.00	4300.00

The resulting ESARR per minute values for the various Erlang loads, assuming that the circuits are used every day of the year for 8 hours per day are:

    0.1 Erlangs,   ESARR/min = 2700/365x48   = $0.154

    1.0 Erlangs,   ESARR/min = 3000/365x480   = $0.0171

    10.0 Erlangs,   ESARR/min = 4300/365x4800   = $0.0024

6.3 . Fiber Optic Transatlantic Cable Requirement

In 1988, a fiber optic cable will be placed in service across the Atlantic Ocean. It will carry 7500 64-kbit/s digital circuits on several digital carriers. By the use of digital channel multiplication equipment (CME) combining speech interpolation and 32-kbit/s low-rate encoding, it will be able to achieve a channel multiplication ratio of 5, thus yielding a total of 37,500 circuits. The capital investment is reported to be $335M for the cable and its beach heads and an additional $35M for its 70,000 digital CME circuit terminations.

The annual revenue requirement per circuit, using the same 4-year recovery period used in the satellite calculations and assuming 50% traffic fill, yields an ARR of $2392.86 compared with $1861.02 obtained for the medium/heavy load satellite system described in this paper.

7. CONCLUSIONS

Using an advanced satellite design, which incorporates onboard demultiplexing and demodulation, baseband switching, hopping spot beams, and low bit rate multiple carrier TDMA uplinks with stations designed to carry medium to heavy intensity loads ranging from 10 to 100 Erlangs, the satellite circuits exhibit an annual revenue requirement that is quite low and compares favorably with that required for the TAT 8 cable. The fact that the satellite circuits also terminate their circuits near the origins and destinations of traffic, possibly at the premesis of the customer, thus avoiding intermediate terrestrial links and CTs, further enhances their desirability.

The discussion presented in the paper also introduces a light load satellite system with full mesh interconnectivity, using 1.8 m VSAT class terminals intended for roof-top-to-roof-top national services among users, who cannot or do not want to be interconnected by other means. The service is examined for station loads ranging from 0.1 to 10 Erlangs. Some space segment traffic efficiency is sacrificed to use the small, low- cost stations; but this can be compensated for by increasing the number of onboard HPAs and demodulators over that assumed in the baseline spacecraft. The cost for 0.1 ERLANG station users is $ 0.231/minute, and becomes lower for the 1 and 10 Erlang users: $0.094/minute and $0.0795/minute, respectively.

REFERENCES

[1] Bargellini, P.L., The INTELSAT IV Communications System, in: COMSAT Technical Review, Vol. 2, No. 2, (Fall, 1972) pp. 437-572.
[2] Paul, D.K. et al., Undersea Fiber-Optic Cable Communications System of the Future, in: IEEE Journal of Lightwave Technology, Vol. 2, No. 4, (August 1984) pp. 414-425.

[3] Campanella, S.J. and Harrington, J.V., Satellite Communications Network, in: Proceedings of the IEEE, Vol. 72, No. 11, (November 1984) pp. 1506-1519.
[4] CCIR Draft New Recommendation (Working Group 4-A), Document 4/419-E, (October 14, 1985).
[5] Fong, F.K., et al., Space Division Unmanned Spacecraft Cost Model, 5th ed., in: Space Division/ACC Report on SD-TR-81-45, (June 1981).

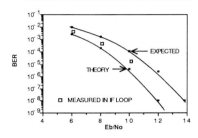

FIGURE 1. QPSK MODEM PERFORMANCE

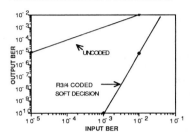

FIGURE 2. RATE 3/4 SOFT DECISION FEC

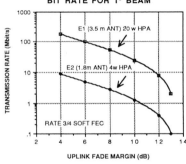

FIGURE 3. UPLINK TRANSMIT BIT RATE FOR 1° BEAM

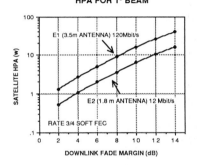

FIGURE 4. DOWNLINK SATELLITE HPA FOR 1° BEAM

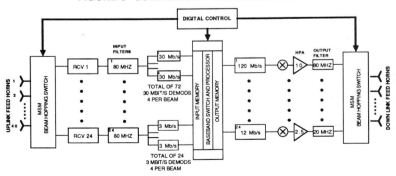

FIGURE 5 COMMUNICATIONS SUBSYSTEM

FIGURE 6. FREQUENCY PLANS FOR MEDIUM/HIGH AND LIGHT INTENSITY ROUTES

FIGURE 7. TRAFFIC BURST STRUCTURES

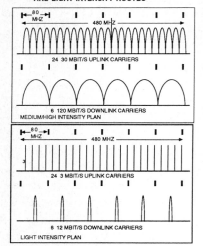

a) HIGH INTENSITY SERVICE

MODULATION = QPSK
FRAME PERIOD = 16 ms
BURST RATE = 30.208 Mbit/s UP and 120.832 Mbit/s DOWN
SYMBOLS/FRAME = 241664 UP and 966656 DOWN
TRAFFIC CHAN (32Kbit/s R3/4 CODED) = (32/2)x16x(4/3)
= 344 SYM (MOD 8)

GUARD	CBR	UW	CTRL	TRAFFIC
8	128	12	12	344xN

← 160 SYM →

SAME FOR UP AND DOWN

b) LOW INTENSITY SERVICE

MODULATION = QPSK
FRAME PERIOD = 16 ms
BURST RATE = 3.072 Mbit/s UP and 12.288 Mbit/s DOWN
SYMBOLS/FRAME = 24576 UP and 98304 DOWN
TRAFFIC CHAN (16 Kbit/s R3/4 CODED) = (16/2)x16x(4/3)
=176 SYM (MOD 8)

GUARD	CBR	UW	CTRL	TRAFFIC
4	32	12	12	176 (SCPB)

← 60 SYM →

SAME FOR UP AND DOWN

FIGURE 8. DAMA CIRCUIT MULTIPLICATION RATIO

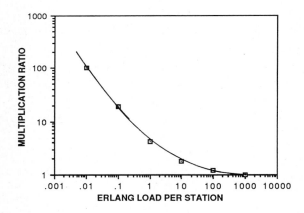

INTEGRATION OF A PAYLOAD ENHANCED NETWORKING SATELLITE (PENSAT) AND TERRESTRIAL BROADBAND ISDN

G. Pennoni & L. Bella,
European Space Agency, ESTEC Noordwijk, The Netherlands.

1. INTRODUCTION

In the field of telecommunications, the satellite currently plays a primary role in providing long distance links between networks at regional and intercontinental level. However, the dramatic reduction of the cost per kilometer per channel now offered by optic fibres will destroy, or at least substantially reduce, the role of the satellite as a "cable in the sky", since a comparable reduction in satellite transmission cost is not foreseeable. To revitalise the appeal of satellite communication, therefore, the European Space Agency has been developing the basic elements of a "switchboard in the sky" system.

2. CABLE & SWITCHING

The cable (or any other transmission media) and the switching elements are almost always treated as two seperate entities, requiring different skills and playing a complementary role in the PSTN. A closer look shows, however, that "cable" and "switching" are not just complementary, but exchangeable terms.

To take a simple example: we want to build a telephone network among N users. We have two basic ways of fulfilling our commitment: a "meshed" (Fig. 1a) or a "star" (Fig. 1b) network, in which each line represents two channels (i.e. one voice circuit).

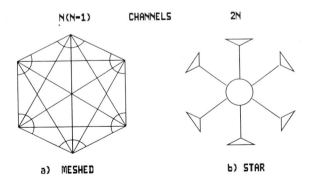

Fig. 1 - Network Architectures

Everybody agrees that the two networks are functionally identical; the first is built only with cables and a cable selector per user, and the second with cables and a centralized switching element.

It is also apparent that in case (a) N(N-1) cables (i.e. channels) are necessary, but only N are used simultaneously, while all 2N cables and the switching elements are permanently busy (i.e. utilized) in case (b). Hence, the switching element can replace (N^2 -3N) cables, and in this sense they are functionally equivalent. Despite the impressive evolution of transmission and multiplexing techniques (from a single voice channel on a copper wire to many thousands on a single fibre optic) the European telephony network of type (a) with about 120 million subscribers would have required about 144×10^{14} channels, with an average length of many thousand kilometers!!

This macro economical example demonstrates that it is primarily switching (first manual, then electromechanical and now digital) that has made the worldwide PSTN as we know it today feasible.

In practice, today's national and international network is organized in a rather complex hierachical structure of both star and meshed elements linked by exchanges.

3. SWITCHING SATELLITE

The satellite community recognized early on that spacecraft in geostationary orbit were ideally placed for real-time demand assignment traffic routing between Earth-based users. The SPADE system introduced by INTELSAT was performing real-time switching in the frequency domain as early as 1972.

A real-time switching function can also be implemented in time and space domains. NASA, to our knowledge, was the first to propose a satellite (ACTS) equipped with a baseband on-board routing processor using the well-known Time-Space-Time switching structure of ground-based digital exchanges.

ESA, in conjunction with the CSELT telecommunications research laboratory, has developed the baseband prototype of an advanced satellite system [1], [2], [3], [4] called TST/SS-TDMA. The main attraction of this system, designed to operate with a spot-beam coverage scenario, lies in its ability to interconnect large/medium Earth stations, small business users and even mobile terminals, as described in another paper in this Workshop [5]. Traffic routing is performed by a switchboard placed on-board the satellite and controlled in real-time by the ground Master Control Centre. A feature of this system is that the "call" can have a capacity that is any multiple of 64 kbps, so that not only telephony but also n x 64 kbps services, 2.048 Mbps videoconferences or higher bandwidth requirements can easily be accommodated.

While developing this "magic box", which is designed to comply with CCITT ISDN (narrowband) recommendations and hopefully also with the future broadband

ISDN requirements, ESA has recognised that the system's novel features make the usual means of comparing "cables in the sky" with cable on ground on the basis of distance break-even point obsolete. A study contract has in fact been placed with the ESCO [6] (a consortium of European PTT's) to assess the optimal integration of this system with the European ground ISDN. This study has already highlighted a number of potential advantages of using satellites of this type integrated with the ground network:
- higher ERLANG/CIRCUIT efficiency than equivalent ground network
- ability to cope with traffic fluctuations (holidays, etc.)
- ready to carry wideband traffic
- remote/low traffic area coverage
- easy point to multipoint connections
- optimal solution for carrying traffic generated by special events (Olympic Games, exhibitions, natural disasters)
- spare capacity in orbit for recovering faults on the ground or delaying investment
- diversity/service protection

The study has yet to produce quantitative results, but some qualitative assessments are discusssed in a companion paper presented at this Workshop [7]. A comprehensive detailed model of the European network, one level below the International Centres, is under development and various network alternatives with/without satellite will be traded-off.

In the following we will present some results of a similar exercise performed in house on a simplified national network model in order to gain better insight into the potential advantage of integrating the satellite system with the ground network.

4. GROUND NETWORK

A network typical for a large European country around the year 1995 is sketched in Figure 2, where the District (DC), Compartment (CC) and International (IC)' layers are fully meshed. The IC layer is also fully meshed with the IC layers of the other European countries We will analyse the national traffic that is offered to the DC layer by the local exchanges. The routing rules are shown in Figure 3a where numbers 1 - 4 are the routing priority choices and the traffic distribution among the hierarchical choices is show in Figure 3b.

In our model (which approaches the Italian network in 1995 [8]) we assume total interdistrict traffic in the busiest hour of 200,000 Erlangs; 50% is closed inside 10 large District Centres (LDC's), 25% is closed inside small District Centres (SDC's) and 25% is exchanged between LDC's and SDC's.

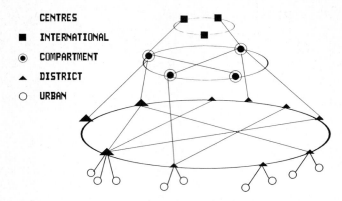

Fig. 2 - National Ground Network (1995)

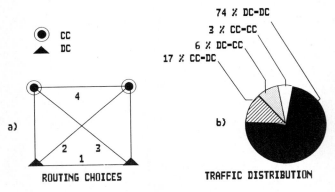

Fig. 3 - Ground Network Routing

The traffic distribution for this simplified but still rather representative situation (for 0.001 blocking probability) is listed in Table 1 while Table 2 gives the main characteristics of the digital exchanges.

TABLE 1 - TRAFFIC DISTRIBUTION IN THE GROUND NETWORK

Link	No Links Bi-Directional	ERLANG Per Link (Offered)	ERLANG Per Link (Carried)	CIRCUITS Per Link	CIRCUITS Total	E/C
LDC - LDC	45	2222	1644	1665	74925	0.99
SDC - SDC	4005	12.4	9.2	12	48060	0.76
LDC - SDC	900	55.5	41.2	47	42300	0.87
LDC - CC	100	725	644	652	62500	0.98
SDC - CC	900	51.7	45.9	52	46800	0.88
CC - CC	45	135	131	144	6480	0.91
					281065	

TABLE 2 - DIGITAL EXCHANGE CHARACTERISTICS

EXCHANGE	No	ERLANGS		TERMINATIONS	
		Per Exch	Total	Per Exch	Total
LDC	10	25000	250000	51470	514700
SDC	90	1666	150000	4116	370440
CC	10	5875	58750	12496	124960
			463940		1010100

5. INTEGRATED NETWORK

A possible integrated network (not necessarily optimized) is shown in Figure 4. In this model all the traffic (100,000Erl) among LDC's is routed via ground as before, but the traffic among SDC's (50,000Erl) and between LDC's and SDC's (50,000Erl) is routed via a cluster of four satellites (plus one spare).

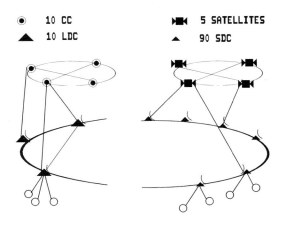

Fig. 4 - Integrated Network

The traffic distribution for this structure (for 0.001 blocking probability) is listed in Table 3. Table 4 gives the main characteristics of the ground digital exchanges and those of the satellite network that include large and small earth stations (LES, SES), Master Control Centre (MCT) and satellites.

TABLE 3 - TRAFFIC DISTRIBUTION IN THE INTEGRATED NETWORK

Link	No Links Bi-Directional	ERLANGS Per Link (Offered)	ERLANGS Per Link (Carried)	CIRCUITS Per Link	CIRCUITS Total	E/C
LDC/LDC	45	2222	1644	1665	74925	0.99
LDC/CC	100	580	514	534	53400	0.96
CC/CC	45	67	65	76	3420	0.85
					131745	
LDC/SAT	10	5000	4995	5096	50960	0.98
SDC/SAT	90	1666	1665	1772	159480	0.94

TABLE 4 - INTEGRATED NETWORK CHARACTERISTICS
Ground network

EXCHANGE	No	ERLANGS Per Exch	ERLANGS Total	TERMINATIONS Per Exch	TERMINATIONS Total
LDC	10	20000	200000	41570	415700
CC	10	2937	29370	5994	59940
			229370		475640

Satellite network

EARTH STATION	No	HALF CIRCUITS Per Station	HALF CIRCUITS Total	ERLANGS
LES	10	5096	50960	--
SES	90	1772	159480	--
MCT	1	--	--	100.000

	No	CHANNELS	TRANSPONDERS
SATELLITE	4+(1)	210440	120

6. COST COMPARISON

Table 5 provides a cost estimate of the two networks based on parameters taken from [6] for the ground network, while for the satellite the figures provided in [9] have been extrapolated to account for ISL and 64 kbps voice channel (i.e. ADPCM and DSI are not considered to increase the bandwidth efficiency).

The costs per circuit/termination are based on the assumption of an average circuit length of 366 km [8] and 140 Mbps data rate, with a fill factor of 26/30.

TABLE 5 - COST ESTIMATE

Pure ground network

COST UNIT	No.	COST/UNIT (K ECU)	TOTAL (K ECU)
Circuit Transmission	281065	4.6	1292899
Multiplexing/Circuit	281065	0.2	56213
ERLANGS Switched	458750	1.5	688125
Circuit Termination	1010100	0.384	387878
			2425115

Integrated network

Ground Network

COST UNIT	No.	COST/UNIT (K ECU)	TOTAL (K ECU)
Circuit Transmission	131745	4.6	606027
Multiplexing/Circuit	131745	0.2	26349
ERLANGS Switched	229370	1.5	344055
Circuit Termination	475640	0.384	182645
			1159076

Satellite Network

COST UNIT	No.	COST/UNIT (K ECU)	TOTAL (K ECU)
ESCD	90	1772	159480
ELCD	10	5096	50960
MCT (100 RE)	1	150000	150000
Satellite	4+1	150000	750000
			1110440
		TOTAL :	2269516

Since the total cost of the integrated solution is 2,425,115 K ECU against 2,269,516 K ECU for the pure ground network, we are tempted to conclude that the integration is advantageous considering that, with a cheaper network cost (that can be furthermore easily dropped using ADPCM and DSI techniques), we can also achieve the inherent benefits listed in Section 3. However, this rather approximate cost exercise is neither intended to look for a threshold of convenience between satellite and ground network, nor to come to any optimised solution, but rather to learn how to take advantage of the switchboard in sky concept.

7. LEARNING

The switchboard-in-the-sky concept increases the satellite circuit efficiency by reducing the per Erlang cost, and compares favourably with ground networks in the case of relatively high numbers of earth stations (100).

It compares better if we take it one step higher in the hierarchical ground network since, at European level, the traffic volume and the network architecture are similar but the distances are far bigger.

The highest immediate benefit from this system concept will probably be achieved by going one step further down in the network hierarchy, to the local exchange, or even to the user's premises. In fact when increasing the number of (smaller) earth stations, since for a given total amount of traffic the station to station link capacity decreases with the square of the number of stations further switching and transmission costs can be saved.

This perspective is really attractive because telematic traffic, estimated to be no more than 20-30% of the total traffic, can be handled by a satellite system that can offer to a widespread business community access to ISDN and B-ISDN. This in turn implies that the on-going replacement of non-ISDN ground exchanges could be delayed, since the ISDN need for residential users is still to be demonstrated.

8. RECOMMENDATIONS - PENSAT

The Agency has developed a prototype switchboard-in-the-sky concept and is in the process of finalizing with EUTELSAT and some European PTT's, a double hop experiment to validate the system, by using an EUTELSAT satellite and placing the "on-board" routing processor on ground.

A number of studies are also now in progress to provide the Agency with the insight necessary to lead a more ambitious development phase. The Agency is currently engaged in finalising the Payload Enhanced Networking Satellite (PENSAT) Programme with an estimated budget envelope of 50 M ECU devoted to developing, by 1992, the engineering models of all of the system elements.

The European network operators are invited to start considering the possibility for integrating PENSAT into their networks.

REFERENCES

[1] Alaria G.B., Pennoni G., "An SS-TDMA satellite system incorporating anon-board Time/Space/Time switching facility; Overall system characteristics and equipment description", Proc. ICC 84 Amsterdam, The Netherlands, 14 - 17 May 1984.

[2] Pennoni G., "A TST/SS-TDMA Telecommunication System: From Cable to Switchboard in the Sky", ESA journal 1984 Vol.8, No 2.

[3]   Alaria G.B., Destefanis P., Guaschino G., Pattini., Porzio Giusto P., Pennoni G., "On board processor for a TST/SS-TDMA Telecommunication System", ESA journal 1985 Vol. 9 No 1.

[4]   Colombo G., Pennoni G., "Advanced Onboard Processing for User Oriented Communication Systems", Proc. ICDSC - 7 München 12 - 16 May 1986.

[5]   Bella L., "Satellite Switching for mobile communications: new problems and perspectives", this Workshop.

[6]   ESCO, "Integration of Advanced Communications Satellite Systems and Terrestrial Networks" ESTEC/Contract No. 6781/86/NL/DG.

[7]   Bartholomé P., Casas J., "An integrated space/terrestrial network for Europe", this Workshop.

[8]   Zanetti L., de Padova S., Giacobbo S., Rossi S., Trimarco U., "An economical analysis on the introduction of a domestic satellite system in a developed national telecommunication network", Proc. ICDSC - 7 München 12 - 16 May 1986.

[9]   Campanella S.J., Pontano B.A., Kao T., "Economics of Multiple Beam Satellites", Proc. ICDSC - 7 München 12 - 16 May 1986.

SATELLITE SWITCHES

S. JOSEPH CAMPANELLA*

Satellite communications systems are far more than pipelines in the sky. From their advantageous location, 22,300 miles above the earth's equator, they view one third of the earth's surface in a single glance. This creates the opportunity to switch space segment circuits to connect any pair of users within view no matter where they are--on the ocean, in the sky, in a desert or in the middle of a large city. Satellites must exploit this unique property to reach their full potential.

1. INTRODUCTION

Satellite systems are switching systems with a unique capability which is their own and is not shared by terrestrial communications systems no matter how huge their capacity is. That capability is summed up in one term-- multiple access--. Because a satellite enjoys a perch far above the earth at an altitude of 22,300 miles, it can view at a single glance one third of the surface of the earth and consequently make and break contact between any number correspondents residing in the coverage. This is done by the relatively simple expedient of sending a radio signal to the satellite from where the signal is repeated back to earth for reception by anyone else. There is no need to dig a ditch, string a line on poles or lay an undersea cable for the correspondence to take place. This is the real power of satellites---the ability to make and break contact with anyone on earth without the need to find a cable path over which to establish the message route.

2. CIRCUIT MULTIPLICATION BY DAMA

When subscribers present light traffic loads --between 0.1 and 10 Erlangs-- to a system, the number of conventional preassigned trunks needed to carry the load with a specified grade of service, expressed in terms of probability of busy out, significantly exceeds the Erlang load. For example, to carry a load of 1 Erlang with a grade of service of 1% requires 5 preassigned trunks. This is a ratio of circuits to Erlangs of 5:1. However, if the load of many such light load users is aggregated into a single pool of a hundred or more Erlangs then ratio of circuits to Erlangs closely approaches 1:1 which is far more efficient. This property is used in local telephone switches that serve a community of subscribers each of whom presents a light load. The aggregate load of hundreds of such subscribers, which may amount to several tens of Erlangs, can be concentrated on a relative few trunks.

The same property can be realized for a community of users distributed over the surface of the earth who share a single pool of satellite circuits. When a user wishes to make a call, he siezes a circuit from the satellite pool, uses it for the duration of the call and on completion of the call returns it to the pool for use by another. By this means the number of users can far exceed the number of satellite circuits when the Erlang load presented be each is light. Figure 1 shows the ratio of circuit ports served in an entire system to the number of circuits in the pool as function of Erlang load. This is called the circuit multiplication ratio. For a group of users, each presenting average load of 1 Erlang, the ratio is 4.5:1. If the average Erlang load is less, the ratio increases, and if greater, it is less. Thus, in a system providing a pool of 400 satellite circuits, the number of circuits served for an average Erlang load of 1 Erlang is 1800 and 400 users (earth terminals) can share the system.

This multiplication is realized for any distribution of the users in the satellite coverage in a satellite's beam by switching the space segment circuits to individual users on demand. This is a property that cannot be duplicated on such a wide coverage scale by terrestrial facilities be they microwave or fiber cable. In the opinion of the author, it is one of the two principle advantages of satellite systems over terrestrial systems. The second is the great reliability of satellite systems once installed and their attending freedom from the outages caused by man and nature.

Two methods of implementing the control system for performing the switching were identified:

---

*The author is with COMSAT LABS, Clarksburg, Maryland 20871, USA.

FIGURE 1. DAMA CIRCUIT MULTIPLICATION RATIO

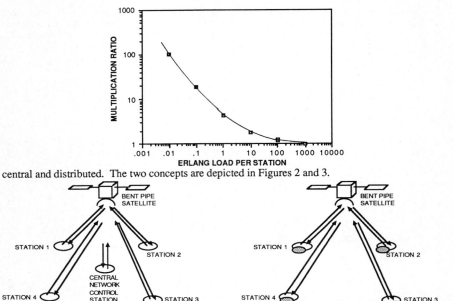

central and distributed. The two concepts are depicted in Figures 2 and 3.

FIGURE 2. CENTRAL NETWORK CONTROL

FIGURE 3. DISTRIBUTED NETWORK CONTROL

3.1. Central Controlled DAMA

In the central control version of the DAMA system, illustrated in Figure 2, a network controller is located at one station and all other stations send their requests for connection or disconnection to the central station. The central network controller keeps a map of all of the current connections and can assign channels for carrying a new circuit without conflict. It sends the choice to both the originating and destination stations over a return control link and the circuit connection is completed. The central control method has the advantages of avoiding assignment conflicts and needing only one network control facility--and perhaps a second one for redundancy-- but the disadvantage of adding double hop propagation delay to the call setup time.

3.2. Distributed Controlled DAMA

The distributed control method, shown in Figure 3, requires that each station directly seize a circuit from the pool. To do this without seizing a circuit being used by some other station, each station must be equipped with a map of the current connections through out the system. This is equivalent to having a central network control facility at each station. To place a call, a station siezes a circuit from the pool of unused circuits by referring to its map. Every circuit comprises a pair of channel frequencies which are automatically assigned upon seizure. The originating station sends a message to the destination station informing it of the choice. Both stations then initiate transmission on the assigned frequencies. Reception of each station's transmission by the other station constitutes acknowledgement. The call digits are forwarded over the established link to complete the signaling. When the call is completed by detection of the on hook condition at either station the circuit is cleared by sending a disconnect message.

Because there is a one way propagation delay of approximately 0.27s between the time that a station executes a seizure and the time it is entered into the memory maps of all other stations, there is a small probability that two stations will attempt to seize the same circuit. This can be resolved by the expedient of a station paying attention only to the first seizure message received. The second reception is ignored, consequently no acknowledgement is made and the second attempt is aborted.

The advantage of the distributed network control method are reduced time delay in call connect and disconnect. The disadvantages are: creation of more opportunities for connect and disconnect errors due error in updating the assignment maps; more complicated protocols and the cost of a network controller at every station in the system.

4. TDMA DAMA

With TDMA, demand assignment is performed in terms of channels that are defined as time slots in the traffic bursts of a TDMA frame rather than frequency assignments of SCPC carriers. The protocols associated with circuit connect and disconnect are similar and can be controlled either by central or distributed network controllers.

TDMA presents a wide variety of alternatives with regard to implementing the DAMA channels as illustrated in Figure 4. To assure that bursts are maintained at their minimum length it is

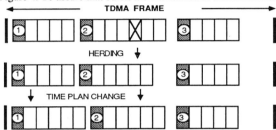

FIGURE 4. HERDING AND TIME PLAN CHANGE

necessary to keep all active channels grouped at the beginning of the burst. This is called herding, and is implemented by always transferring the last channel in a frame to the time slot vacated by a channel that has just been disconnected. Control of a TDMA DAMA system of the type described above involves the reassignment of channels to slots within traffic bursts and the reassignment of the traffic burst locations within the frame. This is a job that requires the accumulation of the demands of all users at a single location and is best accomplished by a central network controller.

TDMA systems are critically dependent on maintaining traffic burst positions in the TDMA frame. To do this, burst position and channel assignment information is presented to all users in a Burst Time Plan which is broadcasted to all stations from the network control center. Stations cannot arbitrarily change their traffic burst positions and channel assignments within the burst without coordinating the change with corresponding changes that must be adopted by all other stations. Otherwise, the bursts would accidentally overlap and no one would know where to find their channels in a burst. To accomplish this, a burst time plan coordination change procedure is adopted though out the network. This is done by establishing a time mark that is recognized by all stations. Each station adjusts its instant of change by taking into account its distance from the satellite. At the satellite this causes all stations to change to the new time plan on the very same TDMA frame. This is called the BURST TIME PLAN COORDINATION PROCEDURE and is an essential part of the protocol to accomplish switched service in a satellite TDMA system.

	CHANNEL A		CHANNEL B		CHANNEL C	
	DEST ADDR	TRAFFIC	DEST ADDR	TRAFFIC	DEST ADDR	TRAFFIC
	32	480				
	← 512 →					

FIGURE 5. DESTINATION DIRECTED CHANNELS IN A TDMA BURST

A novel variation on the method of establishing the channels within a burst was introduced in Satellite Business System's (SBS) TDMA system. The method used a destination directed packet concept illustrated in Figure 5. In the SBS TDMA system each channel packet consists of a 16 bit

R1/2 coded destination address header carried in 16 symbols of a QPSK modulated carrier followed by 480 bits of 32 Kbit/s channel data carried in 240 symbols of the QPSK modulated carrier-- each frame is 16ms long. Each channel packet has a traffic carrying efficiency of 480/512 = 94%. The 16 bit address can steer the channel packet to any of 2^{16}= 65536 ports at any terminal in the system. As a result, the position of a burst in the frame is inconsequential and a station can assign its burst any place in the frame without informing a central network controller. This greatly simplifies the task of the network controller. All the network controller needs to do is assign to each burst its position and length in the frame. To do this it simply needs to be informed of the number of channels each station is currently carrying and assign to each sufficient capacity to accommodate its current traffic plus a small allowance to accommodate one or two new calls. It must administer burst time plan changes on a frequent schedule to keep up with the traffic changes.

5. PACKET DAMA

Packet satellite systems have become very popular in recent years as a means for transmitting computer data using Ku band transponders with E.I.R.P. above 38 dBw. The most prevalent form of the system is the node-hub configuration in which the hub station has a relatively large aperture (5.5m) antenna driven by a 50 w HPA and the node stations are small aperture (1.8/2.5m) antennas driven by 1 to 4w HPAs. These are referred to by the acronym "VSAT" for very small aperture terminal. Typically node stations cost approximately $10000. These are popular for data communications between many outlying sites that need to communicate only with a central facility, such as automobile dealers, department store and grocery chains, rapid mail delivery services, etc.

These systems use random access TDMA/TDM with multiple carriers and transmit data packets to accomplish connections. A typical arrangement is illustrated in Figure 6. The hub station transmits a high bit rate carrier--for example at 256 Kbit/s-- that is actually in TDM form. This hub carrier is segmented into time slots that carry destination addressed packet messages to the node terminals. This establishes a frame reference for the system by virtue of its timing. The node stations synchronize their transmissions to the receive side timing established by reception of the hub TDM carrier, apply a correction to account for their distance from the satellite and send their packets in a time slot on one of several lower burst rate --for example 56 Kbit/s--carriers.

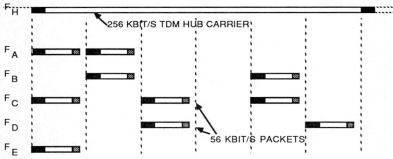

FIGURE 6. RANDOM ACCESS TDMA PACKET SYSTEM

The alignment of the packet transmissions to the low burst rate carrier time slots is performed by cooperative feedback from the hub station via messages directed to the nodes. The node stations do not observe any assignment discipline for access to the node-to-hub carrier time slots. Rather, they randomly send their packets on the assumption that the occupancy is so low that the probability of collision with the packet transmission of another station is very low. The method theoretically has a traffic fill efficiency of 1/e; only half of this value is currently being achieved in actual operation. The data packet transmission is performed in accordance with one of the established packet transmission protocols such as LAP (X.25), SDLC and HDLC. [1]

6. MULTIBEAM INTERCONNECTION

The methods thus far described apply to satellite systems that are fully connected in single beams. However, multibeam operation has been introduced in the INTELSAT IV and even more extensively in INTELSATs V and VI. Multibeam operation has important advantages that become even more attractive in future satellite systems. They are frequency reuse, high uplink G/T and high downlink E.I.R.P. However, multiple beams divide the space segment into separate, disconnected communities of users and new disciplines must be introduced to reestablish connectivity [2].

One method to connect multiple beams is to incorporate static frequency selective cross strapping in which each beam is connected to the others via onboard filters. However, the method requires that the spectrum for each beam be divided into as many segments as there are beams. Thus, frequency cross-strapping works fairly well when a small number of beams are involved--less than 10-- but can fragment the spectrum for systems involving many more beams. Frequency cross strapping also has the disadvantage of inflexibility due to the need to fix the structure of the cross strapping plan before the launch of the satellite and the inability to change it after launch. This impediment could eventually be eliminated by introducing variable bandwidth filters on board.

Another way to achieve the connectivity among multiple beams is by onboard memoryless dynamic switching either at a microwave IF or at digital baseband or switching with memory at digital baseband. These methods are discussed in the following. Because the methods involve time domain switching, they are much more flexible in terms of adjusting the capacity of the routes among the beams and more easily implemented on board the satellite.

The individual beams of narrow beam multiple beam systems may be operated as fixed or hopping. For a fixed beam to be used efficiently, it must be filled with traffic originating and destined to its earth coverage region. If the beam cannot be filled in this way, it can hop to a number of coverage areas to accumulate sufficient traffic to fill it. Thus, it can be visualized that future multiple beam systems may use a combination of fixed and hopping beams. Of course it is obvious that time domain switching is most suitable to systems that incorporate hopping beams.

Methods to accomplish the time domain interconnectivity discussed above are expanded on in the remainder of the paper.

7. MICROWAVE SWITCH MATRIX SS/TDMA

A microwave switch matrix (MSM) operates as space segment switch, for connecting upbeams to downbeams to route traffic bursts transmitted from a community of earth stations to be received by the same community of earth stations. A simplified illustration of a MSM switch

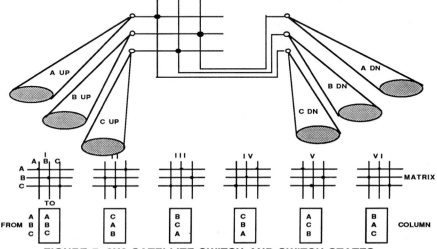

FIGURE 7. 3X3 SATELLITE SWITCH AND SWITCH STATES

interconnecting 3 upbeams and downbeams is illustrated in Figure 7. In this figure, the up and down beams are shown as separate beams but really overlay one another. The beams are interconnected by sets of row to column connections, called switch states, which are maintained for a period of time sufficient to route the traffic assigned to the state. To accommodate all of the traffic presented to a system, a sequence of different switch states occurring in a periodic frame is needed. For N beams, the maximum number of switch states needed in the sequence to accommodate the traffic with a traffic fill efficiency of at least 95% can be shown to be N^2-2N+2.

At the bottom of the figure are shown all possible single point upbeam to downbeam switch states for a 3x3 switch. Switch states are shown in both a matrix and columnar form. The columnar form is best to illustrate sequences of switch states which comprise switch time plans. The first state shown provides the connections A to B, B to C and C to A, the second A to C, B to A and C to B and so on.

7.1. TDMA with Satellite Switching

Satellite switched operation requires a synchronized TDMA ground segment. The combination of a satellite switch and the associated network of TDMA stations is called SS-TDMA. This network is actually a distributed time-space-time switch in which the TDMA transmit and receive buffers form the time domain portion of the switch and the MSM space domain portion. A traffic burst is transmitted from a TDMA earth station residing in a particular beam at a time such that it arrives at the satellite during the switch state interval in which the MSM connects the upbeam to the desired destination downbeam and at a particular time in that interval assigned to its traffic burst. Other stations in the same beam with traffic to the same destination downbeam will transmit at different times in the same switch state. The same traffic station will transmit traffic to a different destination downbeam by transmitting at another time in the frame such that its burst arrives at the satellite during the time interval that a switch state connects the upbeam to the second restination downbeam. This process is repeated for each destination downbeam and for all network stations.

7.2. Satellite Switch Time Plan

A sequence of switch states of various durations and connectivities constitute the satellite switch time plan (SSTP). Algorithms using bin packing theory have been developed to dynamically create optimum switch state and associated TDMA traffic burst time plans for operating demand assigned SS-TDMA systems. Transmission, MSM switching and reception all take place at the TDMA frame periodicity.

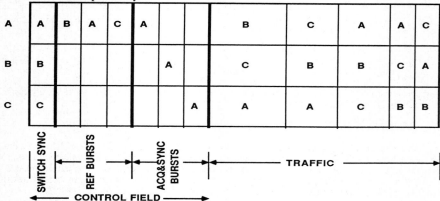

FIGURE 8. SS-TDMA SWITCH STATE SEQUENCE FOR CONTROL AND TRAFFIC

An SSTP for a 3x3 beam system is illustrated in Figure 8. It comprises a synchronization field and a traffic field. The first state of the synchronization field provides loop back connections to the originating beams which are used to establish synchronization between the satellite switch and a station that is designated as the TDMA reference station. The next states in the synchronization field provide for distribution of reference bursts and location of traffic station synchronization bursts.

A subsystem referred to as the acquisition and synchronization unit (ASU) is used to accomplish synchronization between the TDMA reference station and the onboard MSM. It operates by sending a special burst to the satellite that contains a metering segment. This burst is permitted to scan the frame until it is located in the special loop back time slot. This constitutes a coarse location. Once it is so located, the metering segment is aligned with the trailing edge of the loop back time slot such that the metering segment is bisected. This location can be maintained within a fraction of a symbol period of the QPSK modulation.

Having thus synchronized the reference station with the MSM, the reference station can send reference bursts to all of the downbeams by using the reference burst distribution states. Traffic stations receive these reference bursts, use them to establish receive side synchronization and to receive traffic burst time plan and position control information from the reference station.

7.4. Traffic Terminal Operation.

By adding an appropriate delay relative to the instant of reception of the reference burst, the traffic station sends a synchronization burst back to the reference station in a switch state established for this purpose. The reference station observes the time of arrival of this burst and, if it is out of position, it sends a correction back to the station. This process is called cooperative feedback and goes on continuously once a traffic station has accomplished the initial acquisition of synchronization. The initial value of delay used by a traffic station during acquisition is calculated by knowing the approximate distance between the earth station and the satellite.

Once the reference station has been synchronized to the satellite MSM and the traffic stations to the reference station, the reference station can disseminate burst time plans to the stations and carry traffic in the same way as is done for any TDMA system. There is of course the requirement to synchronize changes in the BTPs and SSTPs and this is done by establishing a superframe and synchronizing the changes of both BTPs and SSTPs to this superframe. [3]

7.5. SS/TDMA Traffic Efficiency

Overhead, such as preambles and guard times, associated with each of the traffic bursts, must be accounted for in determining the traffic efficiency of a TDMA system. When the system carries bursts that contain several tens of channels so that the ratio of symbols devoted to overhead to symbols devoted to traffic is small, TDMA is relatively efficient. However, if each burst carries a small number of channels , then the technique becomes less efficient. This is particularly a problem for a network in which the traffic from each station must be split into smaller, separate bursts to be sent to multiple beams. Because of this, SS/TDMA is normally used in networks carrying medium to heavy trunks involving less than 100 terminals.

One way to improve SS/TDMA efficiency for small traffic loads is to use a very long TDMA frame. This lengthens the traffic portion of the burst while the preamble and guard time remain the same, thus improving efficiency. This has been done in the case of the ITALSAT which proposes to use a 32ms TDMA frame period. Even though ITALSAT is baseband switched rather than microwave switched, it still has the same problem. In ITALSAT this expedient makes four channel per burst operation efficient and even permits efficient single channel per burst operation.

8. BBS/TDMA AND HOPPING BEAMS

Narrow beams on satellites provide increased uplink G/T and downlink E.I.R.P. and consequently make it possible to operate with VSAT type earth terminals which achieve full mesh connectivity. However, the narrow beams restrict earth coverage and some means must be introduced to increase the coverage. This can be done by hopping one or more narrow beams among earth coverages, dwelling in each for a time sufficient to collect or deliver the traffic.

Because of the hopping, TDMA is a natural transmission means to carry the traffic between the earth stations and the satellite. The method is referred to as baseband switched TDMA (BBS/TDMA). The beam hopping is controlled by an onboard baseband processor that also permits traffic channels contained in traffic bursts on the uplinks to be disassembled and reassembled into new traffic bursts for transmission on the downlinks. An additional advantage of the method is elimination of the fragmentation of the uplink traffic bursts into smaller less efficient bursts, that resulted for SS/TDMA, to reach destinations in the multiple beams.

A baseband circuit switch for an N beam system is shown in Figure 9. TDMA traffic bursts sent in the hopping upbeams are demodulated and if necessary decoded and the channels loaded into uplink memory locations during the TDMA frame of arrival. There is a separate memory for

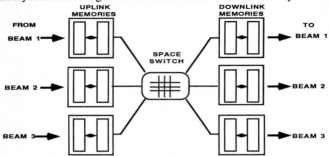

FIGURE 9. ONBOARD BASEBAND PROCESSOR SWITCH

each beam. Each uplink memory actually consists of two sections that alternate their function in a ping-pong manner from frame to frame. The uplink memory section that is loaded in the first frame is connected to the input ports of a NxN space switch and the contents of all uplink memories are loaded into the appropriate downlink memories in the next TDMA frame. Each downlink memory is also of a ping-pong design. On the third frame, the downlink memory sections containing the rearranged channels are played out in the form of downlink transmissions with the channels grouped according to destination beam dwell in each beam. There is a latency of two frames in passing through the switch. Because of the latency, a channel may suffer a transport delay of as little as 1 TDMA frame or as great as 3 TDMA frames.

A network control center and reference station is needed to control the system. The control center controls reference station and traffic station burst acquisition and synchronization with the satellite baseband processor. To do this a method using cooperative feedback directly between earth stations and satellite is employed. Once synchronization is accomplished, the traffic stations place circuit requests over satellite orderwire links to the network control center. The network control center develops traffic burst time plans for all stations of the network and corresponding onboard baseband processor switch and beam hopping instructions. It issues these to the network using a synchronization procedure that assures that the onboard switch routing and beam hopping change occurs at the start of the same TDMA frame in which the traffic burst time plans change.

9. CONCLUSIONS

Satellite systems inherently are distributed switching systems that extend over the entire field of view of the satellite. Judicious use of this property is becoming increasing important in the design of new systems that will be competitive in the face of increasing competition from light guided terrestrial systems. Proper use of switching satellites permits significant multiplication of the space segment capacity for light and medium intensity traffic load users (0.1 to 10 Erlangs). Satellites can efficiently connect such users located in widely distributed regions where terrestrial facilities are too costly or not available at all.

REFERENCES:

1. Tanenbaum, Andrew S., Computer Networks, Prentice Hall, INC., Englewood Cliffs, New Jersey, 1981.
2. Campanella, S.J., Pontano, B.A. and, Dicks, J. L., COMSAT Technical Review, Vol. 16, No. 1, (Spring 1986) pp. 207-237.
3. Campanella, S.J., Colby, R.J., "Network Control for TDMA and SS-TDMA in Multiple-Beam Satellite Systems", Fifth International Conference on Digital Satellite Communications, Genoa, Italy, March 1981, Proc., pp. 335-343.

SATELLITE SWITCHING FOR MOBILE COMMUNICATIONS: NEW ISSUES & PERSPECTIVES

L. BELLA - European Space Agency - ESTEC - Noordwijk - the Netherlands

ABSTRACT

Communication satellites for fixed services were introduced earlier than satellites for mobile services and still retain the major share of the market. As such, they may highlight new issues to be considered for the emerging large mobile market such as interworking with ISDN and adaptability to variable service requirements. This paper describes the basic architecture and techniques of a mobile satellite system able to match these requirements through on-board processing which is compatible technologically with medium term flight goals. Prototype implementation of the main baseband elements has been started and will lead to comprehensive tests and experimentals.

1. INTRODUCTION AND BACKGROUND

The deployment of communication satellites for fixed services in the last twenty years has been a story of successful achievements, as witnessed by everyday practice of intercontinental telephone and TV communications and by worldwide growth of communication satellites organisations, such as INTELSAT. This story with a happy ending has recently been stopped by the tremendous progress of optic fibres, the breakeven distance for which they are cheaper than satellites increasing day by day.

The satellite community, however, is showing a remarkable reaction to their awakening from the beautiful dream of continuous, everlasting growth of communication satellites as "cables in the sky" proving more effective than "cables on ground". Instead of reacting, they have redoubled their efforts taking the viewpoint that if "space cables" alone do not pay enough, "space networks" will pay more [1].

New techniques have been developed for the satellite's new role. High gain spot-beam antennas are accounting for a remarkable reduction in earth-station size and cost, thus penetrating to lower hierarchical levels in the ground network. On-board processing can enhance future satellites by giving them all the capabilities of earth switching, efficiently interconnecting several classes of terminals. Against this background, the ACTS programme has been started in the USA and ESA has developed a prototype of the TST/SS-TDMA system [2].

The short history of fixed satellite services may provide useful insight for the development of future mobile satellite systems.

In the past, a few thin "space cables" have been successfully linked to mobile terminals, most of them large and accommodated on big ships. The resulting system (i.e. INMARSAT) has been very successful, although past satellite technology limits the overall volume of traffic that can be carried by the system and the number of earth stations.

Huge new opportunities [3] are now appearing for mobile satellite communications: growth in maritime communications, extension to aeronautic terminals and, most important of all, introduction of land mobile terminals. The latter, however, seems to mirror the dilemma of fixed services: here again a strong terrestrial competitor, the cellular radio network, is emerging and may pre-empt satellites.

One of the possible answers from the satellite side is described in this paper, namely a mobile communication satellite system based on high-gain spot beams interconnected by an on-board routing processor. This system is orientated towards intercommunications with terrestrial ISDN and to integration with fixed satellite services. The latter might be implemented either by flying two payloads on the same satellite or, more probably, through Inter-Satellite Links (ISL) between fixed and mobile satellites. In both cases fixed-service earth stations may be re-used for mobiles, thus achieving economy of scope and consistent reductions in the length and cost of terrestrial "tails".

2. NEW ISSUES IN MOBILE SATELLITE COMMUNICATIONS

The above considerations led to the awarding of an ESA contract [4] aimed at investigating and implementing prototypes coherent with the basic concept that the future of mobile satellite communications depends largely on their ability for constituting a network (e.g. wide terminal diffusion, routing capabilities, service flexibility, easy internetworking, etc.) As corollaries, several specific issues can be identified and can be roughly divided into three types: transmission issues, networking issues and flexibility issues.

Transmission issues are certainly critical and have been treated by several other works [5],[7]. They are not specifically discussed in this paper, although some relevant assumptions are made (e.g. high-gain spot beam antennas).

Networking issues may be summarised as follows: the mobile satellite system should be the ideal extension of the overall ground network, so that any mobile terminal can offer to the user the same narrowband service availability provided by cable-connected terminals. This target would put mobile satellites

in line with terrestrial cellular network, but implies various requirements:
- Compatibility with ISDN, which in turn requires standard digital links, plesiochronous synchronization, out-of-band signalling methods in line with CCITT No7 systems, digital voice and packet data handling capabilities.
- Co-operation between fixed and mobile payloads, thus substantially increasing the number of "system landing points" as mentioned above.
- Complete routing capabilities (mobile-earth station, mobile-mobile, earth station - earth station), with potential point-to-multipoint communications.

Flexibility issues affect two aspects of the mobile satellite system:
- Adaptability to changing service and traffic requirements, in the case of both long-term modifications (e.g. transition period after system set-up) and short-term fluctuations (e.g. daily or seasonal variations).
- Like ground ISDN, the ability to handle several services with variable bandwidth up to 16 Kbit/s, such as:
    (i) digital toll quality voice, vocoded according to emerging standard (e.g. 16 K/bits). The same channel can be used for bulk data.
    (ii) low-rate data communications (e.g. telex) at various rates. The same channels can be used for highly compressed voice communications.
  (iii) packet data communications.

3. BASELINE CONFIGURATION

The present ESA contract [4] will focus on techniques able to satisfy the above requirements and will include the implementation of laboratory protypes sufficiently representative of the baseband elements, although much simpler than any pre-operative equipment. The baseline architecture, sketched in Fig. 1, is centred around a Mobile Service Satellite (MOSSAT), connected through high-gain spot beam antennas to a large number of mobile terminals and a few earth stations.

Intersatellite links to fixed-service satellites (FIXSAT) are not mandatory although they are immediately compatible with MOSSAT on-board processor.

The latter is able to route on-demand any incoming flow from any up-link to any down-link, through proper switching, so that connections for any service (i)-(iii) can be established throughout the resulting integrated earth-satellite network, as:

- mobile-ground user connections through MOSSAT satellite and
    o   MOSSAT Earth Station only (user A) or
    o   FIXSAT satellite plus FIXSAT station (user B) or
    o   MOSSAT Earth Station plus connected ISDN (user C) or

o   FIXSAT satellite plus FIXSAT station plus connected ISDN (user D)

- mobile-mobile connections (different spots).

  The ability to implement this ambitious architecture relies on a number of key technical elements, which drive the current development effort.
  Mobile access requirements are critical and the proposed techniques play a key role in making them compatible with ISDN and extensive networking for all services (i)-(iii). The basic access schemes are summarized below and are shown in Figure 2 in so far as they concern mobile terminals:
- The mobile up-link requires 3 different schemes for services (i)-(iii), all implemented by the same equipment. Furthermore, alternative synchronization techniques are being investigated for service (i), which implies continuous transmission.
  The possible solutions considered for service (i) consist of SCPC access at 16 kbit/s synchronous with on-board clock and asynchronous SCPC access at 9.6 kbit/s. The former technique matches the bit rate of cellular radio network (with possible easy conversion into narrowband TDMA at n x 16 Kbit/s) and copes directly with on-board switching and down-link interfaces to ISDN. The synchronization is achieved by looping back from the satellite to each terminal information on phase difference between the terminal clock as received by the satellite and on-board clock [6]. 9.6 kbit/s asynchronous access has so far been considered [7] the preferred technique for satellite systems prior to ISDN and has been implemented in prototype multicarrier demodulator and modulator [8], which can be usefully integrated with the proposed prototype. This scheme is made compatible with digital network requirements by means of pulse stuffing/destuffing and bit filling/defilling techniques.
  Service (ii) access is called "microchannel" and consists of a low-rate TDMA with open loop synchronization. In this case the basic carrier (at 16 kbit/s or 9.6 kbit/s) is organized in frames (Frame I, duration = 512 ms) and any mobile terminal can transmit only during corresponding time slots (period = 64 ms) of successive frames, so that the same equipment may be used for connections at various data rates.
  Service (iii) access follows a Slotted Aloha scheme, which utilizes the same frame and slot format used by microchannels.
  This Slotted Aloha access is of primary importance because it is also used by signalling packets to set-up and clear connections for services (i)-(ii).
- TDM links at medium rate (e.g. 256 Kbit/s) are the basic choice for the mobiles down-link. Each TDM multiplexes, in a consistent manner, any kind

of combination of services (i)-(iii), plus information and service channels necessary for mobile synchronization, signalling, housekeeping, etc. The TDM link is organized in Frames, (Frame II, duration = 0.5 ms), Multiframes, and Superframes.
- TDMA scheme at high-rate (i.e. 32,768 Kbit/s) is used for earth-stations up-linking with the same Multiframe/Superframe structure used for the mobile down-link.
- TDM access at high rate (i.e. 32,768 k/bits) is used for the earth-station down-link, with the same structure as the up-link TDMA.

The above flexible strategy copes with a versatile scheme for an on-board switching processor, which relies on a single hardware structure able to handle all services (i)-(iii).

This processor receives up-to-date connection maps from the controller, whose simplified implementation is co-located with the switching processor, as in Figure 1. Medium term technology developments, however, will probably not allow sufficient room on-board for an operative controller, which will therefore be located on the ground at a MOSSAT Master Station and connected to the on-board switching processor through dedicated channels.

This processor is tailored to fit the network functions necessary to manage the system and to make it compatible with ISDN, in terms of:
- Management of connections within MOSSAT, including mobile channels, on-board routing and fixed stations links.
- Interfaces between MOSSAT and corresponding terrestrial facilities (i.e. digital 64 kbit/s channels, Circuit-Switched Data Network, Packet-Switched Network) plus co-operation with ISDN exchanges connecting landing earth stations to end users.

A set of modules handle levels 2-4 of a mobile signalling system presently under study on the basis of both CCITT No.7 system (MUP or ISUP) and procedures used in cellular radio networking. Another set of modules handles CCITT No.7 signalling system, as required by ground network, while a common processor handles and co-ordinates the overall management required by each service. Moreover, the module is in charge of overall housekeeping and of specific mobile procedures such as mobile-to-spot localization and handover in the case of a change of spot.

4. IMPLEMENTATION, TESTS AND POSSIBLE EXTENSIONS

The ESA contract currently in progress [4] will result in laboratory prototypes of the on-board switching processor and the baseband equipment for mobile terminals, with a view to re-utilizing the TST/SS-TDMA ground terminals [2] as much as possible. This approach will save development effort for fixed earth-stations and will lead to mobile satellite prototypes ready to interworking with future generations of fixed satellites.

The immediate result to be expected is the end-to-end prototype integration at baseband level, with radio links replaced by cables, and comprehensive testing of system and equipment capabilities (e.g. multi-service routing, synchronization, inter-networking, etc.).

Another straightforward step is the integration of the above equipment with multicarrier demodulator and relevant modulators developed in the framework of another ESA contract [8]; as a result realistic transmission conditions will be incorporated (e.g. through satellite-channel simulation).

Two areas of further activity have already been envisaged: introduction of new techniques and experimental activities.

Preliminary investigations are being conducted about multipath processing (already assessed for cellular radio system), possible use of various frequency bands, performances of mobile TDMA versus SCPC access.

On the other hand the "Double Hop Experiment" presently planned [9] for TST/SS-TDMA might be fruitfully extended to the prototype mobile system. In this case, field experiments would be carried out using "double hops" via an existing transparent satellite, still maintaining the "on-board processor" on the ground. The result would be validation of the mobile system both under stand alone conditions, and under conditions of co-operation with fixed-service satellite system prototypes.

Satellite Switching for Mobile Communications   71

Figure 1
System Architecture

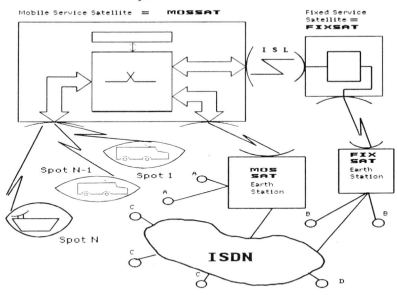

Figure 2
Mobile Access Schemes

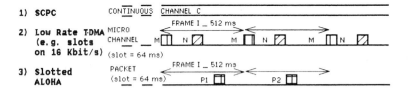

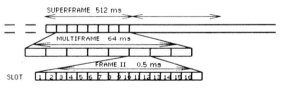

REFERENCES

[1] R. Lovell, C. L. Cuccia "Global interconnectivity in the next two decades - a scenario" - 11th AIAA Conference, San Diego.

[2] G. Pennoni, G. B. Alaria. "An SS-TDMA Satellite System Incorporating On-Board Switching Facility; Overall System Characteristics and Component Description" - ICC 1984, Amsterdam.

[3] J. D. Kiesling, "The United States Mobile Satellite Service" - ESA Workshop on Land Mobile Services by Satellite, June 2-3, 1986.

[4] ESA Contract No. 7080/87/NL/JG(SC). LABEN - ISEL - TELESPAZIO.

[5] R. Rogard "A Land Mobile Satellite System for Digital Communications in Europe" - ICSDC 1986, Munich.

[6] T. Izumisawa, S. Kato, T. Kohr, "Regenerative SCPC Satellite Communications Systems" - AIAA 10th Communication Satellite Systems Conference.

[7] ESTEC Contract No. 5484/83/NL/GM(SC). TELESPAZIO - CSELT - ANT - DORNIER

[8] ESTEC Contract No. 6497/85/NL/IW(SC). ANT

[9] D.B.H.P. (87)2. Plan for the preparation of the Double Hop Experiment. Source: ESA - XRF/181/LB/ld.

E.C.S.: INTEGRATION IN THE EUROPEAN NETWORK AND INTERFACING WITH
THE ITALIAN PUBLIC TELEPHONE NETWORK

D. De Rosa   (*)
B. Vendittelli ($)

1. INTRODUCTION

Telecommunications satellites, used in the past almost exclusively on intercontinental networks, have been recently introduced on shorter distances, becoming competitive with traditional transmission means (cables and radio links) also in regional and domestic areas. This evolution is due mainly to fall of costs resulting from recent fast technological developments; but also important is that some facilities of the latest satellite systems make them apt to provide a number of special services (Mobile Radio services, Videoconference, Point to Multipoint data transmission, etc..) for which there is a growing interest in many potential users.

However the nature and the features of satellite communication systems are so different from those of the traditional transmission means that their adoption requires a careful estimation of their capability to be integrated in the existing networks and of any problem arising. In a more general view the introduction of satellite systems should be referred to the context of a network evolving towards structures oriented to a total integration both of techniques and of services. From this point of view, two significant moments can be distinguished, in relation to the technical characteristics of the network where the satellite will be introduced; these two moments are respectively the present, characterized by the contemporaneous presence in the network both of analogue and digital techniques, and a not far future, when the satellite will find a fully digital homogeneous network.

In the following, the problems resulting from the introduction of a satellite system in the present public switched network will be mainly discussed, taking into account that this network is now composed by switching and transmission means technically not homogeneous. Specific reference will be made to the ECS system which has become operative at the beginning of the last year; in particular the interfacing aspects of ECS with the terrestrial network will be examined and then the possible consequences on the quality of service and the compatibility with the signalling systems will be analyzed. A brief description of the solutions adopted in Italy will also follow. Before the conclusions, a short look at the situation foreseen for the digital context in the near future will be given.

Most of the considerations included in this paper directly descend from the experience gained by the authors on occasion of the ECS interfacing with the national network, therefore the opinions here expressed plainly reflect the point of view of people working in a Company responsible for long distance terrestrial network and services.

2. THE ECS SATELLITE IN THE EUROPEAN NETWORK

The ECS can be considered the first satellite for telecommunications effectively operative in Europe. It has been implemented specifically for carrying both, speech calls coming from the public switched network and

---

(*) ASST-Ispettorato IV zona, ROMA
($) ASST-Direzione Centrale G.R.S.I., ROMA

data traffic generated by large users of the business category.

The engineering and the implementation of ECS system have been cared by EUTELSAT, an agency created to support the development of satellite telecommunications in Europe. All CEPT Countries interested in the use of the space segment, including Italy, participate to EUTELSAT with different percentages.

Three satellites will be available in the final configuration, but presently only two of them are operative. The telephone service has started in the month of march 1986, when the first route via ECS in Europe, linking France to Austria. was set up.

An important contribution to the engineering of the ECS system has been given by numerous Italian industries such as Aeritalia, Fiar, Galileo, Gte, Selenia. Italy has put into service its first circuits via ECS in July 1986 on the direction Rome - London.

The chief features of the ECS system, with regard to the services offered and to its capacity, have been synthetically presented in the table 1 below.

Referring to the telephone services it was decided that each Country supporting EUTELSAT should connect one or more of its Transit Centers (CT) to ECS system through one earth station specifically chosen. For Italy, it was decided to route by satellite part of the traffic originated by the CT of Rome, providing the access to the satellite through the earth station of Telespazio situated in Fucino, which is about 100 kilometers from Rome.

The ECS system uses exclusively digital transmission techniques and, moreover, to achieve a better utilization of the space segment, the earth station has been equipped with Digital Speech Interpolation devices (DSI) which practically allow the duplication of the international telephone channels with respect to the satellite channels.

It is just the introduction of sophisticated circuit multiplication equipments, DSI type or others based on different principles, as will be shortly illustrated in the last part of this paper, that has given a great contribution to lower the price per channel of the system and to make satellite convenient also in regional areas. Moreover, the introduction of ECS is particularly appreciated because it makes immediately available, on long distances, digital transmission resources that terrestrial means do not yet provide. This fact will certainly accelerate the process of digitalization in the European regional network and an expansion of some special services, particularly those requiring "end to end" digital connectivity, will result.

TLC PUBLIC SERVICES	modulation: PULSE CODE MOD.
	access type: TIME DIVISION MULTIPLE ACCESS
	capacity: 12000 CIRCUITS
SPECIAL SERVICES (SMS)	modulation: SINGLE CHANNEL PER CARRIER
	access type: FREQUENCY DIVISION MULTIPLE ACCESS
	capacity: 300/570 CHANNELS (64 Kbit/s equivalent)
TV BROADCAST. SERVICES	covered area: EUROPE AND MEDITERRANEAN BASIN
	capacity: 2 TV CHANNELS

T A B L E 1: principal features of the ECS system

3. ECS TO TERRESTRIAL NETWORK INTERFACING

A perfect integration of satellites in Public Switched Telephone Networks will be possible in the near future when networks will be almost entirely digital. At the present ECS has been inserted in a network where analogue techniques are still dominant, so A/D conversion must be performed wherever non-homogeneous equipments have to be interconnected.

The different elements which must be interfaced with each other and with the satellite system, in all possible combinations, are fundamentally the following:

- Analogue switching systems
- Digital switching systems
- Analogue transmission lines
- Digital transmission lines

Figure 1 below shows all the possible combinations in the case that International Transit Exchange is at a certain distance from the earth station, as usually occurs. The figure also shows A/D converters wherever they are necessary.

While, obviously, an all digital connection is the optimum (case D), for quite a few years to come in many Countries one of the other solutions will necessarily be adopted. Particularly, case C is very costly as it needs A/D converters at both the ends of terrestrial link, but wherever digital exchanges have been introduced before digital transmission lines it will be the only solution possible.

A transmultiplexer, which can convert an FDM supergroup (60 channels) in two digital 2Mbit/s groups (2x30 channels), can be conveniently used to interface analogue lines with the earth station. The same equipment can be used in case C on the switching center side. In case B interfacing with the transit center, analogue side, can be achieved using a normal PCM Multiplex.

Switched networks will use both analogue and digital components for several years to come, their interfacing will result much more complex and costly in the case mutual aid paths between satellite and terrestrial routes have to be implemented.

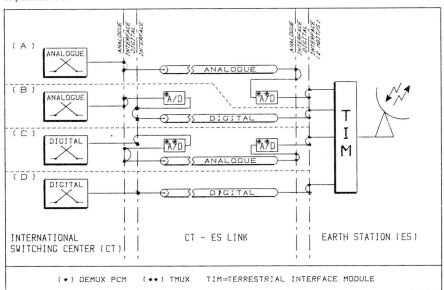

FIGURE 1 - Earth Station to CT interfacing

4. SPEECH QUALITY AND SIGNALLING CONSIDERATIONS

4.1. SPEECH QUALITY - Intrinsic quality of digital satellite links is certainly comparable with quality of digital terrestrial links. However, long propagation delay and undesirable effects due to digital speech interpolation may give the telephone users a feeling of poorer quality. It is well known in fact, that propagation delay causes echo phenomena, which are particularly annoying in telephone conversations. While echo has negligible effects on terrestrial regional connections, due to short delay time (just a few ms), so that there is no need to neutralize it, for intercontinental links echo effects are considerable and neutralization is obtained installing echo suppressors at both ends of the connections. Echo suppressors operation is based on the introduction of additional loss in the echo path: this gives sufficient improvement on links presenting a transmission delay up to about 100 ms but is not fully satisfactory for satellite connections with typically 250 ms delay. Users, accustomed to traditional terrestrial connections, generally feel the difference.

As was mentioned before, ECS adopts Digital Speech Interpolation of channels with a gain of about two (240 channels earth side contending 128 satellite channels). This function is realized by the Terrestrial Interface Module (TIM) part of which is the DSI equipment.

Rather than information on how DSI accomplishes interpolation we want here to put in evidence that it can affect voice channel quality by determining occasional clipping of the initial portion of the speech burst, or, in case of a high number of active channels an increase of quantization noise can occur.

It is clear that great efforts have been made to make satellite systems competitive with terrestrial means and this was accomplished without affecting quality significantly even in the worst working conditions.

4.2. SIGNALLING ASPECTS - The choice of the signalling system to be used for the ECS was made aiming to simplify the integration of the satellite in the existing network and to reduce interworking needs between circuits using terrestrial transmission means and those routed via satellite; so the best choice seemed to be the adoption of a system already in use. The signalling system most widely used in Europe was surely the CCITT R.2, so it was decided to adopt for the ECS this system, although it is not optimized for being used on transmission means characterized by long propagation delay. We will try below to inspect the capability of the R.2 system to support the introduction of the satellite in the network, mainly considering the following aspects: the speed of call routing, the presence of the DSI and the influence of echo suppressors on signalling.

The R.2 system provides, for routing information, a multifrequency code which uses 2 in band frequencies chosen in a group of 6 (inter-register signalling); each signal is sent in a full compelled sequence, so its sending is stopped only when a backward acknowledgment signal is received. This means that a complete compelled cycle is long at least four times the transmission delay Tp. For the longest terrestrial links the Tp time is comparable with the operation time of sending and receiving devices (about 10-20 ms) so that the total time required to send a dialling digit is generally lower than 200 ms. The situation greatly changes when a link includes one satellite jump; in this case, as already said, the transmission delay is about 250 ms, so, it takes more than one second to send a digit. The most significant consequences of this longer time for sending routing information are: the busy time of registers at both the ends of a circuit are prolonged so requiring a revision of their number, the rate of the total busy time not charged during a call rises causing a decreasing of circuits efficiency and, at last, the post dialling delay can increase greatly, in some cases, so inducing the subscriber to release a call going to be completed. All these phenomena need to be continuously monitored to avoid a fall of the service quality due to the presence of the satellite.

Regarding the effects of DSI on signalling, it may in principle occur that, some routing signals which sometimes are transmitted as in band frequency pulses, could be clipped by the interpolation device if a situation of overload is present; however, in service observations have shown that this event is very rare. On the contrary, for the line signalling used by R.2 for calls supervision it is necessary to by-pass the DSI device, because line signals are sent using an out of band continuous frequency (analogue version) or two bits per channel in the time slot 16th (digital version), so they cannot be handled by speech interpolation devices. This problem was solved, at the moment of the DSI engineering, creating a special channel not subjected to interpolation, able to carry all line signals relevant to the speech circuits handled by the same DSI device. Moreover it must be noted that, if the switching center is analogue and a direct encoding of analogue line signal is operated, some internal time-outs must be modified in the exchange, to match the longer transmission delay of the space segment; nevertheless, in order to achieve homogeneous conditions, both for circuits routed by ground and by satellite, it is better to provide a line signalling conversion, from the analogue to the digital version.

Referring at last to the presence of the echo suppressors it is clear that they are not compatible with the compelled sequence of inter-register signalling; in fact, during the routing phase, in band signals are sent at the same time in both the directions, so these echo control devices are induced to introduce on each way the same loss which they normally insert when a double talking condition is present. It is therefore necessary to disable the echo suppressors along all the time the routing information is sent; this disabling function can be easily performed by the exchange when the echo control devices are situated just beside the switching center.

Taking into account the considerations above made it can be said that the integration of a satellite system in the existing network requires attention to some specific quality and signalling problems for which different solutions are possible. The point below shortly describes the solutions adopted in Italy for interfacing the ECS system to the national switched network.

5. THE SOLUTIONS ADOPTED IN ITALY

The public telephone network in Italy is based on 21 compartment centers which represent the nodes at the highest hierarchical level; they are associated with large cities and are almost fully meshed. At the lower level, there are 231 district centers, which are generally interconnected through the hierarchical route via the compartment centers. The long distance transmission network between all compartment centers is operated by the Azienda di Stato per i Servizi Telefonici (ASST), which is also responsable for all traffic with European and Mediterranean Countries. This traffic is handled by five international centers (Rome, Milan, Verona, Turin and Palermo) connected through the intercompartmental network. Of these Rome, with Milan the most important center, was chosen to route traffic via satellite.

The International Transit Center of Rome switches all traffic between Southern Italy with European and Mediterranean Countries. Traffic exchanged between the European Network and the Intercontinental Network also is routed through the CT of Rome which is connected for this purpose with the intercontinental exchange of Acilia (Italcable).

Two exchanges are presently available in the CT of Rome: an analogue Face Standard "Metaconta" exchange and, very recently, a digital ITT 1240 exchange. However the first satellite circuits to be put in service in July 1986 were connected to the Metaconta exchange, while the digital one was not yet operative.

While the exchange to be connected to ECS was analogue, a digital terrestrial link was already available between the CT of Rome and the earth station in Fucino, so their interfacing required an A/D converter as shown in case B of figure 1.

The Metaconta exchange works with R2 system, analogue version, adopting a tone-on-idle line signalling technique, one way for direction, and requiring a separate wire for Interruption Control (IC) information to distinguish the changes of line state due to signalling from line interruptions. Furthermore the release sequence is based on a temporization which takes in account line transmission delay and would therefore need to be modified for satellite circuits.

ECS system supports only R2 signalling in the digital version which uses two bits for line signals, not affected by line interruptions and thus not needing ulterior information nor temporization in the release sequence.

It is clear that besides an A/D voice channel converter, a conversion of analogue R2 signalling, on the exchange side, to R2 digital on the line side, was necessary operating according to CCITT Reccomandations.

The same PCM equipment used for interfacing the exchange to the line performs both voice channel handling and line signalling conversion; it presents exchange side 6 wires (not including an IC wire) per channel. Its capacity is of 120 telephone channels in four 2Mbit/s digital groups.

Also an adeguate number of echo suppressors were installed as shown in figure 2 below which represents the terrestrial section with the indication of all the equipments in the chain.

At the other end of the terrestrial link the 2Mbit/s digital groups enter directly the TIM module. Each TIM module accepts up to 240 channels on 10 digital primary groups and presents satellite side only 128 channels.

Channels used for data transmission can be allocated in the same groups of speech channels but they are handled separately by the TIM and are not interpolated (DNI).

In figure 2 the IC wire is not connected because it is used to signal eventual interruptions of the transmission line only if an analogue section in FDM is present. The echo suppressor is disabled during register signalling directly by the Metaconta exchange on a separate wire (ESC).

ECS interfacing with the national network was realized as described above because the digital exchange was not available at that time; the solution is however temporary and presently a small number of circuits are directly connected to the new digital exchange ITT 1240 now in service. This switching

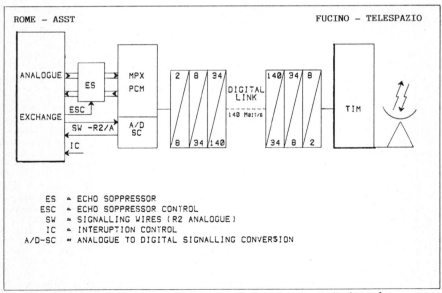

FIGURE 2 - ECS interfacing to the national network - phase 1

system accepts directly 2Mbit/s groups and digital version of R2 signalling system so that no conversion is necessary.

At the moment circuits with London, Reims, Nicosia and Lisbon are already in service on the new exchange. Within a short time the four digital groups connected to the Metaconta exchange will be transfered to the ITT 1240. Within the month of November new routes are planned with Denmark, Finland, Norway, Luxemburg and Yugoslavia. Presently 110 circuits via ECS are in service and by the end of the year they are planned to become 660.

Moreover two different 140 Mbit/s digital links between Rome and Fucino are now available so that alternative routing in the terrestrial section is possible in order to improve overall availability.

6. SATELLITE INTEGRATION IN THE ISDN

Before going to the conclusions, it may be interesting to look through the situation foreseen for the near future, when the satellite will be placed in the context of a wholly digital network dealing with both voice and data services.

Although circuits in the digital environment generally will use end to end four wire links, echo could be still present to some extent, due to phenomena that cannot be cancelled such as acoustic coupling at the subscriber site. So the echo control function, in such a context, is still necessary and will be exclusively performed by extended use of "echo cancellers" devices which are much more sophisticated and efficient than traditional suppressors. The echo canceller operation is based, in principle, on the subtraction from the signal received of the component due to the echo phenomena, providing that it can be exactly estimated. The echo canceller performances have been already positively tested on ECS wholly digital circuits. It has been certified that they can cancel multiple echoes and also cope well with very long terrestrial extensions. Moreover they are not sensitive to the background noise and do not generate clipping or other troublesome phenomena, typical of traditional suppressors especially during double talking situations; so, an appreciable increase of the speech quality can be expected.

Regarding signalling, it can be assumed that the system extensively used in such homogeneous digital environment will be the CCITT N.7, or, in some domestic areas, a national version of the same. In any case signalling systems will be common channel type, so, links carrying signals can be routed independently of relevant traffic channels. Nevertheless, for needs of the circuits routed by satellite, it is possible to use signalling links passing through the same route; in this case it is expected that the signalling procedures should not suffer for the overall transmission delay. However a double jump through satellite should be avoided and, moreover, when the same signalling link serves both circuits bundles routed by ground and others routed via satellite, it is advisable to set up this link using the terrestrial transmission means, avoiding so any constraint for the traffic routing choices. In this way the presence of the satellite should not produce any trouble with respect to the signalling needs.

It remains to speak about the possible effects produced by the presence of DSI devices and similar. Various types of Circuit Multiplication Equipment (CME) have been developed to be used on the most expensive transmission systems (submarine cables and satellites); besides DSI techniques, already mentioned, some Low Rate Encoding (LRE) methods have been developed that allow to transfer the voice signal using the bit rate 32 Kbit/s and also lower (16 Kbit/s), without appreciable loss of speech quality. DSI and LRE techniques can be used together so achieving a circuit multiplication factor (gain) typically equal to 4 or more. But, just the circuit multiplication equipments, always present in the satellite transmission systems, could produce some problems for their integration in a network able to carry both voice and data services. In fact CMEs cause a loss of transparence of the transmission network, so obliging the exchange to perform a preliminary check of the transmission resources available, before starting the routing of a special call. Normally, when an exchange has to route a call, it only needs to know whether or not an outgoing circuit is available; on the contrary, when a CME is present, it is possible that a call may find a spare circuit from the exchange to the CME but no available channel between the two CMEs at the ends of the satellite trunk could exist. This situation can occur especially for service requests different than speech calls. CME, in fact, is generally able to route speech calls, also in overload conditions, but the same does not occur when the call to be routed requires, for instance, a full 3.1 kHz band (in band 9600 bit/s data call) or when a 64 Kbit/s bearer service is requested by the exchange. The situation becomes even more complex when two or more exchanges are linked to the same CME so sharing all the available transmission resources. In order to overcome the outlined problem, it is necessary to exchange some amount of information between the switching centers and the location of the CME, so that all interested users can be continuously informed about the resources available for each type of service. It is also possible that the echo control function may be incorporated in the CME; in this case some additional information should be sent to the CME in order to disable the echo canceller device when a 3.1 kHz or 64 Kbit/s bearer service is requested by the switching center. So, it is clear that, the full integration of a satellite system including a CME, in a network offering bearer services in addition to speech calls, requires the specification of a set of signals to be exchanged between the switching center and the transmission system and the relevant procedures should also be defined.

7. CONCLUSIONS

An extended terrestrial network is present today in Europe and a large spare capacity is still available, however the choice made by Eutelsat seems to be successful both for the technical solution proposed and for the particular present network situation; in fact the demand of long distance digital transmission means is quickly growing but it cannot yet be satisfied using the terrestrial resources available, because analogue technique are still dominant. On the contrary, digitalization of switching centers is coming fast, so just the timely introduction of ECS has allowed to set up some wholly digital CT to CT links.

However, the integration of the satellite in the existing switched network requires a careful estimation both of quality and of signalling aspects. Referring to the present, speech quality seems to be not quite satisfactory chiefly because of echo phenomena; but as soon as echo cancellers will be extensively used, this problem will be overcome. As for signalling, some events affecting routing can be observed when using associated R2 system as today; no difficulties are foreseen for the future adoption of common channel systems as CCITT N.7. Moreover, in order to take account of both speech and data services in the same network, a full integration of satellite systems in switched networks requires the implementation of technical specifications regarding the exchange of information on resources availability between switching centers and satellite systems, supposing that these make use of CME devices.

Satellite Integrated Communications Networks
E. Del Re, P. Barthelomé and P.P. Nuspl (eds.)
© Elsevier Science Publishers B.V., 1988

INTERNETWORKING OF VSAT AND MSAT SYSTEMS FOR END-TO-END CONNECTIVITY

K.M.S. MURTHY and D.J. SWARD
Telesat Canada
333 River Road
OTTAWA K1L 8B9
CANADA

A number of technical and economic issues related to the use of VSAT bypass technology for the MSAT backhauling network are discussed in this paper.

1. INTRODUCTION

In recent years the two areas of satellite communications which have been gaining major thrust from leading satellite industries and institutions are: 1) mobile satellite (MSAT) systems, and 2) very small aperture terminal (VSAT) communication systems. The driving forces behind these developments are: technology advance, de-regulation, and market demand.

MSAT which can be called "VSAT-on-Wheels" have several features in common with VSAT. Both these technologies have made successful use of developments in antenna and RF technology, LSI based codecs and modems, low bit rate voice codecs, multiple access protocols, packet transmission and switching, terminal protocol standards, microprocessors and software. While VSAT takes the satellite communication to fixed user premises, MSAT takes it to mobile user and they both exploit inherent satellite strengths viz., universal coverage and networking flexibility. They both offer economic communication and information services, with greater level of cost control.

A typical mobile network involves communication between mobiles and vehicle operators and between vehicle operators and their customers. These links can be established either through all-satellite networks or a combination of satellite and terrestrial networks. A fully matured mobile network will have fixed and mobile satellite network for end-to-end communications. However, some applications may require terrestrial network interconnection. Therefore, internetworking and interfacing of mobile and fixed satellite networks are of paramount importance in the development of mobile services.

This paper addresses some of the technical and economic issues related to internetworking of mobile (MSAT) and fixed (VSAT) systems. In Section 2 the overall MSAT network is discussed. A "tripple star" structure embedded in the end-to-end mobile network is discussed in the context of land mobile and aeronautical applications in Section 3. An analysis of interconnection requirements are presented in

Section 4. Section 5 discusses the use of VSAT for the backhaul networking and Section 6 compares the economics of VSAT backhauling with terrestrial services.

2. MSAT COMMUNICATION NETWORK

The major elements of an end-to-end MSAT communication network (see Figure 1) are: 1) satellite system, 2) mobile earth stations, 3) fixed earth stations, 4) interface/backhaul network, and 5) auxilliary systems.

2.1 Satellite System

The current baseline satellite system will have communication transponders operating at L-band (1.5 - 1.6 GHz) and Ku-band (11 - 14 GHz) frequencies. The system will generate nine 2.7° beams covering Canada and the United States. The first generation Canadian and U.S. satellites are likely to be similar and will back each other up. The total capacity of the satellite will be distributed non-uniformly among the nine coverage areas.

2.2 Mobile Earth Stations

The mobile/remote user earth stations can be divided into three types: 1) vehicle mounted mobile terminals, 2) portable terminals, and 3) remote user fixed terminals. These terminals will operate at L-band frequency and support voice, data and image transmission services. Depending on the application, mobile terminals may have different antenna types, capabilities and cost. A typical mobile channel will have 5 KHz spacing and will support one analog or low bit rate digital voice or one 2.4 kbps data. Other channel spacing and data rates are possible.

2.3 Fixed Earth Stations

The fixed earth stations (viz., network control station, satellite control station, gateway station, and base station) form the backbone of the overall network and operate at Ku-band frequency. A <u>centralized</u> control station (CCS) will control the satellite system and the communication network and the <u>distributed</u> gateway and base stations will link the mobile and fixed users. The satellite control system will perform the usual satellite tracking and control functions, station-keeping maneuvers, reconfiguration of the backup satellite, etc.

The network control system performs two major functions, (1) resource allocation and (2) service management. The resource allocation to authorized users is done by demand assigned multiple access (DAMA) subsystem. This handles events and activities of fairly short duration and perform functions related to the assessment, co-ordination, allocation, monitoring, maintenance and control of satellite resource. The network management subsystem (NMS) manages the overall communication network and service. The NMS's functions include efficient network utilization, maintenance of customer directory and user statistics, billing and administration, monitoring and reporting of the health of network elements, aid resource allocation, aid and supervise overall satellite control and DAMA control systems, aid and monitor backhauling facilities. In general, NMS manages events and activities of longer duration or those occuring less frequently.

2.4 Interconnect/Backhaul Network

The interconnect and backhaul network is the last segment of the mobile to fixed-end link. The interconnection function include switching (circuit, packet and/or message switches), signalling, multiplexing, protocol conversion, storage (voice, alphanumeric and coded messages for paging and scheduled broadcast applications). While the network control center will support all the above functions, the gateway stations will only support limited switching interconnect, protocol conversion and multiplexing functions. Depending on the type of interconnection established, the backhaul network may perform only the transmission function or both the transmission and switching functions. The backhauling facility could be satellite, terrestrial or a combination of both.

2.5 Auxilliary Systems

The auxilliary systems are the ones which act as a source of or sink to part of the information transmitted between the mobile and fixed-end user. The vehicle based auxilliary systems include position determination (Loran-C, GPS, etc.), monitoring, measurement, alarm and control subsystems. Also, systems offering information and agency services may belong to this category. These systems may be directly or indirectly connected and may require occasional or frequent access to main mobile network.

3. THE TRIPPLE STAR TOPOLOGY

The topology of the end-to-end MSAT communication network has a complex "tripple star" structure as shown in Figure 2. The figure depicts the topology of a typical land mobile application viz., the trucking industry.

Usually the medium and long haul public carriers will control the load distribution and routing operations from a fleet management center (FMC). The FMC of each operator will be linked individually to either a base or a gateway station. Another layer of communication network exists between the shippers (S1, S2, ...Sn) and the FMCs. Note here, the concept of layers in Figure 2 is used for identification purpose are not based on any Standards. The shippers can choose one or more of the carriers depending on their cargo type, destination, cost, etc. Many private carriers (e.g. a retail chain) operate their fleet much the same way and in which case (S1, S2, ...Sn) will represent a group of retail stores. In some cases S1, S2, ...Sn will also represent a host of public or **government agencies (e.g. fueling stations, motels, restaurants, banks, customs, safety** control agencies, etc.) who may communicate to mobiles and/or their operators.

There is yet another layer of communication taking place between FMC and the corporate branch offices (CBO1, CBO2, ... CBOn). Although not in direct control of the fleet operations, the CBOs will provide related services such as local pick-up, goods storage, vehicle meaintenance, etc. and hence may require communication links to FMC or direct to mobiles. Further, the CBO's may require facilities for intra-corporate communications.

As can be seen at the top of Figure 2, mobiles derive position or other information from the auxilliary systems such as Loran-C, GPS, etc. and will report to their operators either on a scheduled or on a demand basis. Between these four layers (L1 - L4 in Figure 2) of communication networks there is an imbedded "tripple star" structure. The communication links between layers L3 and L4 can be called as the "main" or "backbone" MSAT network and those between layers L1, L2, and L3 can be called as the "secondary" or "backhaul" network.

3.1 MSAT Aeronautical Communication Network

Aeronautical communication network is perhaps the most complex of all mobile communication systems both on-board the aircraft and on the ground (See Figure 3). Communications to and from a commercial aircraft can be broadly divided into three categories: 1) air traffic services, 2) airlines operations, and 3) public correspondence.

The ATS will provide air traffic control (ATC) and advisory services to aircrafts flying in the airspace overlying the country. The airspace is divided into a number of flight information regions (FIR). Area control center (ACC) within a FIR is responsible for providing air traffic control, advisory and alerting services to aircraft flying in the region. The ATC services on the airport and in the airspace in its immediate vicinity is offered by airport control towers (TWR). Flight service system (FSS) usually located at airport provides specialist services and advice to assist the pilot in the planning and conduct of his flight. Area control center gathers aircraft position information (through voice or data) as well as from primary and secondary radar surveillance (RS) centers within the FIR. ACC's will frequently communicate with other ACC's to transfer the control of the aircraft while moving to other ACC region. The communication network architecture for airlines operations is similar to the trucking industry discussed earlier. Booking agencies (BA) and corporate branch offices (CBO) will communicate with airlines operations center.

4. ANALYSIS OF INTERCONNECTION REQUIREMENTS

The MSAT backhaul network is a very complex network requiring interconnection to public/private, switched/ non-switched, data/voice, satellite and terrestrial facilities. A number of technical, economic, operational and regulatory factors are to be considered in establishing appropriate interconnection facilities. To analyze some of the issues involved, a typical transportation industry communications can be divided into five categories, viz.: 1) public correspondence, 2) vehicle operations, 3) corporate communications, 4) agency services, and 5) distress, safety and control services.

Some of the general requirements/features of these services which will influence the choice of interconnection facilities are summarized in the table below. (Legend: * Primary, + Secondary, - Occasional/Rare).

FEATURES	PUBLIC CORRESPONDENCE	VEHICLE OPERATIONS	CORPORATE COMMUNICATIONS	AGENCY SERVICES	SAFETY/CONTROL & EMERGENCY
Voice-Interactive	*	*	*	-	*
Voice-Messaging	-	*	-	-	+
Data-Interactive	+	*	*	*	*
Data-Messaging	-	+	+	*	*
Data-Broadcast	-	+	*	*	+
Data-Collection	-	+	*	-	+
Inter-Communication	no	yes	yes	maybe	yes
Intra-Communication	no	yes	yes	yes	yes
Delay Tolerance	low	high	high	high	low
Priority Required	low	high	medium	low	very high
Privacy Required	low	high	high	low	very high
Call Origins	random	random/scheduled	random	random/scheduled	random/scheduled
Reliability Required	low	high	high	low	very high
Net Availability	low	high	medium	low	very high
Storage Required	no	yes	yes?	yes	yes
Standards Required	yes (PSTN,PDN)	yes & no	yes	yes & no	yes
Access To/From Vehicle	direct	vehicle operator	limited	vehicle operator	direct
Interconnect Point	gateway	gateway, base, NCC	gateway, NCC	gateway, NCC	gateway, base, NCC
User Cost Affordability	low	low/medium	medium	medium	high
VSAT Suitability	not quite	yes	yes	yes	yes
Justify VSAT	no	yes	yes	yes	yes
Double Hop - An Issue	not quite	no	no	no	no
ISND Compatibility Req.	yes	no	no	no	no
Early Entry Potential	medium	high	low	medium	high

5. VSAT BYPASS FOR MSAT BACKHAUL

A typical VSAT network can meet most of the backhauling/interconnection requirements outlined in Section 4. However, for public correspondence applications, user access to public switched telephone or data network may be required. VSAT's can handle a wide range of interactive voice, data and image services. The adaptive access protocols employed in many commercial VSATs can efficiently handle traffics with message length, terminal speed, and terminal activity each varying about three orders of magnitude. Typical VSAT aperture diameters range from 1.2 m to 1.8 m and can transmit up to 128 kbps continuous data and can receive several times this rate. Commercial VSATs can support a wide range of terminal protocols and can meet competitive terrestrial delay constraints.

Such a VSAT network can effectively replace the three layers (e.g., L1A, L1B, and L2 of Figure 2) of the communication network. The VSAT hub can be co-located with MSAT hub and the nodes of each layer equipped with on premises VSATs. The network architectures of Figure 2 can be simplified with the use of VSAT network as shown in Figure 4.

The backhaul communication requirements in many cases are ideally suited for VSATs. Consider the case where an FMC collects messages from the shippers, formats and passes them to mobiles. In this case, the total volume of traffic from FMC to mobiles will be equal to or higher than the total transmitted to FMC from the shippers. This situation is similar to one large user generating more than 50% of the traffic and a large number of small users generating less than 50% of the traffic. If these users are allowed to use a Slotted-Aloha channel, the resulting throughput may increase to around 50% to 70% which is much higher than the 37% theoretical maximum for Slotted-Aloha.

6. EXAMPLE: VSAT VS. TERRESTRIAL ALTERNATIVES

In this section, the economics of using VSAT for backhaul is compared with terrestrial facilities with specific examples. Let us consider two-way data messaging service between land mobiles and vehicle operations center (FMC). Besides VSAT, Telecom Canada's Dataroute and Datapac terrestrial facilities can be used for the backhauling. Dataroute is a dedicated facility whose cost increases with the line speed and the distance of separation of the nodes. Datapac is a packet switched facility whose cost depends on the number of packets, number of calls, node location and in some cases, total access time.

Let us assume that the MSAT and VSAT hubs are co-located and the message transfer between FMC and the MSAT-NCC is done in a store and forward (SAF) mode or in real time. Typical message is 32 characters long. The cost per mobile per month as a function of fleet size is plotted in Figure 5.

For the range of parameters considered, VSAT cost is practically insensitive to fleet size and message traffic. In addition, distance- insensitive cost, network flexibility are some of the assets of VSAT. Within the constraints of the assumptions, it may be suggested that a total cost of $400 - $600/month Canadian be used for a VSAT node for the type of application considered here. This amounts to about $10/month/mobile for small fleets (less than 50 mobiles) and about $2/month/mobile for large fleets (more than 250 mobiles) for the message backhauling. The advantage of VSAT will increase with the increase of per node traffic since the space segment is only a fraction of the total cost.

7. CONCLUSION

The use of VSAT network for MSAT backhauling has been discussed from the interconnection and economic points of view. The architecture of MSAT communication network was discussed in the context of land and aeronautical mobile applications. The issues and requirements of backhaul internetworking have been identified. A VSAT bypass for the MSAT backhaul has been suggested and analyzed. The economics of VSAT backhauling had been compared with terrestrial facilities and some numerical results were presented.

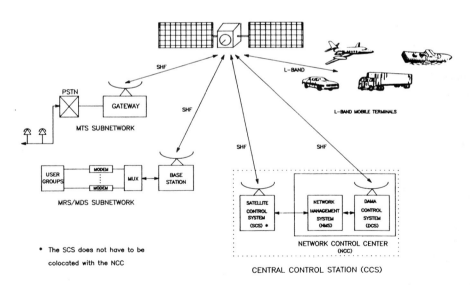

FIGURE 1 MSAT NETWORK CONFIGURATION

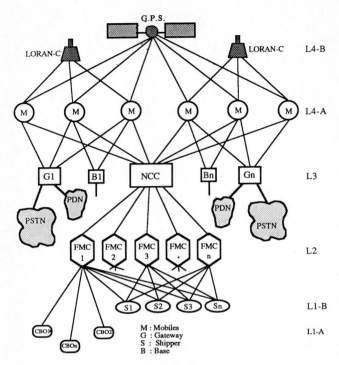

FIG.2 MSAT COMMUNICATION NETWORK: LAND MOBILE APPLICATION

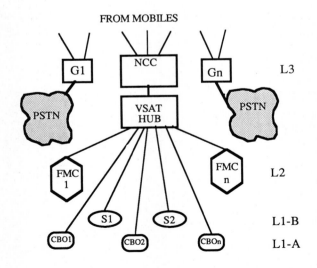

FIG. 4 MSAT NETWORK WITH VSAT BYPASS
(COMPARE WITH FIG.2)

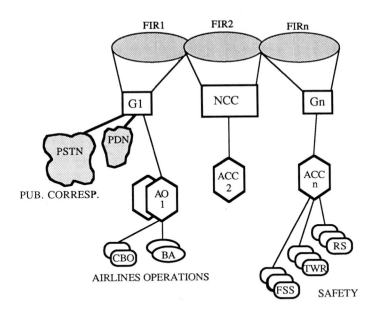

FIG.3 MSAT COMMUNICATION NETWORK: AERONAUTICAL APPLICATION

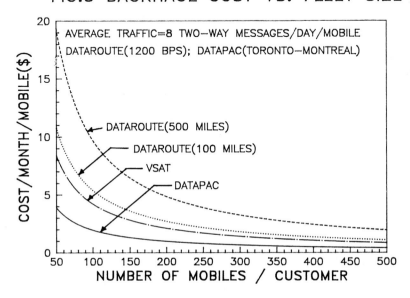

WIDE AREA NETWORKING BY INTERCONNECTING LANs AND MANs VIA SATELLITE

K.M.S. MURTHY and D.J. SWARD
Telesat Canada
333 River Road
OTTAWA K1L 8B9
CANADA

1. INTRODUCTION

Resource sharing and information exchange management within a local environment are two key benefits which motivated the use of local area network (LAN) by the business and other communities. The commercial benefits to LAN users will enhance several fold if these two capabilities are extended to access facilities available at a much wider area viz, regionally, nationally or internationally. In fact existing metro area networks (MAN) already provide such benefits to some extent to metro area users. The LAN users could benefit by accessing MANs, or remotely located information and computing services.

This paper proposes a satellite based wide area networking (WAN) concept to link MANs and LANs. In Section 2, the need for internetworking of LANs and some of the problems of the existing interconnection facilities are outlined. Satellite based WAN concept and inter-LAN traffic assessment issues are discussed in Section 3.

2. LAN INTERNETWORKING ISSUES

2.1 Market Trend

Organizations use multiple types of LANs to meet special application, security, organizational growth and maintenance requirements. Market studies indicate that the North American LAN market will grow by about 20% annually for the next five years. Although office LANs will continue to remain as the largest percentage of share, factory and medical application LANs are expected to experience highest growth over this period. About 39% of the installed LANs are within a building floor, 27% within a building and 34% outside the premises.

About 30% of the business communication takes place at inter-corporate level and about 20% at the inter-site intra-corporate level. This means nearly 50% of business communication is with the agencies outside the premises.

2.2 Limitations of Interconnection Facilities.

Lack of established Standards for the LANs as well as the interface units has made the task of interconnection very complicated. Further, analysis of applications indicate that the message length, terminal activity and data rate of user terminals each vary up to three orders of magnitude. Therefore, the LAN and inter-LAN traffic will vary even over a wider dyamic range.

Currently private lines are most commonly used to physically link local facilities to wide area network. Private circuit charges are fixed and are not traffic-adaptive. Other disadvantages of private lines are; rising costs, delays in acquiring facilities, and inability to reroute traffic around faulty link or nodes. Although dial-up services to some extent can be cost effective, they have inherent security drawbacks. In addition to speed constraints, terrestrial facilities are not economic or flexible for interconnection, reconfiguration and growth adoption.

2.2 Backbone Network to Information Services

Providing LAN user the capability to access resources and services available nationally and internationally, will bring into the user enormous computing and communication power. In due course these facilities will form the primary vehicle to convey information services to the end-user. This will aid the decision process and enhance the productivity at all levels. Thus, formation of a wide area network by interconnecting LANs and MANs will provide the end-user the capability to utilize enormous communication, computing and information services. The value of such a powerful network will increase as more and more information and computing services become accessable. It is the authors' view that such a powerful, flexible, traffic-sensitive backbone network can be realised using satellite technology.

3. SATELLITE-BASED WAN

3.1 Network Concept

The configuration of a satellite-based WAN is shown in Fig 1. First, multiple LANs at a location can be interconnected with appropriate gateways or bridges. They inturn can be linked to the nearest MAN by cable, fibre obtic, or microwave digital radio links. The MANs of different locations are then linked through a high bit rate satellite link. LAN clusters which do not have or require immediate access to MAN can have on premises low bit rate satellite earth stations.

Wide Area Networking

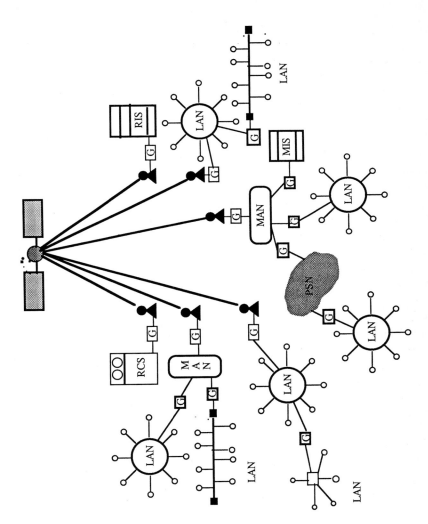

FIG. 1 WIDE AREA NETWORKING VIA SATELLITE

Computing and information services centres can be linked to the network with appropriate gateways, satellite resource allocation and network management can be controlled by a Hub Station (centralized) or distributed among the network nodes.

A number of factors will determine the type of satellite network required. These include, the number and type of LANs, MANs and terminals, terminal activity, application and the overall traffic characteristics. Satellite network can be designed to maximize flexibility and minimize transmission costs by appropriately segregating the inter-LAN traffic on the basis of message type (datagram, stream, etc.), traffic volume, delay tolerance and bit integrity (low-error, error-free). A TDMA type system with traffic-adaptive access protocol capability will be suitable. Currently there are systems (very small aperture terminal-VSAT network) which will support random access Aloha, reservation Aloha, demand or pre-assigned virtual channels at different bit rates. Message transfer delays for delay-sensitive traffic can be minimized either by additional coding, or by separate high performance low bit rate channels or low bit rate terrestrial feed-back channels. A detailed innovative satellite network design can be done with the knowledge of a well defined traffic model. However, this is not pursued here.

Perhaps the most important element of a flexible satellite-based WAN is the earth station-to-earth station gateway and the associated interface processor. This gateway, at the satellite end, should be able to handle broadcast type satellite protocols, operate at wide range of bit rates, and tolerate differential delays. The gateway or the interface processor should be able to segregate messages and adaptively route through appropriate satellite facility. On the other side, the gateways should be universal to interface with diverse LANs and MANs.

3.2 Inter-LAN Traffic Assessment.

The knowledge of inter-LAN traffic is very important to evolve an efficient and innovative satellite network (See Fig 1). Failing to locate any measured data in the open literature about the inter-LAN traffic, we have attempted to make a parametric assessment. The inter-LAN traffic is a function of a number of parameters; LAN speed, cable length, message length, number of active devices, average peak-hour device activity, application etc. An inter-corporate traffic of 20% and intra-corporate traffic of 30% have been assumed for the traffic outflow based on a market survey. The LAN parameters assumed are: 1) Type = Ethernet; 2) Cable Length = 0.5KM; 3) Capacity = 10Mbps; 4) Channel Efficiency = 89%; 5) Packet Length = 1024; and 6) Average Device Activity = (1% - 20%).

In Figure 2 the peak hour average inter-LAN traffic based on the assumptions above is drawn as a function of device activity for both the inter-corporate and intra-corporate cases.

4. CONCLUSION

In this paper, the need for internetworking of LANs and some of the problems of the existing interconnection facilities were discussed briefly. A satellite based wide area network comcept and inter-LAN traffic assessment issues were also discussed.

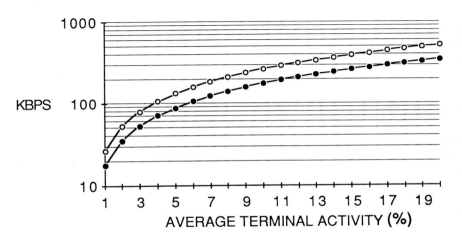

SESSION 2

INTERSATELLITE LINKS: GEO-LEO

Chairman: G. Mica *(ESA/ESTEC, The Netherlands)*

ESA's UNDERTAKINGS IN THE FIELD OF OPTICAL
INTERSATELLITE/INTERORBIT LINKS:
PROGRAMME, CAPACITY AND LINK CHARACTERISTICS

M.Wittig, G.Oppenhaeuser
European Space Research Centre
ESA/ESTEC
Keplerlaan 1
2200 AG Noordwijk, The Netherlands

Abstract

Intersatellite links allow to optimise network interconnectivity or to provide new sevices in a more favourable manner. These links can be established either at RF frequencies in the Gigahertz region or at optical frequencies. The advantage of optical frequencies is the available large bandwidth and the ease of suppression of external interferences due to extreme narrow beamwidth. Potentially optical system can be made smaller, ligther and are consuming less electrical power when compared with their RF counterparts at comparable capacity. However, there are complex technical and technological problems to be solved before such a system is mature for operational application. These problems are mainly related to the pointing and tracking system, the optical characteristics of the various elements and the reliability of the laser diodes and the required detectors.
ESA, in the frame of the PSDE programme (payload and satellite development and experiments) is developing an optical communication system by means of which data links will be established either between two geostationary satellites or between geostationary satellites and a low flying satellite.
In this paper the design concept of the system will be described and the expected link characteristics will be given. Advantages and problems related to various modulation schemes will be explained. Bit errors in an optical link will not have the well known distribution caused mainly by thermal noise effects in the link, but will occur as burst errors caused by excessive pointing errors. The statistics of such burst errors will be described and the effect of coding will be shown.
Finally the programme planning of PSDE will be outlined and an outlook on future developments of optical communication will be given.

I **Introduction**

Optical communication with semiconductor laser diodes was identified as a promising candidate for a European data relay satellite DRS. But system feasibility has not been demonstrated yet. Therefore significant development effort is necessary for delivery of flight-hardware.
For a launch of DRS in 1996 the semiconductor laser diode communication system should meet the following schedule:

      1988   Proof of concept
      1989   Components qualification
      1993-94   In-Orbit demonstration of optical payload
      1996   Operational DRS

For a diode direct detection system based on GaAlAs laser diodes operating in the wavelength region around 800 nm all basic components are available today.
ESA started within its payload and spacecraft development and experimentation program (PSDE) the development of an optical payload based on todays technology for in-orbit demonstration in 1993. SILEX (Semiconductor Intersatellite Laser EXPeriment) was selected as baseline for the in-orbit demonstration mission for optical payloads. This in-orbit demonstration mission scenario is shown in Fig.1.
The maximum data rate between the GEO satellites - called the intersatellite (ISL) link - is 120 Mbit/s. For the link between LEO and GEO spacecraft - called the interorbit link (IOL) - the maximum datarate in the return link, i.e. from LEO to GEO is four times 60 Mbit/s whereas for the forward link (from GEO to LEO) only 1 Mbit/s is foreseen. In order to serve different users (at LEO) and to ensure a good link quality of the experimental system the datarate is variable in such a way that possible datatrates are integer fractions of 120 Mbit/s namely 60, 30 and 15 Mbit/s.
The blockdiagramme of the optical payload for the GEO-2 spacecraft is shown in Fig.2. The blockdiagrammes of GEO-1 and LEO terminal are very similar to GEO-2.
The telescope diameter of the LEO terminal is 20 cm and for the GEO-2 35 cm is choosen whereas for GEO-1 a final selection is not made yet, but a telescope diameter around 20 cm or below is foreseen. The optical transmitter power for SILEX is assumed to be 30 mW. As a consequence of this power level limitation the transmission capacity in the return link (i.e. from LEO to GEO-2) is limited to four times 60 Mbit/s.
On LEO a multichannel transmitter assembly consisting of four twofold redundant laser diode assemblies for 821, 827, 833 and 839 nm nominal center wavelength is foreseen. The modulation format is 4-PPM. The wavelength multiplexer consists of a combined filter/beamcombiner assembly. At GEO-2 receiver the same approach is used as wavelength demultiplexer. The wavelength division mulitplexing scheme is described in detail in Ref.1.
The data detector for each channel consists of two redundant Si-APD's followed by a low noise transimpedance amplifier.

For the IOL forward link a threefold redundant diode laser transmitter is located on GEO-2 and a twofold redundant receiver front end on LEO.
The GEO-2 terminal will be capable of operating with ground, with a LEO satellite and with GEO-satellites positioned east or west of GEO-2.
For acquisition and tracking two different CCD-matrix sensors are applied. For compensation of satellite motions and for tracking of a relative moving target a fine pointing assembly (FPA) consisting of two orthogonal arranged mirrors driven by electromagnetic torquers is used. The fine pointing assembly is further used for generation of a search pattern during acquisition phase.
For acquisition GEO-2 illuminates the probable location of LEO or GEO-1 with a seperate beacon laser. The beacon laser is a laser diode array assembly with a beam divergence of 850 microrad operating at 860 nm wavelength. Both terminals generate a scan pattern independent from each other. The displacement between GEO-2 and LEO or GEO-1 line of sight is measured by the acquisition CCD-matrix and used for correction with the FPA.

The interface to the feeder links requires the data to be transmitted in baseband NRZ-format, the output from optical receiver delivers NRZ-baseband format as well as the associated clock. The optical transmission system is a transparent data channel. The link performance is compiled in Table 1.

II Data Transmssion Performance

The selected direct detection schemes requires intensity modulation of the laser radiation. With semiconductor lasers this modulation scheme is easy to achieve by modulation of the injection current of the laser diode. The simplest modulation scheme is the so called on-off-keying (OOK): the laser output power is switched on or off by the information which is to be transmitted. The bit error rate is a function of received optical signal power, of optical receiver front end (RFE) performance, of received background power level and of detection threshold. If this threshold is set into the center between signal level of received one and zero (which is the usual methode in microwave transmission/detection systems) the bit error rate (BER) as a function of received optical signal power plotted in fig.3 is obtained. The assumed receiver front end consists of an avalanche photo detector followed by a low noise transimpedance amplifier. But the noise after the optical detector is a function of received optical signal power. This is contrary to microwave or RF systems were the noise is independent of received power level (thermal noise). It can be expected that the choice of center threshold is not optimum for detection of optical signals. With an optimum threshold, calculated using the different noise power levels for a one and a zero the BER as function of received power level with the same RFE is also shown in fig.3. To achieve a BER of 10E-6 2.6 dB less optical received power is required with this optimal detector. But the question of realization of such an optimal detector is still open.

A second modulation scheme is considered here. It is the M-ary pulse position modulation (M-PPM). The idea behind this modulation scheme is that a laser is used which is able to produces very short pulses and a peak power much higher than the cw-power. Therefore the receiver has a better chance to discriminate a received pulse against the unavoidable background power. This is of advantage for the high data rate LEO-GEO-2 link in which the GEO-2 receiver sees always the earth radiation (albedo) as background power. The coding of a serial data stream into a M-PPM signal is simply the transformation of a block of N bits into its decimal equivalent M=2E N and transmission of a logical one in the corresponding time slot. The required transmission bandwidth increases exponential with M and limits therefore M for data rates in the Mbit/s region . For 4-PPM the BER as a function of received optical power with the same RFE as for OOK is also shown in fig.3. To achieve a BER of 10E-6 a received power 4 dB below the required value for OOK with optinal threshold is required. Here a maximum likelihood detector (ML) is assumed. Such a data detector was not used up to now. But in frame of PSDE two different versions of ML-detectors are considered and a breadboarding is foreseen to evaluate its performances.

Another very important point assumed for calculation of BER=f(P) for 4-PPM was that the power transmitted of the laser diode per time slot is 4 times the cw-value,e.g. 4 * 30 mW = 120 mW. It is known from experimental investigations (Ref.2) that not all types of laser diode and not all elements of the same type would operate properly under this condition. In addition it must be investigated wether the increased peak power does not degrade the operational lifetime of the laser diode to unacceptable low values.

Finally the BER using forward error corection coding (FEC) with a very short code of 22 message bits and a 2 error correction capability is plotted in fig.3 for the three modulation/detection schemes discussed above. With FEC a reduction of required received power about 2 dB is possible.

III Special Aspects of Spatial Tracking

One advantage of optical intersatellite/interorbital links from the operations point of view is its extremely sharp beam which allows to communicate with neighboured satellites without interference. But to obtain this advantage one has to pay a price for a precise beam pointing system which have to compensate the motions of the satellite, because the attitude instability of satellites is several orders of magnitude higher than the optical beamwidth. The impact of satellite motions on optical intersatellite data links was investigated by computer simulation. In this approach a tracking control loop with a very high signal to noise ratio was assumed and the BER as a function of observation time was obtained. The improvement of system quality using FEC was also investigated and it was found that short block codes, having a great amount of redundancy, can reduce the BER during a large part of the observation time. But if the BER is worse than a threshold, which depends on the code choosen, an improvement of BER is not possible. Due to the duration of reduced BER in the order of milliseconds, caused mainly by the bandwidth of the beam pointing control loop, one talks of burst errors. Fig.4 shows the result of a simulation for the SILEX IOL return link. The transmitted power is 30 mW, the data rate is 60 Mbit/s. For the LEO a mean pointing error of 0.5 microrad and a rms value of 0.3 microrad is assumed, and for the GEO a mean value of 0.5 microrad and a rms value of 0.25 microrad. This plot gives the first 1000 samples from the total length of simulation of 20000.

Another interesting question concerns the behaviuor of tracking system if the tracking loop has a limited signal to noise ratio. Fig.5 gives the pointing error in one axis influenced by sensor noise as a function of observation time for a signal to noise ratio of around 15 dB, which leds to an instable tracking system. With a received signal power 2 dB higher than in the previous case a stable behaviour is obtained. As a result of this simulation it must be pointed out, that the tracking sensor/control loop noise plays a very important role in optical intersatellite/interorbit links and can override the effect of satellite vibration/motion on link quality if the tracking link budget degrades below a value depending on system parameters, like tracking sensor characteristic, detector noise performance, loop bandwidth,etc.

V Schedule

PSDE/SAT-2 will carry the GEO-2 optical payload of SILEX. The launch date for SAT-2 is mid 1993.
The phase A study of SILEX is just finished and phase B2, which contains mainly breadboarding of critical items, will be started in May 1987. Mid 1989 starts the phase C/D which will be completed by the launch of SAT-2 in mid 1993 which is the start date for phase E, the in-orbit experimentation phase.
As an operational LEO spacecraft SPOT 4 will be equipped with an optical payload for experimental and preoperational use. This optical payload will be developed and manufactured by ESA and handed over to CNES for operation on SPOT 4.
As GEO-1 SAT-1 of PSDE will be one candidate with a nominal launch date in 1992.

IV Outlook

In order to provide the possibility to embark more advanced component/transmission schemes than those actually planned, it is foreseen to connect to the baseline package an add-on channel. This add-on channel could operate at an alternative wavelength using alternative laser sources and/or may use modulation schemes other than direct modulation. Detailed concepts for this add-on channel do not exist yet and it may deliberatly be kept open for a while in order to allow addition of advanced techniques rather late in the program.
That the SILEX concept is the rigth baseline for optical intersatellite/interorbit data links was proved by the announcement of several laser diode manufacturers of high power laser diodes. In SILEX a transmitter output power of 30 mW is assumed and led to limitation of LEO-GEO2 return link data rate of 4*60 Mbit/s. With higher optical output power this limitation can be avoided.
If a 1 W laser diode with a single lobed far field pattern will become available the transmission of 120 Mbit/s with a margin of more than 3 dB between two GEO spacecrafts, which are separated by 140 degree will become realistic with the present baseline of SILEX ISL.

References

/1/ B.Laurent, J.L.Perbos:
Design of a wavelength division multiplex/demultiplexing system for optical intersatellite links.
SPIE Proceedings, Los Angeles, 1987

/2/ M.Roux:
Evaluation of laser diode candidate for optical links.
ESA Round Table Discussion on Optical Space Communication.
2. April 1987, Summary Report

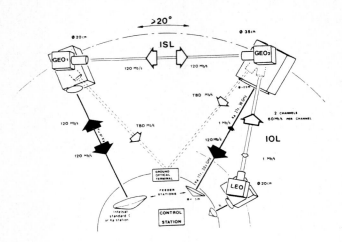

Fig.1: SILEX Mission Configuration.

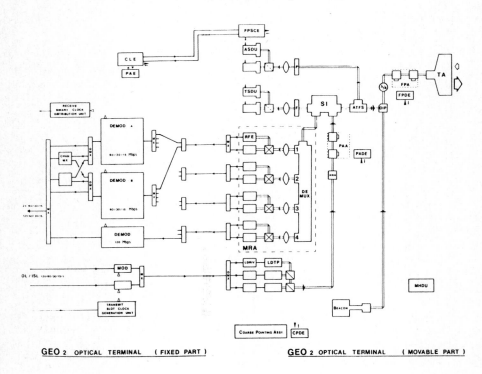

Fig.2: GEO-2 Optical Terminal Blockdiagram.

Link Parameter	LEO -> GEO	GEO -> LEO	GEO -> GEO	
AVERAGE TRANSMIT POWER	.03	.03	.03	WATTS
MAX. DATA RATE	60	1	120	MBPS
WAVELENGTH	821	870	870	nm
and	827	and	821	nm
and	833	or	827	nm
and	839	or	833	nm
		or	839	nm
MODULATION FORMAT	4-PPM	4-PPM	4-PPM	
TRANSMIT ANT. DIAM.	20	35	35	cm
RECEIVE ANT. DIAM.	35	20	tbd	cm
BIT ERROR PROB.	10-6	10-6	10-6	
MARGIN	3.6	3.2	tbd	dB

Table 1: SILEX Communication Link Parameters.

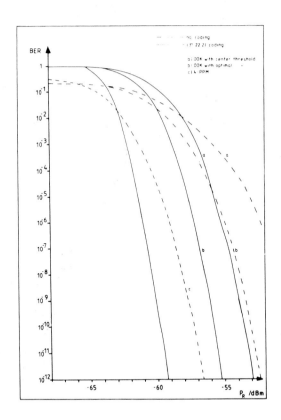

Fig.3: Bit Error Rate as function of received power for various modulation/coding schemes.

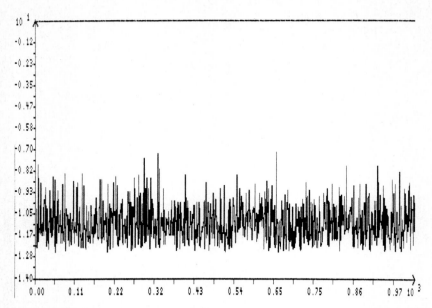

Fig.4: Logarithm of Bit Error Rate for SILEX IOL as function of time.

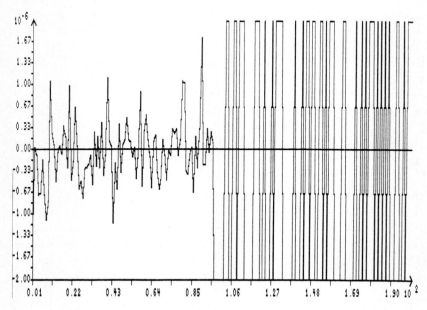

Fig.5: Telescope pointing error in radian for SILEX IOL as function of time.

GEO - GEO ISL PROJECT OF ESA/EUTELSAT/INTELSAT

D. DHARMADASA [1], O. MILLIES-LACROIX [2], R.A. PETERS [3].

1 INTRODUCTION

Intersatellite Links promise a number of advantages to satellite system users. Recognizing this, ESA, EUTELSAT and INTELSAT, initiated discussions in 1985 to establish a framework of cooperation between the three Organizations, with the end aim of demonstrating an in-orbit experimental optical Intersatellite Link. The culmination of those discussions was the identification of an experimental programme consisting of four phases. The first phase of this programme has been agreed by the three organizations. This phase of work is now in progress and it consists of mission and system studies.

This paper summarizes the overall programme and also highlights some of the relevant system and technical considerations.

2 WHY ISLs ?

A major rationale for introducing ISLs between operational satellites is that ISLs would reduce the number of earth stations required for complete system interconnectivity, when the system involves two or more satellite. Also when more than one system is involved, the user requirements for access to both systems could be economically facilitated through the availability of ISLs.

---

(1) EUTELSAT, FRANCE
(2) ESA, FRANCE
(3) INTELSAT, USA

A consequential outcome of this is the widening of the effective coverage provided by the systems involved.

Besides increasing the effective coverage, ISLs also bring an increased flexibility in selecting the orbital locations of the satellites, allowing in some cases higher elevation angles than would otherwise be the case. This is of particular importance for small dish installations at customers' premises in urban areas. Furthermore, the flexibility in the choice of orbital locations would ease frequency coordination problems and would permit a choice of system parameters to minimize mutual interference between FSS/FSS and BSS/FSS. For example, in Figure 1 is shown the situation that could exist for mutual interference between Region 2 BSS/FSS and Region 1 (European) BSS/FSS. The BSS to FSS interference arises from the sidelobe radiation from the high power BSS of Region 1 through the sidelobe of the FSS receive antenna of Region 2. It is clear from Figure 1 that having the flexibility of moving the Fixed Service Satellite away from the Broadcast satellite will enable such interference to be minimized.

3 Joint ESA/EUTELSAT/INTELSAT project

As mentioned in the introduction, the three organisations have identified a four phase programme for the joint ISL project as follows :

Phase I : Mission and system studies (1987-1988), to quantify the advantages of optical ISLs.
Phase II : Critical payload hardware development (1988/1989), and detailed definition of associated satellites.
Phase III: Flight model optical terminal development (and installation on 2 satellites).
Phase IV : IN-ORBIT Demonstration

ESA/EUTELSAT/INTELSAT have agreed on phase I. In June 1987, a contract was placed by ESA with a consortium led by Telespazio and including MATRA, Selenia and CISE, to perform the detailed studies necessary to complete phase I. In this phase, the most value-added applications of ISLs will be identified on the basis of the present status and expected evolution of satellite networks as regards traffic volume, traffic type and space and earth segment configurations.

The improved system efficiency with respect to frequency-orbit resource utilization, connectivity, new satellite configuration, the roles of on-board baseband processing, earth segment configuration, single earth station access to domestic, regional and international systems, avoidance of double hop, routing strategy and network protection will be studied. Critical aspects related to systems design will also be dealt with. Then an experimental system limited in scale and in scope, but sufficiently advanced as regards system and technology, will be proposed as a proof-of- concept experiment for an optical ISL using semiconductor laser diodes. In the proposed experimental system, the ground segment is expected to use existing equipment and transmission standards to the maximum extent possible.

The details of the Phases II to IV have yet to be defined.

4 Possible Time Frame

EUTELSAT is in the process of procuring its new generation of satellites, EUTELSAT II. INTELSAT will issue shortly an RFP for the INTELSAT VII series of satellites. Therefore, the earliest operational ISL may be feasible in the post INTELSAT VII and post EUTELSAT II generations of satellites or possibly later units of EUTELSAT II/INTELSAT VII. According to the present plans of the two organisations this would set a date of mid-1997 as the earliest possible date for an operational ISL.

Therefore, bearing in mind both the long lead time for the procurement of satellites and the need to have some in-orbit experience before including an ISL package on an operationel satellite, ESA, INTELSAT and EUTELSAT have agreed that the joint optical ISL in-orbit experiment should start in 1993, or soon thereafter.

5 Objectives for Operational ISLs

An ISL would form only a part of an overall connection and, therefore, its performance and other attributes must be compatible with the requirements laid down by the various regulatory bodies for such connections.

For data and voice transmissions the end-to-end performance objectives are defined by the CCITT. The recommendations for satellite link (Hypothetical Reference Digital Path - HRDP) performance are made by the CCIR to conform with the overall CCITT objectives. The performance of an ISL must be compatible with these objectives. Relevant aspects are discussed below.

(a) **Long Term Performance**

A satellite network containing an ISL will inevitably form a part of the ISDN (Integrated Services Digital Network). The CCITT Recommendation G821 defines the error performance objectives for international ISDN connections as well as the portion attributable to a satellite HRDP.

The CCIR have converted the satellite HRDP allowance of Rec. G821 into the following form which is more suitable for satellite system design (CCIR Recommendation 614):

"The bit error ratio at the output of a satellite HRDP operating below 15GHz and forming part of a 64 kbit/s ISDN connection should not exceed during the available time the values given below:

- 1×10^{-7} for more than 10% of any month;
- 1×10^{-6} for more than 2% of any month;
- 1×10^{-3} for more than 0.03% of any month."

It should be noted that in converting from CCITT Rec. G821 to CCIR Rec. 614, the assumption has been made that the bit errors occur randomly (i.e. no bunched errors) and that they would have a Poisson distribution.

Whilst this may be true for Gaussian noise induced errors in networks not using error correcting techniques (Forward Error Correction - FEC), for those networks using FEC this assumption is not valid. This is particularly the case for the INTELSAT/IBS and EUTELSAT/SMS/SCPC networks, which use 1/2-rate convolution encoding, and the INTELSAT 120 Mbit/s TDMA system which uses 7/8 BCH (Bose Chauduri Hocquenghem) code, to enhance performance.

The modifications that may be necessary to the CCIR Rec. 614 when different FEC codes are used needs to be determined during Phase I of the project. In this analysis it would also be necessary to take into account any other bunched error causing mechanisms peculiar to ISLs such as burst errors due to mis-pointing.

(b)   Availability

The CCIR recommendation on performance objectives for a satellite HRDP applies when the link is "available". A link is considered not available when the information BER excess 10^{-3} for more than 10 seconds or if the digital signal is interrupted (i.e. alignment or timing is lost).

The allowable unavailable-time due to non-propagation effects is 0.2% of a year (CCIR Rec. 579). A proportion of this 0.2% may be attributable to the ISL for solar conjunction. Solar conjunction (when the sun will be shining directly on the detector of the ISL) could give rise to a total loss of the ISL and will contribute directly to the unavailability allowance. Solar conjunction will occur twice a year and last about one minute each time. While it may be possible to track during solar conjunction, it is unlikely that an optical ISL could maintain a communications link during this time. Because two spacecraft are involved, solar conjunction will lead to an outage of at least 4 minutes per year. A longer outage may result from imperfect baffling.

(c)   Orbital Separation

Transmission delay is a critical element for speech. Also for data, especially for situations where ARQ is used, this could be a significant design constraint.

For speech, CCITT sets a transmission delay limit of 400 ms. In order to meet this limit the orbital separation between the two satellites connected by an ISL needs to be less than about 60°.

6 Some critical aspects of optical ISLs and brief technology review

The design of systems involving high capacity duplex ISL's for transmission of either high data rate digital or wideband analogue signals has to deal with many technological constraints. These constraints will mainly influence the following parameters:

- transmit power capability
- receiver sensitivity
- performance of the Pointing Acquisition and Tracking (PAT) subsystem
- optical quality
- device reliability

Most recent R&D and studies have investigated various technological options, for IOL (inter orbit link : LEO-GEO) as well as for ISL (GEO-GEO) system. For medium term applications, the most promising approach appears to be the use of GaAlAs semiconductor laser technology, with direct detection and wavelength division multiplexing (WDM) schemes. For longer term applications InGaAsP semi-conductor lasers are attractive, either in systems using direct detection or heterodyne detection.

Therefore, in the first phase of the project, the three following options will be considered:

- direct detection, WDM, GaAlAs (wavelength~0.8 μm)
- direct detection, WDM, InGaAsP (wavelength 1.3 - 1.5 μm)
- heterodyne detection, InGaAsP.

The most technically mature approach now is to use semi-conductor lasers of wavelengths of either 0.85 μm or 1.3 μm with direct detection and baseband intensity modulation of the laser through the injection current. The heterodyne techniques are less mature.

7 Conclusions

Some possible advantages of ISL's for operational use which led to ESA, EUTELSAT and INTELSAT to agree to conduct a joint programme to demonstrate the practicability of optical ISLs have been described. The programme plan summary and some details of its initial phase has also been given. Finally the criteria to be taken into consideration when designing operational ISLs and a brief summary of the status of technology developments which could influence the studies conducted under the joint programme have been highlighted.

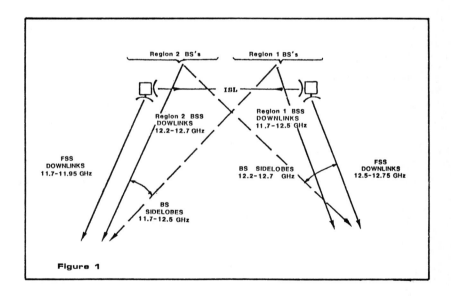

Figure 1

DATA TRANSFER OVER AN INTEGRATED COMMUNICATION NETWORK
FOR AN ORBITAL INFRASTRUCTURE

L. ZUCCONI and C. SMYTHE - UNIVERSITY OF SURREY - Guildford - UK.

ABSTRACT

A major feature of the European Orbital Infrastructure is the communication network; this facilitates the exchange of information between space and ground through interconnection of the various elements of the infrastructure.

The aim of this work is to provide an overview and an independent synthesis of a global scenario which is to be discussed within ESA, and which will contribute to the gradual implementation of the integrated communication network.

This paper provides an overview of a procedure for data transfer over the space to ground link of the communication network.

1. INTRODUCTION

At the ESA Council meeting at Ministerial level held in Rome in January 1985, a long term plan was approved for space activities in Europe until the year 2000 (and beyond).

This plan focusses on, and supports, a basic space infrastructure with a view to European autonomy, and centres on transportation using the Ariane-4 and 5 launchers, the Hermes spaceplane, participation in the Space Station Programme with COLUMBUS, and the preparation of future Earth observation and communication satellite systems.

An essential part of this infrastructure is the communication network, which will enable the exchange of information between space and ground via the various elements of the infrastructure. The communication network is based on the use of Data Relay Satellite Systems (DRSS) which will satisfy communication requirements between Low Earth Orbit (LEO) user satellites and ground.

The data generated within the LEO user satellite are received by the earth terminal of the control centre via the DRS and are routed through the control centre network to the end user (directly or via ISDN public networks).

The data to be transferred can be subdivided into continuous data (i.e., video, voice) which requires minimum end-to-end delay, and into computer data which requires minimum error rate and less contraints in terms of delay.

2. DATA TRANSFER

This description of data transfer comprises two sub-topics, the first dealing with the problem of establishing an access procedure over the space-to-ground link, and the second dealing with end-to-end data transfer (this end-to-end data transfer includes data communication networks both on ground and on board, i.e, LAN's and/or WAN's). For each of these sub-topics, the options are described and, where possible, more emphasis is given to the more adequate solution.

2.1 Access to the Space/Ground Link

Two methods of accessing the link are analyzed:
- Synchronous TDM (PCM) used in telephony (fixed-frame, fixed allocation of time slots to each channel, etc.).
- Logical Multiplexing (transmission resources are allocated according to traffic requirements).

The physical link over which the information is transferred shall multiplex data originated by N sources and addressed to N destinations.

Four major evaluation criteria are considered for trade-off analysis between the above mentioned methods:

. Synchronization technique; a classical synchronous TDM network relies on a centrally distributed clock such that the TX and RX are locked together at bit level. Logical channel multiplexing transmission relies on sync pattern recognition.

. Efficiency in terms of throughput; in synchronous TDM the number of bits is counted, and in a specific instant the destination of data is known (i.e., there is no need for in-band signalling and no overheads per information unit transmitted, on the other hand, out of band signalling is needed). In the case of logical multiplexing, in-band signalling is required since there is no correlation between the time clock and the data destination (this causes an increase in overheads).

. Efficiency in terms of flexibility; where synchronous TDM synchronisation assumes that the time slot made free by one user is taken by another on the basis of statistical network traffic. This assumption works well in a large community of users (e.g., PTT users), but in fact would greatly limit efficiency in an environment where user data types are heterogenous, and the number of users is limited.

NOTE: Heterogenous means a slot allocated to a type of data (computer transaction), which cannot be easily reallocated to another type of data (e.g., video data). Statistical allocation of available slots does not work as well as for a small number of users.

Implementation of priority scheme; logical multiplexing allows for a very flexible solution to this problem by allocating a bandwidth on request to high priority users, thus maintaining nominal efficiency.

With synchronous TDM this is only possible through rigidly allocating a higher number of slots within the available bandwidth.

When comparing the advantages and draw backs of these two proposed baselines in terms of user requirements, it appears that the logical multiplexing approach will offer better performance, and has the advantage of being compatible with existing link access procedures used in space systems for scientific satellites (Packet Telemetry Standard).

The present Packet Telemetry Standard at the level of a link access procedure has the following main features:

. Frames have a fixed length, and consequently a fixed repetition rate, thus simplifying sync pattern recognition.
. A logical channel (virtual circuit) identifier, which allows for real statistical multiplexing between frames.
. Each logical channel carries the data either of one single user (dedicated) or of a group of users (shared). Therefore, when a user is not active, the logical channel can be skipped, or reallocated to other users (thus providing the advantages of flexibility and optimum bandwidth usage).
. Control information carried in the frame header allows for sequence control of all frames received, and of frames per logical channel.
. The frame error control field, implemented by the polynominal CRC and/or Reed-Solomon coding, is able to ensure the BER on the link is suitable for the required grades of services most required.

This basic structure is adequate for the variety of classes of data already defined. Examining each basic class of data will give us the following details:
. Video high rate (user has large bandwidth requirement), may be allocated one dedicated logical channel with appropriate priority in order to meet the maximum allowed access delay.

. Audio (low rate and possibility of fragmentation of information); digital audio channel requires a nominal bandwidth of 64 Kbit/s, and assuming a transfer frame length of 10,000 bits, the number of bits per frame varies according to the channel data rate.

From an efficiency point of view, it would be ideal to wait for 10,000 bits to be generated in order to fill up one frame. However, it is reasonable to assume that the delay between successive audio samples should be in the order of 20 to 50 us, and therefore additional fragmentation of the audio information will be required. The result is that in each frame, a small number of audio bits will be available for transmission, thus decreasing efficiency. A possible solution to this problem is the sharing of the frame data field by audio channels and other users. This implies some difficulties, mainly in the area of signalling (channel allocation/deallocation), and in influencing traffic which shares the virtual channel with audio.

As a final consideration, we may say that both solutions (dedicated or synchronous insert) are technically feasible; the selection criteria are mainly linked to the frame repetition rate and the maximum acceptable audio sample length.

In the case of asynchronous user data (computer data), users require a low data throughput compared to high rate payload, and therefore, in oder to efficiently use the link, users should share one single logical channel; this of course requires the implementation of higher level multiplexing. In the present Packet Telemetry Standard, this is supplied by the Source Packet, and the issue of adequacy of the source packet for such a function being applied to new users is addressed in some detail below.

2.2 Transfer of Data over the Interconnected Environment

The method of accessing the link from space to ground and vice versa previously described is based on the assumption that all sources and destinations of data on board and on ground were physically local to the two end points of the link (LEO terminal and earth terminal).

This assumption was made in order to simplify the global scenario of the communication network. In fact, the scenario is rather more complex because on-board users are interconnected to the LEO terminal, via an on-board LAN, as well as to ground users, via the decentralized control center network (where interconnection may be established via LANs, or via public wide area networks).

As a consequence, we can state that end-to-end data transfer takes place over interconnected sub-networks (in this respect the space-to-ground link is considered a sub-network). Refer to Figure 1.

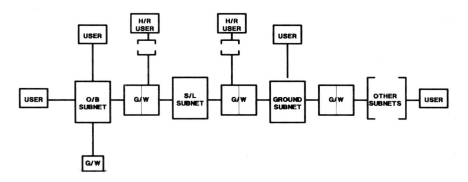

Figure 1 SPACE-GROUND NETWORK

G/W, GateWay
H/R USER, High Rate User
O/B SUBNET, On-Board Sub-Network
S/L SUBNET, Space Link Sub-Network

The data communication technique to be adopted for transferring information through this environment depends (among other things) upon the characteristics of the data generated by end users. The majority of end users on the Low Earth Orbit terminals are computer-based payloads with significant autonomy, requiring large amounts of software and therefore transfer of files. This type of data is of a "burst" nature (asynchronous with respect to accessing the network), and is based on interactive transactions between users. Hence, a transmission based on a data structure "packet" seems to be preferable to the synchronous distribution of data.

Given the structure of the interconnected environment already described, it is necessary to provide a mechanism capable of handling data transfer of this type.

Two basic options can be considered for the implementation of this function:

(1) Data transfer takes place entirely via protocol; each user data unit is formatted in such a way as to provide full addressing capability throughout the system (datagram-like approach).

(2) Data transfer is completely handled by management; in this case the user needs only to be known at the local network interface, and the communication management function facilitates delivery of data to the final destination.

There are advantages and disadvantages to each of these solutions:

(1) A fully data-driven protocol (e.g., the ISO 8473 Internet protocol) implies a large overhead on each data unit transmitted (low efficiency), and provides a low grade of service in terms of reliability. In fact, such a system does not guarantee correct sequencing of the delivery of data units (without omission or duplication), and can also, in the event of link breakdown, create an unknown situation on the state of transmission that is the users responsibility for recovery from the instant of link interruption. All this is, of course, true in the case of connection-less service (datagram).

(2) The transfer by management solution certainly simplifies the highlighted problem but does imply a very heavy processing capability within the internal elements of the network. This can be a problem for the processing power required by the on-board processing unit, especially due to the relatively high data rates involved.

A possible alternative is a compromise between the two, and involves the establishment of an end-to-end logical link over which a relatively simple network protocol may be run according to requirements.

In such a case the logical path for data transfer is initially established by means of a connection request phase (signalling), and then data transfer takes place by means of data packets which can themselves use either a connection-less, or a connection oriented, protocol according to the identified need. The basic principles of this solution are inspired by the switched virtual circuit service of the CCITT X25 recommendation; this solution combines the low overhead obtained and reliability in the transfer by management solution, with the relatively simple protocol handling of the data driven approach.

The solution described above holds for those classes of data which:

. Can stand an asynchronous type of service by the network with an apriori random access delay.
. Produce data at rates which can be effectively handled by the present generation of data networks available both on ground and on-board.

There are some classes which, for either or both of the reasons listed above, are excluded from this classification:

. Audio data which requires asynchronous or quasi-synchronous transfer; this cannot be easily supported by present public data networks, and therefore requires special care in transmission.
Video data with synchronized audio also falls into this category.
. Payloads (including video cameras) whose generated data rates are far too high to be handled by any on-board LAN or ground network; it is worth mentioning here that a SAR (Synthetic Aperture Radar) processor produces data in the order of up to 300Mbit/s, but multi-band optical IR Thermatic Mappers will go above 500 Mbit.

For these data classes, the interconnected environment described above can be simplified by making two basic assumptions:

(1) On-board, the physical connection between the user and the LEO terminal is via a point-to-point dedicated link (i.e., no on-board LAN).

(2) On ground, data transfer is assumed to terminate at the ground terminal (or at the user site when direct data distribution is considered).

Under this hypothesis, the problem is reduced so as to provide fast and simple access to the space-to-ground link.

The selected solution is to provide such user data with direct access to the data field of the Transfer Frame, where each high rate user is allocated one dedicated virtual channel (from an efficiency point of view, this is justified by the high rates produced), so that Virtual Channels can be efficiently filled.

The requirements of synchronous transmission over the link are fulfilled by the implementation of a priority scheme; in this way virtual channels carrying synchronous data gain immediate access to the link when ready.

3. CONCLUSION

Given the major evaluation criteria as suggested earlier in the paper, the conclusion is that such a procedure provides a flexible and efficient space telecommunication network capable of satisfying user requirements.
In particular it would provide: minimum transmission delay, user access direct or via ISDN, flexibility in bandwidth-to-service allocation, further extensions.

4. REFERENCES AND BIBLIOGRAPHY

. Consultative Committee for Space Data Systems (CCSDS), Space Station: Application of CCSDS recommendations for Space Data System Standards to the Space Station Information System Architecture "Green Book".

. Telescience Requirements, Interactive Payload Operations on Columbus, Report of Telescience User Team (ESTEC), 1987.

. Union Internationale des Telecommunications, Comite Consultatif International Telegraphique et Telephonique, Tome III-Fascicule III.3, Tome VIII-Fascicule VIII.3, Tome VIII-Fascicule VIII.5.

. Honvault, ESA, Future Integrated Space Data System, AIAA/NASA International Symposium on Space Information Systems in the Space Station Era, Issue 22-23 1987, Washington, D.C.

. Kwei, TU, J.H. Johnson, Space Shuttle Communications and Tracking System, Proceedings of the IEEE, Vol.75 No 3, March 1987.

. G.A. Beck, The NASA Advanced Communications Technology Satellite (ACTS).

. G. Maral, M. Bousquet, Satellite Communications Systems, Wiley Interscience Publication.

ACKNOWLEDGEMENT

The authors wish to thank Mr. A. BODINI, ESA/ESTEC, for the support provided in this work.

OPTICAL COMMUNICATIONS LASER DIODE INTERSATELLITE LINKS

A. ARCIDIACONO
Advanced Systems Dept.

S. DE VITA
Marketing Dept.

SELENIA SPAZIO S.p.A. - SPACE SYSTEMS DIVISION
Via Saccomuro, 24 - 00131 ROMA

In the last two decades satellite optical communications have been analyzed by communications theorists and by device designers. In fact most of the hardware development efforts have been devoted to guided light systems and only a few experiments have been conducted on unguided optical channels.
In recent years a new interest has arisen in optical satellite communications for high data-rate links. This interest has been focused on Relay Satellite Systems ACTS (NASA), DRS (ESA) and DRTS (NASDA) where a connection must be established between a Low-Earth-Orbiting (LEO) platform or satellite and a relay satellite in geostationary orbit and then transmitted to the ground station operating at an overall data rate of up to 500 Mbit/sec.
The use of an optical intersatellite link (ISL) between two geostationary satellites for the future INTELSAT and EUTELSAT Satellite programs have been also taken into account. In order to follow these new interests in optical communications terminals in its PSDE program which will test relay and intersatellite optical communications performance and systems degradations.
The scope of this paper is to analyze the critical areas in optical communications systems with particular emphasis on the modulation/demodulation techniques.
The design of a 4-PPM or a QPM demodulation with an analyses of the system implementation problems at high data rate has been carried out. A special emphasis has been given to Maximum Likelihood (ML) demodulation techniques, slot and frame synchronization techniques have been done. An attempt was made to define a strategy in order to obtain frame synchronization tracking without using unique words along data patterns.
Finally, for the optical ISL the use of coding was suggested to relax link budget requirements. The use of coding could in fact be useful reducing problems due to pointing subsystem instabilities. The use of on board soft-decoding without system bandwith expansion is also foreseen.

1. INTRODUCTION
The design of a laser communication system, when operating as a satellite-to-satellite (geostationary) links, has to be defined to meet the following leading requirements of low weight and volume, small power

consumption, long lifetime, long range operation (15.000 - 72.000 Km), high data rate.

Starting from the first system requirement a large effort must be made to reduce masss and volume of our optical terminal without affecting system performance. For this reason the fundamental goal will be the reduction of the telescope diameter in order to reduce both its volume and control mechanism mass and volume. A further advantage reducing telescope diameter is obtained in terms the induced effects of Noise Equivalent Angle (NEA) which affects burst error statistical distribution /1/.

The reduction of telescope diameter together with all other requirements listed above will lead our design approach in the direction of a modulation technique able to minimize the required SNR to obtain the design bit error rate (B.E.R.) (normally fixed at 10-6).

Obviously a large number of trade-offs must be performed in order to optimize the transmitter and receiver design requirements.

Among this trade-off /2/ /3/ we have focused our interest on modulation techniques and their possible implementation at high speed (up to 120 Mbit/sec) of a complete MODEM.

The use of coding techniques have also been taken into account even though for the implementation of the PSDE terminal have not been considered to relax communication system complexity.

In section 2 we will show results about communications subsystem coming from a design effort performed along Semiconductor-laser Intersatellite Link Experiment (SILEX) Study. This study was performed by Selenia Spazio under ESA contract with MATRA Espace as prime contractor.

Among the responsability of Selenia Spazio there was the definition of an optical payload for a gestationary satellite (GEO1 terminal) /3/.

In the ISL system GEO-1 terminal is the optical payload to be mounted on a geostationary user satellite. It will be able to optically communicate with the master relay satellite and with ground station via an RF feeder down link. For these reasons this payload could be also seen as a standard optical payload for a future ISL network.

2. COMMUNICATION SUBSYSTEM

From a theoretical point of view the leading parameter to optimize the communication subsystem is the receiver sensitivity. This parameter represent the minimum signal incident power that is needed to obtain design B.E.R..

Starting from this assumption we have designed a feasible communication subsystem for the optical ISL as a result of the performed trade-offs herebelow described. Two are the possible optical communication systems:
- Direct Detection Systems
- Heterodyne Systems (or coherent systems (optically)).

In the first case, direct detection, the information waveform modulates the laser intensity and the receiver through a focusing lens, an optical chain and a photodetector demodulates the field intensity. In the second case we have an heterodyne receiver in which the receiver uses a local laser field to mix with the received field prior to photodetection.
In heterodyne systems a strong focal field (relative to the received signal and background power) is added to the received field prior to photodetection. The photodetector responds to the intensity of the combined field (laser plus background plus local field). Since the local laser power is much larger than

the receiver field power we may operate in this case also in conditions of solar conjunction.

If we use these heterodying systems from a theoretical point of view we will obtain better performance than in the case of direct detection systems. But if we consider the real hardware implementation we have problems of laser diode linewidths, frequency doppler due to spacecraft attitude motions and in general problems of phase coherency.

Moreover the received laser power still depends on the alignment between local and received optical fields. If these fields are not perfectly aligned the receiver area is effectively reduced and the received power (at the detector) is decreased. Hence a heterodyne system requires a more delicate receiver design. This fact alone may preclude its use for spaceborne satellite crosslink application. We may conclude that the use of coherent modulation techniques could be of interest but seems to be more a candidate for a second generation of optical intersatellite links.

In the second case we have an information waveform modulated onto the laser field by modulating the transmitted power with two possible system solutions:
- transparent laser crosslink
- regenerative laser crosslink

For the transparent crosslink an RF link is used for uplink and transmission to the satellite allowing the use of one of the possible multiple-accessing satellites utilizing a direct detection optical link for waveform transmission. A frequency conversion can be useful to match central frequency requirements of the RF links and physical requirements for laser diode modulator and detector.

Unfortunately this system architecture strongly affects receiver sensitivity because of the wide bandwidth required for the optical crosslink together with the low level of output power from the laser diode at transmitting side.

For this reason we have devoted our analysis to the regenerative crosslink.

With this system architecture that has been analyzed in a detailed manner, data on the up link are decoded at the satellite prior to retransmission on the crosslink. The easiest method to operate retransmission is simply to intensity modulate bit waveform directly onto the laser, and decode the photodetector output by standard baseband circuitry.

A detailed trade-off among viable modulation formats for direct detection have been done. Three fundamental modulation formats have been taken into account :
1) ON-OFF Keying (OOK)
2) Quadrature Pulse Modulation (QPM)
3) M-PPM (PPM = Pulse Position Modulation; $M = 4,8$)

The first format could be of interest because for the reduced requirements in terms of demodulation and frame synchronization. Neverthless this solution has been disregarded because we need a threshold detector (with outadaptive monitoring of signal level) and we cannot use a maximum likelihood demodulation technique.

A substantial modulation gain can be achieved using PPM format with maximum likelhood detection. In this case savings in terms of energy per bits increases with the PPM order (M) and them from a theoretical point of view a large value for M could be suggested.

Unfortunately we have some problems when M 8:
- receiver bandwidth increases with M because of the shortened pulse width.
- maximum likelihood receiver complexity increases with M because of the number of comparison to be done for ML decoding (6 comparators for M = 4, 28 comparators for M = 8).
- M can be increased as long as the laser peak power can be increased at constant average power (at present the maximum value of M for laser diodes is fixed at 8).

For these reasons and because of the high information rate required in our ISL (from 15 to 120 Mbit/sec switchable) a 4 PPM represents a good compromise between theoretical performance and real technological implementation. Problems arising here because of high speed have been also seen in view of further enhancement in information rate of future programs (300 400 Mbit/sec INTELSAT-EUTELSAT, 500 Mbit/sec, DRS).

Another modulation format (similar to 4 PPM) of great interest is QPM where two information bits are sent by two possible field of polarization. Unfortunately for the PSDE program this kind of modulation format is not applicable because of system requirements in the optical chain at the receiver that does not allow the use of separate polarization. Nevertheless this kind of modulation format could be of great interest for future ISL communication systems when higher information rates will be required. In this technique that the entire communication link is divided into two independent data channels at half data rate, one driving each channel independently giving at the communication system a further modulation gain (in the order of 3 dB) and an intrinsic redundancy in the event of a single modulator failure.

Moreover a 4 PPM demodulator excluding the frame synchronization structure is very similar to the QPM demodulator structure.

4 PPM demodulator system architecture given in the subsequent section is then easily adaptable to the QPM channel when 2 PPM is used on every polarization.

2.1 High Speed 4-PPM Receiver: Receiver Front End

The fundamental purpose of an optical receiver is to detect the light incident upon a photodetector and to convert it into an electrical signal containing the information impressed on the light at the transmitting end. It consists of a photodetector and an associated amplifier along with necessary filtering.

Whereas achieving optimum sensitivity in the design of a receiver is the basic goal, other considerations influence the details of the design, and in fact may result in a practical receiver with less than the optimum in sensitivity. An example of such considerations include achieving a wide dynamic range in terms of signal levels or in terms of switchable data rates. So we intend to obtain a manufacturable and cost effective design looking at a receiver front end which might be optimum at one bit rate (for example at the highest data rate where we have the most stringent requirements) and non optimum at others.

The basic optical receiver is shown in fig. 1 and consists of an Avalanche Photodetector (APD), a transimpedance Low Noise Amplifier (LNA), which is eventually equalized, the main amplifier and a filtering section.

The choice of an APD was dictated by the substantial increase in the sensitivity of the optical receiver due to the internal current multiplication. The photocurrent generated by the detector must be converted to a usable signal for further processing with the minimum amount of noise added. The preamplifier is here defined as the first stage or stages of amplification following the photodetector.

It will thus be the dominant source of noise added to the signal and hence its design will be the leading factor in determining the sensitivity of the receiver. For this reason a LNA is considered in a transimpedance configuration as a current-to-voltage converter (see figure 1).

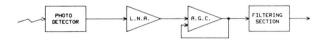

Fig. 1 Low Noise Amplifier

In this case we will have wide bandwidths, a greater dynamic range than using an high impedance amplifier and a noise performance comparable with that of an high impedance front-end.

The most significant parameters which causes compromises in the design of a practical receiver are receiver sensitivity and dynamic range. (9 dB: data rate + 13 dB: distance).

In fig. 2 a detailed block diagram of the RFE is given with details referring to the Automatic Gain Control (AGC) philosophy in order to drive correctly the APD voltage polarization and the subsequent stages (demodulator).

The final element in the receiver front end will be a filter operating on both signal and noise sources. Its main function is to maximize the signal to noise ratio while preserving signal characteristics. The filter will be chosen to reduce the amount of in band and out of band noise.

2.2 High Speed 4-PPM Receiver: Demodulator

At the output of the RFE (a direct detection receiver) we have voltage counts proportional to the incidental energy on the receiver detector with modified (because of APD) Poisson statistics.

As we have already seen at receiver front-end, when the signal level is low it is not convenient to switch circuit elements in order to optimize system performance. For this reason at receiver front -end level we haven't considered any parameter switching.

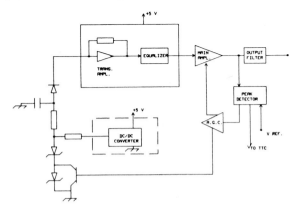

Fig. 2 - RFE Block Diagram

Now at the input of the demodulator a switchable set of filters is considered in order to minimize SNR at the input of the demodulator. With this set of filters with amplifiers it is also useful to have always the same input signal level at demodulator (for every data rate). A system block diagram for the proposed 4-PPM demodulator is given in fig. 3.

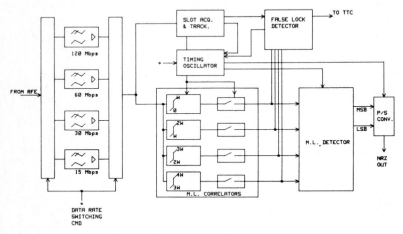

Fig. 3 Demodulator Block Diagram

The demodulator is composed by four ML correlators realized with discrete components because of high speed operation required, a slot synchronizer controlling timing oscillator, a frame false lock detector, a ML detector and a parallel to serial converter for the NRZ output.

A sampling gate is introduced at the output of every integrator in order to give a sufficient time to discharge slot integrators (because of the high data rate). A slot synchronizer is also considered in order to acquire slot timing and to obtain slot tracking during data transmission (see fig.4).

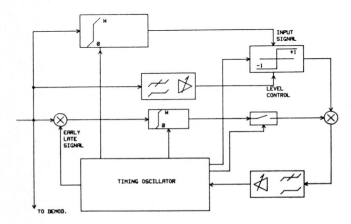

Fig. 4 Slot Synchronizer

An implementation trick is considered in order to overcome problems due to the instantaneous discharge of the integrating capacitor required when we want to implement a slot by slot tracking. Integration for the slot tracking will be done on subsequent slots corresponding to subsequent frames.

In order to solve problems of frame synchronization along PPM patterns there are two possible solutions. The first solution, which is the classical one, may be realized inserting unique words and performing a frame acquisition and tracking via software. Another solution can be considered using signal statistical properties in order to monitor a in-lock condition and to detect out-of-lock circumstances.

Though we do not have any need of using data on board demodulated we cannot use the TDMA synchronization protocols only, because of the difference between QPSK and 4 PPM modulation formats. In fact in the QPSK case we have only one sample per symbol (also when symbol synchronization is lost). On the contrary we may have double or missing pulses inside a symbol frame when a PPM symbol is considered in an unlocked status.

For this reason we will use a false lock detector in order to acquire and to track frame patterns. In the subsequent description we shall assume that slot synchronization has been achieved and there is only the problem to solve frame synchronization. In fig. 5 a detailed system block diagram is given for the false lock detector (including also a part of the slot synch.).

If we compare integrator outputs with a threshold, corresponding to the mean value of the slot energy, it is impossible to detect the evidence of one or more slots along a single frame. Monitoring the events with two consecutive confirmation, a false lock event can be detected. When a false lock event occurs we only have to rotate integrators signal clock enables in order to return to in-lock condition. This condition will be reached when the false lock detector finishes counting consecutive confirmations.

In the case of 4-PPM we are dealing with a SNR of about 15 dB for the ML data demodulation which corresponds to a PD = 0.999 at the input of the adaptive threshold false lock detector. Starting from these values if we use 2 frame confirmations out of 16 frames we will have a first event probability of false lock detection of 0.994.

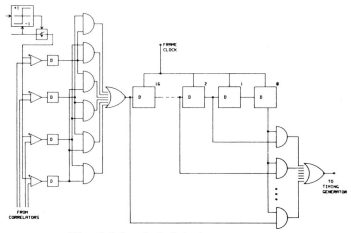

Fig. 5 False Lock Detector

2.3 Coding Techniques for Not-Coherent Modulation

The results obtained along the study performed for ESA /3/ have made clear that, while laser diode optical ISL is technically feasible, every conceivable advanced technology should be used to optimize the link budget. For this reason forward error correction (FEC) seems to be the only viable solution together with the research in the field of more powerful laser sources and photo-detectors.

The design effort toward the use of sophisticated coding techniques and new electroptical components is justified also to finally assess if optical ISLs are real competitors with Millimeter-waves ISL (60 GHz ISL).

In this section we are focusing our efforts toward the use of FEC for an optical ISL. Starting from the assumption that the output voltage of an optical receiver (using APD) is statistically the sum of a discrete count variable and a continuous Gaussian thermal noise variable we may approximate this distribution with a continuous Gaussian density.

Therefore as with the RF satellite channel, random errors together with moderate burst errors would be the error sources to be combatted.
The use of Reed-Solomon (RS) codes seems to be the first obvious solution for the use over the PPM frames since their alphabet, defined over GF (2^m) is naturally matched to the 2^m-ary PPM channel. The complexity and effectiveness of the error correcting algorithm for RS codes is an increasing function of block length for a given code rate. For this reason if we intend to reduce significantly the telescope diameter the use of only RS on ground encoding and decoding (extrapolating from /5/) will give only a slight gain in the GEO terminal mass budget.

The possibility of using short constraint length convolutional codes have also been taken into account by several authors /6/ - /7/ looking at 2^m - ary PPM as a parallel combination of completely correlated Binary Erasure Channels.
Results obtained by Massey are applicable only to the erasure channel where background radiation can be disregarded. This last condition in our ISL is always respected (if we except solar conjunction) and a further effort toward the evaluation of system performance could be interesting also in view of the use of the VLSI Viterbi decoders today under advanced development /8/. A further problem that could arise in this case is represented by the hardware implementation of PPM demodulator operating at twice the information rate if we intend to use a rate 1/2 convolutional code.
In order to overcome this problem, using the same hardware complexity of Viterbi decoders, we may introduce combined coding and modulation techniques using an optical set partitioned PPM format together with a convolutional encoding.

Starting from the previous assumption that the output of the APD is statistically Gaussian asimptotic coding gain have been obtained for a 4-8 PPM interleaved structure /10/.

In particular, if we consider on board implementation with present technology an 8-state 4-8 PPM decoder seems to be feasible with an asimptotic coding gain of about 4 dB over uncoded 4-PPM in the assumption of ideal soft decision.

In order to obtain a further improvement in terms of coding gain the use of combined on board Viterbi soft decoding, for the convolutional encoding or for the Trellis code modulation, with on ground RS decoding is foreseeable

to reach an overall coding gain in the order of 6-7 dB with a low level of redundancy.
Finally it has to be stressed that the introduction of on board decoding does not affect RF up and down links that will look at the optical ISL as in the case of an uncoded system.

4. CONCLUSIONS

As with any spaceborne communications package, the most obvious design trades involve the macroscopic parameters: performance, size, weight and power consumption. Nevertheless, because laser diode communications systems today utilize many new technologies and really a small number of off-the-shelf building blocks, the system designer must also be sensitive to the less tangible parameters such as design and hardware complexity, testability and above all development and production cost competitivness.

Obviously as more such system become operational, less subjective will become the mathematical relationships between the less tangible parameters.
In our system, design requirements for the communications links were: data rate (up to 120 Mbit/sec), required bit error rate (10^6) and range (from 15.000 to 72.000 Km).

These requirements have driven system design together with the specific requirements of an optical communications system: link acquisition and pointing control, transmit beamwidth, satellite attitude motions, requirements for the optical and thermal components.

Starting from these assumptions an ISL terminal has been carried out /3/.
In table 1 an overall performance summary of the designed ISL terminal is given.
Thightly referred to this performance summary are mass and power budget evaluation given in table 2.

Results given in these tables represent a design effort toward the optimization of an experimental optical communications package using technology today available on the market.

The results of the study can be improved looking at the reduction of the optical package size (reducing telescope and pointing assembly dimensions) privileging the communication subsystem and looking for new technologies for the implementation of laser sources and photo-detectors.

REFERENCES

(1) R.A. Peters "Effect of Coding on B.E.R. of an Optical ISL" Proc. SPIE 748, (Los Angeles, Jan. 1987)
(2) "Final Report Silex-A1"(4200) (Matra Espace, Nov. 1986)
(3) "Silex A2: WP 2222/3212" (Selenia Spazio, June 1987)
(4) R. Gagliardi, G. Prati -"On the Gaussian assumption for an optical PPM Channel" (IEEE Tr. on Comm., Jan. 1980).
(5) D. Divsalar, R.M. Gagliardi, J.H. Yuen "PPM performance for Reed-Solomon Decoding over an optical RF Relay Link"(IEEE Tr. on Comm., March 1984)
(6) J.L. Massey " Capacity, cut-off Rate, and coding for a Direct Detection Optical Channel" (IEEE Tr. on Comm., Nov. 1981)
(7) R. Gangopadhyay, G. Prati. "Performance of convolutional codes in a direct detection optical PPM channel".

(8) A. Arcidiacono, R. Giubilei "DRS transmission system: architectural trade- offs and coding" (1987 Tirrenia International Workshop on Digital Communications, Sept. 1987).
(9) G. Ungerboeck "Trellis-Coded Modulation with Redundant Signal Sets - An overview" (IEEE Comm. Magazine, Feb. 1987).
(10) A. Arcidiacono, G. Prati
"Trellis-coded modulation for an optical PPM Channel" to be published.

GEO1 COMMUNICATION PERFORMANCE SUMMARY

TX PERFORMANCE:

WAVELENGTH	839 ns $\pm$ 2 ns
POLARIZATION	LEFT CIRCULAR
E.I.R.P.	1.67 E8 W/SR within 2 E-6rad
OPTICAL LOSSES	2.2 dB
W.F.E.	LAMDA/10
SPHERICITY ERROR	3.4 E-5 a^{-1}
LASER POWER	5%
EXTINTION RATIO	
MODULATION	4 PPM
BIT RATE	120/60/30/15 Mbps

RX PERFORMANCE:

WAVELENGTH	870 ns $\pm$ 2ns
POLARIZATION	RIGHT CIRCULAR
FIELD OF VIEW	50 E-6 rad
TELESCOPE DIAMETER	20 cm
OPTICAL LOSSES	3.2 dB
MODULATION	4 PPM
BIT RATE	120/60/30/15 Mbps
RECEIVER SENSITIVITY	3.0 nW at 120 Mbps
	1.7 nW at 60 Mbps
	0.9 nW at 30 Mbps
	0.5 nW at 15 Mbps

S/N REQUIRED AT DEMODULATOR   15 dB

	UNITS MOUNTED ON SATELLITE PANELS	UNITS MOUNTED ON MOBILE PLATFORM	SUBTOTAL
COMMUNICATION ELECTRONICS	3.3 KG/20W	2.6 KG/6W	5.9 KG/26W
P.A.T. ELECTRONICS	14 KG/26W	6.6 KG/13.3W	20.6 KG/39.3W
COARSE POINTING ASSEMBLY	4.6 KG/11W	8.9 KG/11W	13.5 KG/22W
AUXILIARY ELECTRONICS & THERM. HW	7.5 KG/14W	4.5 KG/21.5W	12 KG/35.5W
RELAY OPTICS AND TELESCOPE		7.2 KG	7.2 KG
STRUCTURE, SUPPORT AND HARNESS	12 KG	6.5 KG	17.5 KG
TOTAL	40.4 KG/71W	36.3 KG/51.8W	76.7KG/122.8W

TAB. 2 - GEO1 MASS AND POWER BUDGET

COMMUNICATION AND TRACKING DESIGN ASPECTS IN
DRS-LEO USER LINK

BY L. BARDELLI - A. FLORIO
SELENIA SPAZIO S.P.A. VIA SACCOMURO, 24 - 00131 ROME

1. INTRODUCTION

Purpose of this paper is to outline the general design aspects, regarding RF Communications and tracking, for the inter-satellite link between a geostationary satellite of DRS (Data Relay Satellite) type and a Low Earth Orbit (LEO) user satellite. The use of two data relay satellites, suitably spaced along the geostationary orbit, improves the link availability. Particular aspects of these links are the big value of the pointing angular ranges, the doppler effect, the multipath effect and the antenna angular acquisition.

2. GENERALITIES ON DRS-LEO.USERS LINK

The DRS is operating as a relay repeater between the LEO user and the ground stations, increasing the availability of the link toward ground, because of its intrinsic large coverage on the ground and on the LEO users.
The corresponding geometry for the visibility of RF links is depicted in fig. 2.1. Therefore the real time communications capability with the Earth is improved, per each orbit of the LEO user, so avoiding or reducing the necessity of using the onboard data storage and retrieval in successive times.
Table 2.2 reports computed data of link availability and contact time in a single DRS system for different values of the LEO user orbit altitude and then orbital period (computed in an earth frame); the following explanations apply:
- D (%) is the duty cycle, or availability, of the direct link between a LEO spacecraft and a single earth station, just for reference.
- D2 (%) is the allowed link duty cycle with a single DRS, just excluding the crossing of the tropospheric layer in the link path.
- DK (%) is the allowed link availability with a single DRS, compatible with The multipath rejection requirements in EDRS KA band and TDRS KU band (secondary path reflecting on the earth surface); the link availability reduction is exploited also for the limitation of the Power FLux Density (PFD) interference on ground due to the return link LEO-to-DRS.
- Tc (WA) is the contact time with the DRS, per LEO user orbit, corresponding to the duty cycle DK, without considering the acquisition time.
- Tc (SA) is as Tc (WA), but considering a short acquisition time of 0.5 min.
- Tc (LA) is as Tc (WA), but considering a long acquisition time of 4 min.
- It should be noted that the link duty cycle values D2, DK and the contact time values are referred to the worst case of the LEO user orbit plane including the DRS.

A dual DRS system, with two data relay satellites spaced in longitude of about 110 degrees along the geostationary orbit, improves further the link availability in the European zone, achieving also a better flexibility in the communication traffic management.

The relevant frequency plan includes the S band, K band and KA band, depending on the different services and links. Various digital modulation types are involved, for the different links, including BPSK, QPSK, possible use of subcarrier and spread spectrum. The DRS system architecture follows the experience of the US TDRS system, but with the use of higher frequency bands (KA band instead of KU band), higher data rates and more limited dimensions of communication antennas. Particular attention is dedicated to the problems of the LEO user antenna pointing coverage and DRS/LEO user angular acquisition.

3. LEO USER TYPES

DRS will provide its services to different LEO user types, from a point of view of communications, tracking, telecommands, telemetries and, more in general, support for mission operations.
Some relevant examples of different LEO users are the following:
- European Elements of the Columbus Programme, like the Polar Platform and the Man-Tended-Free-Flyer
- Eureca
- Spot 4
- Ers 3
- Future remote sensing satellites
- Hermes
- Ariane 5

Table 3.1 is a summary of max data rate requirments for some significant DRS users.

4. THE DRS COMMUNICATION ARCHITECTURE

The DRS communication reference configuration is being designed to provide simultaneous services to two users (ref. 1) in KA band and two users in S band (possibly compatible with TDRS). The maximum foreseen information rate in the KA band return link is 2 x 500 MBPS by means of channellization in the frequency band 26.2 to 27 GHz (previous baseline), or 2 x 600 MBPS by channellization into the overall band 25.25 to 27.50 GHz (new baseline); an overall number of eight carriers (channels) is foreseen.
Moreover the maximum information rate in the S band return link is 2 x 5 MBPS.
An antenna system possible configuration is constituted by two large deployed reflectors, able to scan up to plus or minus 10° about the nadir, in order to cover LEO spacecrafts orbiting at altitude up to 1000 km.
The steering loss should be possibly minimized, considering that this scan angle corresponds, for an aperture diameter of 2.7 mt, to about 35 HPBWs in KA band and 3 HPBWs in S band.
The communication transponder is of transparent coherent type, as primary option, with the feeder links towards ground in the 20/30 GHz bands, also for what concerns the S band users.

The design objective is to achieve a link performance, at the maximum bit rate, with a system BER 10 (optionall 10) for at least 99% of the time. Whereas the TDRS system architecture follows the centralized data distribution concept, where the data of all the LEO users are bent pipe transmitted to a unique ground station at White Sands and then redistributed to the ground network, the EDRS system architecture is oriented toward the more flexible decentralized concept, where the LEO user data can be directly received at the corresponding user ground station, or toward a mixed approach.

4.1 PREVIOUS APPRAOCH

The previous approach of the DRS communication architecture and particularly of the frequency plan is outlined hereafter.
The maximum information capacity in the return link is 2 x 500 MBPS at KA band and 2 x 5 MBPS at S band; in forward link the maximum information capacity is 2 x 25 MBPS at KA band and 2 x 5 MBPS at S band.
Table 4.1.1 shows the corresponding approach for the DRS frequency bands and max data rates for the DRS-LEO user links.
In the overall foreseen KA band from 25.25 to 27 GHz, the first 100 MHz (25.25 to 25.35 GHz) are dedicated to the forward link, while the last 800 MHz (26.2 to 27.GHz) are dedicated to the return link.
For what concerns the KA band return link, being the allowed bandwidth over the information rate ratio equal to 0.8, and being foreseen a FEC (forward error correction) coding of 7/8 rate for each channel, the QPSK modulation is envisaged without frequency reuse within the RF band.
The QPSK modulation is envisaged for the KA band forward link too.
The feeder links toward ground are in the frequency range 29.5 to 30 GHz in the forward link (up-link) and 19.7 to 20.2 GHz in the return link (downlink), also for what concerns the S band users.
The return link signals of the two users accessing the DRS in Ka band, for a total information rate of 2 x 500 MBPS, are rounted in the same bandwidth of 500 MHz of the down link toward ground, by means of frequency reuse with half channel staggering and polarization discrimination. The previous approach for the DRS frequency plan is shown in fig. 4.1.1, both for the inter-orbit links (IOL) and for the feeder links.

4.2 BASELINE

A new baseline is under investigation/definition for what concerns the DRS communication architecture in general, and particularly for what concerns the DRS frequency bands, max data rates for the DRS-LEO users links and FEC coding type.
In fact the more powerful 1/2 rate FEC convolutional coding is now included in the Ka band return link, together with increased requirement for the information data rate up to 600 MBPS per each of the two users, in concomitance with a larger available RF bandwidth from 25.25 to 27.5 GHz.
Rationale is the following:
- The 1/2 rate FEC coding can provide greater benefits against noise and interferences, with respect to the 7/8 rate FEC coding;
- The increased requirement of data rate from 500 MBPS to 600 MBPS allows a

possible upgrading of the polar platform payloads operation;
- The greater availability of RF bandwidth is depending on the possible use of other frequency band dedicated to the forward link (in the range from 23.12 to 23.55 GHz).

Table 4.2.1 shows the baseline approach for the DRS frequency bands and max data rates for the DRS-LEO user links.

The maximum information capacity in the return link is 2 x 600 MBPS at Ka band and 2 x 5 MBPS at S band; in forward link the maximum information capacity is 2 x 25 MBPS at Ka band and 2 x 1 MBPS at S band.

Optionally the Ka band forward link could be in the range 25.25 to 25.35 GHz, as in the previous approach; in that case the Ka band return link should be in the range 26.2 to 27 (or 27.5) GHz.

The block diagrams of fig. 4.2.1, 4.2.2 depict a possible configuration solution for the DRS forward repeater and DRS return repeater respectively.

4.3 DRS SERVICES

The services to be provided by the DRS to its users are of several types, but can be summarized in the following three types:

- LEO spacecrafts operations support
  It is the provision of telecommands, ranging and housekeeping telemetry transmission service, from/to the DRS system control earth station at K band, to/from the LEO spacecraft at S band and/or K band.
  In addition to conventional TTC services, the accurate localization of LEO spacecraft may also be provided for support to orbit and/or trajectory determination, as possible alternative to the capabilities offered by the Global Positioning System (GPS), for instance during the rendez-vous and docking (RVD) maneuvres.
- LEO payloads communications
  It is the provision of communication links for the transmission of payloads data from the LEO user to the data receive earth station (return link) and for the transmission of telecommands (or other data) from the control earth station to the LEO user (forward link).
- Communications for manned LEO spacecrafts
  It is the provision of communications with persons in space, i.e. on-board of space stations or free-flyers spacecrafts (audio channels, video channels, printer, fax or other data).

5. THE COLUMBUS COMMUNICATION SUBSYSTEM(S)

A particular group of LEO DRS users is constituted by the Columbus Elements, which are demanding many services to DRS. The Columbus project has experienced a continuous and also consistent evolution, since the previous phases of design definition until the actual phase B2 extension.

The reference configuration is changed several times, in connection with the progressive design developments, for what concerns not only the general system architecture, but also the particular communication subsystem(s) requirements and definition.

5.1 PHASE B1 REFERENCE CONFIGURATION

The system reference configuration of phase B1 was composed by:
- The Pressurized Model (derived from the Spacelab Module), either in integrated or in attached option in respect to the US Space Station (orbit altitude 500 Km, inclination 28.5°);
- The Resource Module, which, associated with a Pressurized Module, constitutes the Man-Tended-Free-Flyer (MTFF) coorbiting with the Space Station and having a sun pointing control attitude;
- The unmanned Platform in two options (polar and coorbiting with the Space Station); the Polar Platform is orbiting at 700 Km of altitude and 98.2° of inclination, with sunsynchronous orbit and earth pointing attitude control;
- The Service Vehicle as support for the Polar Platform operations.

The primary reference launch system was the NSTS (Shuttle), but also the compatibility with Ariane 5 was envisaged.

The use of TDRS as data relay satellite in Ku band was considered the primary approach, whereas the use of EDRS in Ka band was considered in a frame of growth capability to be achieved by means of in-orbit reconfigurations substitutions regarding the electronic equipments and the antenna as well.

A single antenna coverage approach in Ku band (reconfigurable in Ka band) was chosen for the Platform, whereas a dual antenna coverage approach was chosen for the Resource Module.

In the dual antenna approach, the two antennas, mounted on-board in opposite directions, are sharing a global spherical coverage in two complementary emispherical zones and are used alternatively, together with their related transponders, depending on the mission time sequence.

The on-board mounting of the two antennas can be suitably chosen and optimized, minimizing the interference with the spacecraft structures, in indipendent way from the attitude control concept.

The two Ku band (or Ka band) antennas of Resource Module are foreseen to include also the S band operation, to provide an S band link capability toward TDRS (or EDRS).

The impact of this S band link, with spread spectrum modulation, on the related S band transponder, was not investigated in phase B1.

An antenna autotracking, by means of RF sensor, provides the necessary pointing accuracy and stability to a such narrow antenna beamwidth in Ku band or in Ka band.

5.2 PHASE B2 REFERENCE CONFIGURATION

Following the general criteria of programmatic readjustments for phase C/D costs reduction, better design definition and design simplification for the concept of commonality between Columbus Elements, a certain number of changes, with respect to Phase B1, were introduced into the baseline reference configuration of Phase B2.

The deletion of the Service Vehicle and of the Coorbiting Platform from the Columbus Programme were part of the major changes; Hermes is the new servicing support vehicle.

The system reference configuration of phase B2 was composed by:
- A four segment Pressurized Module (PM4), in configuration attached to the US Space Station (orbit altitude 500 Km, inclination 28.5°);

- The Resource Module, which, associated with a two segment Pressurized Module (PM2), constitutes the MTFF coorbiting with the Space Station and having a sun pointing attitude control;
- The unmanned Platform in the polar option; the Polar Platform is orbiting at 850 Km of altitude and 98.2° of inclination, with sunsynchronous orbit and earth pointing attitude control.

Ariane 5 has become the primary reference launch system, instead of NSTS, apart the PM4, which has to be launched by the NSTS.
The use of both TDRS and EDRS is envisaged as data relay satellites, without in-orbit reconfigurations/substitutions, even if the Ku band link toward TDRS is not simultaneous with the Ka band link toward EDRS; the switchover between the two links is operated by means of proper telecommands.
A single antenna approach in Ku/Ka band is chosen for sake of commonality, for both Polar Platform and Resource Module, possibly with the same antenna dimensions.
The single antenna on-board mounting/orientation is depending on the type of attitude control and on the availability requirement for the DRS link; in this sense the steerable antenna on-board mounting with its supporting mast is different for the PPF and for the RM.
No particular requirement is envisaged, in this phase, of having an S band link toward TDRS/EDRS.

5.3 PRESENT REFERENC BASELINE

The major changes introduced into the Phase B2 extension (actually in course) reference configuration, with respect to that of the previous Phase B2, are:
- The deletion of OMV (orbital maneuvre vehicle) from Columbus Programme operations support vehiles
- The addition of Eureca B (enhanced Eureca) into the Columbus Programme, as small coorbiting platform
- The requirement of an S band link toward TDRS/EDRS for TT&C purpose
- The link toward NSTS is changed in frequency from S band to Ku band.

The Columbus S band transponder, with transmit and receive functions for TT&C signals, shall be compatible with the pseudo noise spread spectrum modulation (TDRS type) and with the subcarrier standard modulation; the two different operating conditions are not simultaneous.
To increase the commonality between Polar Platform and MTFF, to DRS forward link data rate for the MTFF is changed from 25 MBPS to 1 MBPS (in normal flight configuration).
Table 5.3.1 shows the actual data rates requirements of Columbus Elements as DRS users in Ku/Ka band.
The required data rate for the Columbus Elements as DRS users in S band is 1 KBPS tentatively (TT&C purpose).
A characteristics summary is reported in table 5.3.2, referred to the Columbus project evolution through its phases of design definition.

6. MULTIPATH EFFECT

Due to the secondary route reflecting on the earth surface, as represented in the simplified geometry of fig. 6.1, a multipath effect can arise, with possible degradations in the communication signal transmission, for instance

a propagation fading.
The differential time delay and the RF frequential period 1/ of the fading, having impact on the communication signal modulation distortion, have large variations during the LEO user trajectory, respectively from a fraction of usec to some msec and from some MHz to a fraction of KHz; this is referred to an orbit altitude range from 100 to 1000 Km.
The corresponding fading frequency has large variations too, along the orbit, from a fraction of KHz to some ten of KHz (for a radio frequency of 15 GHz).
A multipath rejection factor of at least 25 dB is needed, so limiting the peak-to-peak ripple to about 1 dB and the fade depth to about 0.5 dB.
In order to get a good rejection factor against the multipath, a certain reduction, even if limited in percentage, of the link availability must be accepted, in correspondance to the visibility edges of the user orbit.
In fact in this way we can obtain important contributions to the multipath rejection, like the decrease of the earth surface reflectivity related to the grazing angle and the antennas angular discrimination (specially for what concerns the LEO user antenna).
On the basis of these criteria the allowed link availability DK of table 2.2 has been computed, referred to a single DRS system, compatible with the multipath rejection needs for the EDRS Ka band and TDRS Ku band.

7. DOPPLER EFFECT

Doppler effect is particularly consistent in the inter-orbit-link.
Considering LEO users at orbit altitude in the range from 100 to 1000 Km and for any orbit inclination, the max value of the doppler effect is about 26 KHz/GHz, while the max value of the doppler effect rate is about 32 Hz/GHz.Sec.
Fig. 7.1 depicts the behaviour of the doppler effect versus time for an user orbit altitude of 1000 Km, in the hypothesis of user orbit plane including DRS.
The curve is referred only to half of the DRS visibility period, while in the other period half the curve is in practice simmetrical in respect to the origin.
Due to the high value of the doppler effect, a service of doppler effect compensation is envisagd by EDRS, as by TDRS.
In similar way as for doppler effect, fig. 7.2 shows the behaviour of the multipath differential doppler effect versus time, in the same conditions of fig. 7.1, with a peak value of 5 KHz/GHz.
The multipath differential doppler can give a contribution of rejection against the multipath effect, at least for a certain percentage of the user orbit.

8. ANTENNA TRACKING SYSTEM

The antenna beamwidths in Ka band are very narrow, not only for DRS but also for the LEO user, in respect to the available pointing accuracies in open loop. Therefore particular needs are identified of antennas autotracking by means of RF sensing, specially concerning the angular acquisition phase; in fact small acquisition times are required, in order to avoid an excessive reduction of the link availability, and a simple/reliable acquisition

procedure is required too.
The steerable antenna of the LEO user has to be compatible with TDRS in Ku band and/or EDRS in Ka band, for transmission of high data rate signals. The RF links are bidirectional (TX and Rx).
The Ku band link with TDRS and Ka band link with EDRS, from an operational point of view, in any case are not simultaneous, but are referred to different periods of mission and, in general, the switchover between the two systems is not in the same orbit.
For a certain number of LEO users, the antenna tracking system operation has to be compatible with the microgravity requirements, imposed by the on-board scientific payloads; this means that the antenna pointing mechanism must not generate excessive disturbance torques on the spacecraft body.

8.1 POINTING COVERAGE

The analysis results of the max antenna pointing coverage, for LEO users at orbit altitude in the range from 100 to 1000 Km, are depicted in fig. 8.1.1 and 8.1.2, corresponding to the user S/C attitude stabilized toward the earth and toward the sun respectively, so that possible commonality aspects can be verified.
The horizontal dotted line identifies the orbit plane of the user. These pointing coverages must be obtained by means of steerable antenna(s) with narrow beamwidth, i.e. HPBW less than 1.5°, for a radiating aperture diameter greater than 1 mt, in any case of link with TDRS in Ku band or EDRS in Ka band.
Let's observe the pointing coverage patterns of fig. 8.1.1, 8.1.2, in order to verify the necessity or opportunity of using a single antenna or a two antennas system.
In general we can see that the cases of Fig. 8.1.2, with AOCS stabilized toward the sun, are more demanding in terms of coverage with respect to the cases of Fig. 8.1.1, with AOCS stabilized toward the earth (360° instead of +/- 115° in the orbit plane).
In particular the case of the polar orbit with the sun pointing AOCS is the most demanding one: in that case it is strictly necessary of making recourse to a two antennas system, sharing the global spherical coverage in two complementary emispherical zones.
The other three cases can be solved easily with a two antennas system too, but we could also solve these patterns with a single antenna, provided that some particular constraints are taken into account.
In particular the coverage of the earth pointing polar S/C can be considered as an overdimensioned emisphere (with an extra angle of 25°), whose symmetry axis is laying in the Orbit Plane and is passing through the earth centre; whilst the coverage of the sun pointing S/C, with orbit inclination of 28.5°, requires an elevation angle of +/- 35° over the orbit plane and has a simmetry axis orthogonal to the orbit plane.
In the case of the two antennas system, the orientation of the two antennas in the on-board mounting can be whichever, from a point of view of coverage, provided that the two emispheres are complementary and possibly with some margin of overlapping, in order to avoid the dead time of acquisition in the switchover from one antenna to the other.
Due to this degree of freedom, the on-board mounting of the two antennas can

be suitably chosen and optimized, minimizing the interference with the spacecraft structures.
Instead in the case of a single antenna, the orientation of the antenna on-board mounting has not this freedom of choice, but it is strictly determined from a point of view of coverage; consequently compatibility problems and interferences with the spacecract structures can arise, so constraining the steerable antenna design design and even the spacecraft configuration.
A particular problem arises when we want to implement an extra angle in addition to an emispheric coverage: in that case the extra angle mechanical steering could request a cumbersome antenna geometry (in terms of dish offset with respect to the supporting mast) in order to provide the necessary clearance with respect to the mast; this problem is more consistent as larger is the antenna diameter.

8.2 THE ACQUISITION PROBLEM

The problem of the acquisition, to establish the link between DRS and LEO user, arises each time the LEO user leaves the eclipse, due to earth obstruction, and so one time per orbit (if the orbit orientation with respect to DRS is such to have the eclipse).
In addition, in the case of dual DRS system, the acquisition must be performed again also in the handover between the two DRS (reacquisition procedure).
TDRS does not transmit any beacon signal for LEO users antenna tracking system, so that the tracking must be performed using as reference the communication carrier of the forward link in Ku band; by the way the acquisition time, generally expected, is enough small (less than one minute), being not required the cross search procedure (search simultaneous from both sides).
Therefore for the TDRS case, the LEO user antenna must acquire and track the communication carrier at 13.775 GHz, being this carrier modulated by data during tracking.
In the DRS case, the LEO user antenna HPBW in Ka band is very narrow, specially for high values of aperture diameter, so that the expected acquisition time in ka band can be consistent (for instance three minutes), being required a cross search procedure aiding the acquisition.
During the cross search phase a scan search is carried out by both sides, i.e. DRS antenna and LEO user antenna, following a suitable procedure or sequence, which can·be complex and long.
Acquisition and handover/reacquisition can be enough long, specially in the cross-search case, also because of the limited antenna pointing mechanism motor speed, due to the operating characteristics of the APM itself and to the constraints imposed by the microgravity requirements (for the LEO user).
In order to avoid the cross search necessity and to limit the acquisition time to a value within one minute, the DRS could transmit a beacon signal in Ku band, so implementing an intrinsic widebeam concept.
In that case the LEO user antenna can acquire and track the beacon signal in Ku band, transmitted by the DRS antenna, with larger beamwidth of the two antennas in respect to the Ka band link.
In this way, for the users that have to be compatible both with TDRS and EDRS, a single tracking approach in Ku band is allowed, simplifying the

tracking system complexity and the antenna design, and improving the performance in terms of acquisition time.
The acquisition time reduction becomes even more important in a dual DRS system, when the reacquisition is required in the DRS handover; in this case a consistent improvement is achieved ragarding the overall link availability. In the single tracking approach the tracking receiver is operating on the same frequency band, but its configuration has to be compatible with a non-coherent operation (envelope detection) for the TDRS link and with a coherent operation (phase lock approach), to limit the necessary beacon signal transmit power, for the DRS link.
The solution of including a beacon signal transmitter in Ku band on-board of DRS may give some complication to the DRS architecture design, but provides the benefit of a design simplification for all the LEO users operating in dual band (Ku/Ka), while improving the acquisition time performance for the links in Ka band.
The required beacon signal transmit power in EIRP is typically in the range 44 to 48 dBW for a possible frequency of 13.6 GHz or 13.9 GHz about, far enough from the TDRS carrier frequency of 13.775 GHz and from the used frequency bands for TVB or special services purpose.
The EIRP value of 44 to 48 dBW, to be better verified and determined, is required for the acquisition phase, where an uncoherent monitor is used for the energy detection, whereas a 10 dB less value (about) is required for the tracking mode.
This widebeam concept (for antenna tracking) with DRS transmitting a dedicated beacon signal in Ku band is in any case still under investigation and subject to general system compatibility verification, before to be confirmed.

11. BIBLIOGRAPHY

(1) D.L. Brow, K.G. Lehart - An European Data Relay Satellite (DRS) System - ESA, may 1985
(2) A. Florio, L. Bardelli - Columbus Communication Subsystem Design. 33rd International Congress on Electronics and 26th International Meeting on Space - Rome, march 1986
(3) A. Kutzer - The European Columbus Programme - IEEE AES, january 1987
(4) F. Longhurst - The Columbus System Baseline and Interfaces - ESA Bulletin n. 50, 1987.

LINK ANALYSIS OF DATA RELAY SATELLITE SYSTEM UNDER 'PFD' CONSTRAINT

T. Tanaka[*], Y. Tsujino

Total System Analysis Lab., System Engineering Dept.
Tsukuba Space Center, NASDA
2-1-1 Sengen, Sakura-mura, Niihari-gun, Ibaragi-ken 305 Japan

In a data relay satellite system, we clarified that, under S band power flux density constraints, the frequency spread technique is essential, especially for the forward link. We calculated the attainable data under the constraints. The results are shown in the graphical resolutions.

1. INTRODUCTION

NASDA has a plan to perform S band Multipe Access (SMA) and Single Access (SSA) inter-satellite communications experiments, using Engineering Test Satellite VI (ETS-VI) which will be launched in 1992. Following ETS-VI, Experimental Data Relay & Tracking Satellite (EDRTS) and Data Relay & Tracking Satellite (DRTS) for practical use, will be launched in 1994 and in 1996 respectively. The three Space Agencies, NASA, ESA and NASDA will probably use the common S band frequency for the low and medium data rate relay purposes. The frequency bands are planned to be 2200-2290 MHz for the return link and 2025-2110 MHz for the forward link.

Under the ground P.F.D. conatraints, ruled in the 28th article of the Radio Regulations by WARC 1979. NASDA began to study about communications techniques and attainable data rates. So, frequency spread spectrum communications [1] were introduced to increase data rate and to keep the constraint.

This paper describes the results of calculations with four figures, classifying combinations between the user spacecraft (S/C) antenna types (omni/directional) and the data transmitting directions(return/forward) into four cases.

2. POWER FLUX DENSITY ON THE EARTH SURFACE

The P.F.D. constraints are described with maximum permissible radiation power on the earth surface to avoid harmful interference in the sharing band, assigned to both space and terrestrial communications services with the equal priority. The related band to the Data Relay Satellites (DRS) system, discussed here, is the S band (1525-2500 MHz). The constraint can be expressed so that

---

[*] Now, he is the Directior of Earth Observation Program Office, Program Planning and Management Dept., NASDA; World Trade Center Bldg. 2-4-1 Hamamatsu-cho, Minato-ku Tokyo 305 Japan.

the ground P.F.D. value should not exceed the value defined in Fig.1. In order to keep the constraint, we must consider the next two cases, shown in Fig.2.

(1) Forward link P.F.D constraint

The configuration is shown in Fig.2(a) with a data beam transmitted from a DRS to a user S/C. The radiation beam bore sight may aim to the earth surface or to out-of-the earth.

(2) Return link P.F.D. constraint

Fig.2(b) shows the return link with a data beam transmitted from a user S/C to a DRS. In this case, part of radiation power also arrives at the earth.

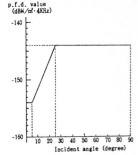

Fig.1 S-band p.f.d. constraint

3. FREQUENCY SPREAD SPECTRUM

The S band P.F.D. is defined as an amount of radiation power density within a 4kHz frequency bandwidth. Therefore, it is changeable, depending on a shape of the frequency spectrum of the transmitted signal. The transmitted signal in this report, is assumed to be QPSK-modulated-digital data. The spectral P.F.D. of the signal which arrives at the earth is expressed as

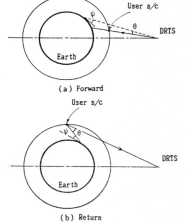

Fig.2 Link Geometry of DRTSS

$$P(f) = \frac{2*S}{f_b} * \left(\frac{\sin X}{X}\right)^2 \quad (W/(m^2*Hz)) \quad (1)$$

where S is total power flux density in one square meter (W/m^2), f_b is a data rate (bps), $X=2\Pi*(f-f_c)/f_b$, and f is a variable of frequency (Hz) and f_c is the carrier center frequency (Hz). This function reaches the maximum at $f=f_c$ and then the maximum P.F.D. value within a 4 kHz bandwidth is

$$S_{df} = \int_{f_c-2kHz}^{f_c+2kHz} P(f)df \quad (W/(m^2*4kHz)), \quad (2)$$

The direct frequency spread spectrum technique [1] is available to a DRS system and the technique is such that original digital data are overlapped with a special code, called psuedo noise code, with each other logically exclusively

ORed. The expression is exactly the same equation as Eq.(1) except for f_b. In this case, f_b in Eq.(1) is corresponding to a psuedo noise code rate, and usually called chip rate and referred to as symbol fp in this paper. f_b is referred to as data rate. The chip rate is usually set $f_p >> 4$kHz and if so, sinX/X almost equals 1 in the bandwidth. Therefore, Eq.(2) is approximated to the following Eq.(3).

$$S_{df} = \frac{2*df}{f_b} * S \qquad (W/(m^2 * 4kHz)), \qquad (3)$$

where df=4 kHz.

4. LINK ANALYSIS

A communication link between a DRS and a user S/C is to be built with E.I.R.P.:P_e(dBW) for a transmitter, G/T (dB/K) for a receiver and the longest range d(m) between a DRS and a user S/C. The attainable data rate is

$$f_b = \frac{P_e * (G/T)}{\left(\frac{4 \Pi d}{\lambda}\right)^2 * C * k} \qquad (bps), \qquad (4)$$

where k is Boltzmann's constant (J/deg), λ is a radio wavelength (m) and C is a necessary S/N ratio to detect digital signals with a proper bit error rate.

On the other hand, at one point on the earth surface, the P.F.D. of radiations made from leakage components of the link is given by Eq.(5). Here, l is the range (m) between a transmitter and the point on the earth concerned.

$$S = \frac{P_e * G_t'(\theta)}{4 \Pi l^2} \qquad (5)$$

where $G'(\theta)$ is a relative transmitting antenna gain from the main beam. The P.F.D. with the 4 kHz bandwidth centering at $f=f_c$ is obtained from Eq.(3) and Eq.(5) as follows.

$$S_{df} = S * \frac{2*df}{f_b} = \frac{P_e * G_t'(\theta)}{4 \Pi l^2} * \frac{2*df}{f_b} \qquad (6)$$

The condition that this communication link fulfills the P.F.D. rule is as follows.

$$S_{df}(\theta) < S_r(\psi) \qquad (7)$$

where ψ is an elevation angle, viewing a transmitting spacecraft from the earth (Fig.2). S_r is a function of ψ, given in Fig.1. S_{df} is a function of θ. From Eq.(7), we get Eq.(8).

$$\frac{P_e}{f_b} < \frac{S_r(\psi) * 4\Pi l^2(\theta)}{G_t'(\theta) * 2 * df} \tag{8}$$

The righthand term of Eq.(8) changes due to the geometric relation between the point concerned and the DRS-user S/C link. There exists a minimum value Pe' and it can be expressed as

$$P_e' = \min_{\theta} \frac{S_r(\psi) * 4\Pi l^2(\theta)}{G_t'(\theta)} \tag{9}$$

Deriving the above Eq.(9), we assume that a DRS antenna and a user's are rightly faced. Substituting Eq.(9) into Eq.(8), Eq.(10) is obtained.

$$\frac{P_e}{f_b} < \frac{P_e'}{2*df} \tag{10}$$

The value of P_e/f_b can be obtained from Eq.(4), and if it does not fulfill Eq.(10), a violation against the P.F.D. constraint might occur at the point cocerned. Adopting the minimum value of f_b is preferable in an ordinary hardware design, within the value range which meets Eq.(10). That is, the chip rate f_p should be given by Eq.(11).

$$f_p > \frac{P_e}{P_e'} * 2 * df \tag{11}$$

From Eq.(4) and Eq.(10), we get the next inequality regarding G/T.

$$G/T \geq \frac{2*df*C*k*\left(\frac{4\Pi d}{\lambda}\right)^2}{P_e'} \tag{12}$$

The righthand term of Eq.(12.) is called critical G/T, written as $(G/T)_c$. Using $(G/T)c$, f_{pc} is given as below.

$$f_{pc} = f_b * \frac{(G/T)_c}{G/T} \tag{13}$$

The necessity to adopt the frequency spread spectrum technique can be classified into the next two cases.

(1) $G/T > (G/T)_c$ (region1) $f_p = f_b$ not necessary
(2) $G/T < (G/T)_c$ (region2) $f_p > f_b*(G/T)_c/(G/T)$ necessary

Drawing a link chart of Eq.(4) with E.I.R.P.=P_e as parameter and putting this $(G/T)c$ on the chart, the chart like Fig.3 can be obtained. However, there does not appear frequently the $(G/T)_c$ in the practical link designs because

SESSION 3

MOBILE SATELLITE COMMUNICATIONS

Chairman: S. Shindo *(NTT, Japan)*

A MULTI-BEAM MOBILE SATELLITE COMMUNICATIONS SYSTEM IN JAPAN

Shuichi SHINDO, Kohei SATOH, Eiji HAGIWARA, AND Kozo MORITA

Electrical Communication Laboratories, Nippon Telegraph and Telephone Corporation.
P.O. Box 8, Yokosuka Post Office, Yokosuka, 238, Japan

SUMMARY

A multi-beam mobile satellite communications system is under study at Nippon Telegraph and Telephone Corporation.
The main objectives of the system are as follows : ① the expansion of the domestic maritime telephone service area. ② the complement of land mobile communication systems. ③ the establishment of an economical satellite communication system.
This system can provide telephone class services, low bit rate message services and 64kb/s class digital non-telephone services.
A new multi-beam system called Dynamic channel Assigned Multi-beam satellite system (DAM system) and compact mobile terminals proposed to realize this system.
NTT participates in the Experimental Mobile Satellite System (EMSS) planned and conducted by the Ministry of Post and Telecommunications, Japan. In this program, evaluations of propagation characteristics, mobile terminals performances and various application experiments are to be carried out.

1. INTRODUCTION

In recent years, mobile satellite communications systems are being investigated in many countries [1-4]. To develop an economical mobile satellite communications system, system cost per channel should be low, and inexpensive mobile terminals are required.

The multi-beam mobile satellite communications system is an advantageous technique which meets these requirements. In multi-beam mobile satellite communications systems, the high satellite e.i.r.p. is achieved by increasing the satellite antenna gain. Therefore, the compact mobile terminals can be used for communication through a satellite.

This paper describes the multi-beam mobile satellite communications system and the mobile terminals studied at Nippon Telegraph and Telephone Corporation (NTT). Then, the experiments using the developed mobile terminals are described.

2. SYSTEM DESIGN OBJECTIVES

Satellite communications are promising for mobile communications. A multi-beam satellite communication system is under study at NTT. It have to achieve following objectives.

(1) Expansion of the domestic maritime telephone service area

The domestic maritime telephone system using the 250MHz band is operated by NTT within a coastal range of 50~100km. There are increasing demands for the service area expansion and the blind zone elimination. In addition, non-telephone services, such as facsimile, are also required. To fulfill these requirements, a mobile satellite system is extremely advantageous, because of its broad service area, slant path propagation characteristics, and channel assignment flexibility.

(2) Complement of land mobile communication systems

The automobile radio-telephone system using 800MHz band, and the paging system using 250MHz band have also been introduced by NTT for use mainly in major cities and along trunk roads. The mobile satellite system may be more economical than a terrestrial system for expanding these services in suburban and rural areas.

(3) Establishment of an economical satellite communication system

In addition to two of the demands above, the cost of the system should be comparable to the domestic maritime telephone system or the automobile radio-telephone system. The following factors are important for this requirement : ① satellite and base station equipments must be shared with other systems. ② very compact and economical mobile terminals should be made possible by using a large scale multi-beam satellite antenna. ③ the compatibility of terrestrial mobile communications equipment should be maintained.

3. THE MOBILE SATELLITE COMMUNICATIONS SYSTEM

3.1 Service Overview

The mobile satellite communications system studied at NTT is schematically shown in Fig.1.

The main functions of this system are as follows :

(1) Telephone and Non-Telephone services

As well as analog telephone services, telephone services below 16kb/s using a high efficiency coder/decoder (CODEC), can be provided through the public switched telephone network (PSTN). Various kinds of non-telephone services, including G-III class facsimiles, communications between personal computers, and Videotex services, can also be utilized with modulator/demodulator (MODEM) through the PSTN.

(2) Low bit rate message services

Low bit rate (≤ 300b/s) message services are effective and attractive for mobile satellite communication systems for achieving an efficient radio frequency use and the compact, low-cost mobile terminal equipments.

These services consist of two service modes, a receiving only mode and a two-way (transmitting/receiving) mode.

The receiving only mode can provide a nation wide paging service with messages by utilizing the broad coverage of a satellite communications system. Since the paging services require the relatively low C/No, the large propagation loss is permissible. Therefore, the link budget design of the paging services is more feasible than that of 8kb/s or 16kb/s telephone or non-telephone services. The paging mode can perform an effective and important role in urban areas where telephone class services cannot be utilized.

In the two-way mode service, various kinds of message communications as well as paging responses are possible. In this system, a media conversion, from messages to voices, is feasible with the aid of the data communication processing system [5].

(3) Digital Non-telephone services

In analog network services (including telephone, facsimile, data services and other non-telephone services), services below 16kb/s are sufficient to satisfy demands in maritime and land mobile communications system.

However, when the development of terrestrial digital network services and demand for moving picture service are taken in account, 64kb/s class non-telephone services might have to be provided in the mobile satellite communications system. Therefore, 64kb/s as well as 16kb/s class non-telephone links are also under study.

3.2 Dynamic channel Assigned Multi-beam Mobile Satellite Communication System

In a conventional multi-beam satellite communication system, radio channels of transponders are allocated to each beam. Because of traffic variation among beams, effective use of satellite signal power is not possible when the HPA output power to each beam is fixed. Traffic loss is also caused by the fixed allocation of radio channels to each beam. The proposed multi-beam satellite communication system[6] called Dynamic channel Assigned Multi-beam satellite communication system (DAM system) can solve these problems.

(1) Operation of the DAM system

Transponder configuration of a 5-beam DAM system and radio frequency channel allocation of SCPC signals are shown in Fig.2 . The operation of the DAM system is as follows :

In Forward Link (base station → mobile), SCPC signals are transmitted from a Satellite Mobile Base Station (SMBS) to a satellite. The received signals at a transponder are divided into five groups and supplied to Multi-port Power Amplifier (MPA). The MPA consists of several unit amplifiers and flexibly shares the total output power among all beams. All the SCPC signals are commonly amplified at each unit amplifier [7]. Then, the MPA outputs are transmitted to mobile terminals in each beam zone.

In Return Link (mobile → base station), a mobile terminals transmits SCPC signals from each beam to a satellite. Received signals independently pass through converters and are combined. The combined signal is, through a high power amplifier, transmitted to an SMBS.

(2) Tolerance of Traffic Variation among Beams

Figure 3 shows the number of allowable subscribers versus the traffic concentration rate to a specific beam, when the total number of channels (50 channels) is fixed. The traffic concentration rate is defined as the ratio of the number of allowable subscribers in a specific beam to the total number of system subscribers. Traffic volume per subscriber and the blocking rate are assumed to be 0.02 erl. and 0.05 erl., respectively.

In a conventional system, the number of subscribers decreases as the traffic concentration rate increases. This means excessive radio channels must be reserved for each beam in order to carry the traffic volume of imbalance subscriber distribution. On the other hand, a DAM system can be equivalently viewed as a single-beam system. The allowable number of subscribers can be kept constant without depending on the traffic concentration rate. For example, compared with a conventional multi-beam system, a DAM system can increase the allowable number of subscribers by 1.3 times even if the traffic concentration rate is 20%, i.e. all the traffic volume is same in each beam. The DAM system is superior to a conventional system in reducing the system cost per subscriber.

3.3 Mobile Terminals [8]

In mobile satellite communications, mobile terminal equipments has to be small-size, light-weight and inexpensive. In addition, it is required to track a satellite and to operate under low C/N levels. To overcome these requirements, two types of mobile terminals have been developed for telephone class services and message services.

(1) Mobile terminals for telephone class services
(a) Configuration

This terminal was designed for telephone class services. The outlook of the terminal is Fig.4. A schematic block diagram of this terminal is shown in Fig.5. Mobile station consists of Antenna unit, Modulator/Demodulator unit, and Control unit. The major parameters are given in Table 1. Non-telephone signals, such as facsimile (≤ 4.8kb/s), are converted to 8kb/s or 16kb/s bit stream, and transmitted through the voice channels.

(b) Antenna unit

To achieve simple satellite tracking, an electronically beam-steerable antenna is developed [9]. Having no mechanical parts, above-deck equipment size was reduced. The size of this antenna is 25cm in height and 40cm in diameter at L band. The radiator section consists of 6 annularly placed microstrip patch antennas. Since a microstrip antenna has broad beam width, beam steering is not necessary in the elevation direction. According to the vehicle direction, however, this antenna switches one element to another in order to point a satellite in the azimuth angle.

This antenna can cover elevation angles in Japanese waters with a gain of 6dBi or more. By exciting 2 elements of this antenna, higher gain can be achieved [10].

(c) Modulator / Demodulator unit

In this modulator/demodulator unit, Offset-QPSK modulation and coherent detection are used. A bit stream of voice signal by APC-AB CODEC, to which 0.5kb/s in-channel signal is added, is encoded at a rate 1/2 convolutional error correcting code. After adding 1 kb/s frame signal, the bit rate to transmit is 18kb/s or 34kb/s for 8kb/s or 16kb/s mobile terminal, respectively.

To perform an error correction decoding of the convolution code and a unique word detection for frame synchronization, LSI chips are introduced [11]. In addition, a digital signal processing technique is applied to the demodulator and the baseband circuits, which reduces the number of components and their size.

(d) Control unit

The control unit has three main functions, location registration, originating and paging. Each control signal is transmitted in the control channel, or by in-channel signal in the voice channel.

(2) Message Mobile Terminals
(a) Configuration

This terminal is designed for low bit rate (≤ 300b/s) message services. Conversion from messages to voice is introduced to communicate between this terminal and an analog telephone. As shown in Fig.6, this terminal consists of antenna unit, modulation/demodulation unit, and display/keyboard unit. A schematic diagram of the transmitting/receiving terminal is shown in Fig.7. The main parameters are shown in Table 2.

(b) Antenna unit

Quadrifilar helix antennas can achieve conically shaped radiation patterns [8]. Since they require no satellite tracking, a simple and compact mobile terminal can be developed. They are also of small size, such as mobile antennas in the automobile radio telephone system.

(c) Modulator/Demodulator unit

In this message mobile terminal, FSK discriminator detection is used, because it is tolerant of received signal fluctuations and its IC chips are commercially available. In addition to the convolutional coding / Viterbi decoding error correction, interleaving is used for burst bit errors.

4. FIELD TRIALS IN EXPERIMENTAL MOBILE SATELLITE SYSTEM (EMSS)

The Japanese mobile satellite program called EMSS (Experimental Mobile Satellite System) using the Engineering Test Satellite V (ETS-V), is being performed. This experimental program is planned and conducted by the Ministry of Post and Telecommunications, Japan. It is to start from this autumn.

NTT participates in this program. The major experiment items at NTT are as follows:

(1) Evaluation of the SMBS and the mobile terminals performance.
(2) Propagation measurements including reflections, shadowing, and multi-path fading.
(3) Evaluation of the signal transmission techniques
(4) Operation of multi-beam mobile satellite communications system
(5) Application experiments to various information services.

The configuration of NTT's experimental system for this program is shown in Fig.8. C band (6/5GHz) and L band (1.6/1.5GHz) are used as feeder link and mobile link, respectively. This system is composed the various types of mobile terminals.

They are used for various types of services such as telephone/non-telephone services message service, and 64kb/s class non-telephone service. Telephone/non-telephone services are available through PSTN. On the other hand, several non-telephone services for ISDN are available through experimental digital switching equipment.

5. CONCLUSION

A multi-beam mobile satellite communications system studied at NTT was described. This system can provide telephone/non-telephone services, low bit rate message services and digital non-telephone services. A proposed multi-beam system called DAM has an effectiveness in dealing with traffic variations among beams. Compact mobile terminals were developed to provide telephone class services and message services for users on the move.

At NTT, the performances of mobile terminals and their communication experiments are planned on the Experimented Mobile Satellite System program (EMSS) with the guidance of the Ministry of Post and Telecommunications, Japan.

ACKNOWLEDGEMENT

The authors would like to thank the Ministry of Posts and Telecommunications and the Radio Research Laboratory for kindly permitting NTT's participation in EMSS and for the helpful guidance during the course of this program. The authors also wish to thank Dr. H. Fuketa and Dr. H. Yamamoto for their advice and encouragement.

REFERENCES

[1] F. Nadori : "An advanced generation land mobile satellite communication system and its critical technologies", NTC, B.1.1. Texas (1982).
[2] J. L. McNally et al. : "Some results of the MSAT phase B studies and their impact on system design", IAF'84, Lansanne, IAF-84-89 (1984).
[3] E. Miura : "Experimental mobile satellite system for communications using Engineering Test Satellite- V (ETS- V/EMSS-C)", IAF'84, Lansanne, IAF'84-87 (1984).
[4] E. Hagiwara, H. Mishima, K. Satoh and N. Yoshikawa : "The prospect of a multi-beam mobile satellite communications system", IEEE VTC'86 (1986).
[5] Y. Muraoka : "Data communication processing system", ECL Tech. Jour., NTT, Jpn., 32, 6, pp.1027-1035 (1984).
[6] S. Nakajima, M. Kawai and F. Yamazaki : "A Proposal on a Flexible Beam-capacity Multi-beam Mobile Satellite Communication System", ICC'87 (1987).
[7] S. Egami and M. Kawai : "Multiport power combining transmitter for multibeam satellite communications", Trans. IECE Japan, Vol.J69-B, 2, pp.206-212 (1986).
[8] T. Sakai, H. Komagata, N. Terada and E. Hagiwara : "Compact mobile terminals for multi-beam mobile satellite communications systems", Globecom'87, to be published.
[9] N. Terada, K. Satoh and F. Yamazaki : "Compact Mobile Antennas for Mobile Satellite Communications", VTC'87 (1987).
[10] T. Hori, N. Terada and K. Kagoshima : "Electronically steerable spherical array antenna for mobile earth station", 5th Int. Conf. on Antennas & Propagation, ICAP'87 (1987).
[11] K. Enomoto, et al. : "Study about LSI for TDMA Equipment", Report of Tech. Group, IECE Japan, SAT84-36 (1984).

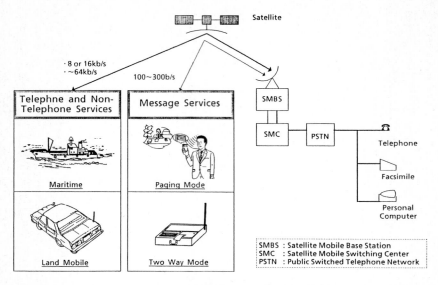

Fig. 1 Mobile Satellite Communications System.

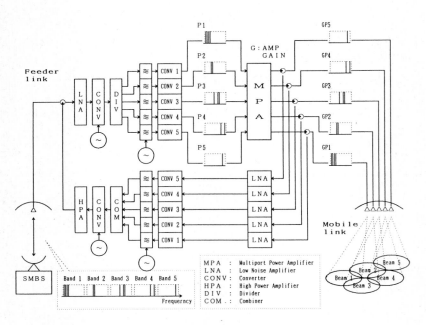

Fig. 2 A Dynamic channel Assigned Multi-beam Satellite System.

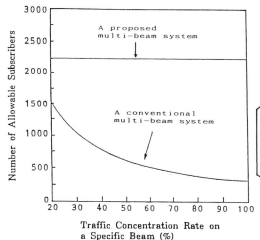

Fig. 3 The Number of Allowable Subscribers versus the Traffic Concentration Rate.

Conditions
Traffic volume per subscriber : 0.02 erl.
Blocking rate : 0.05
Number of channels : 50

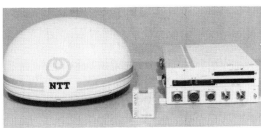

Fig. 4 Mobile Terminal for Telephone Class Services. (antenna unit and main unit.)

Fig. 5 Schematic Diagram of the Telephone Class Mobile Terminal.

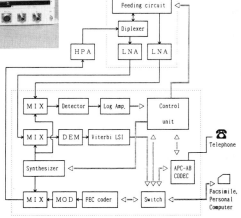

Table 1. The Major Parameters of Telephone Class Mobile Terminal

antenna type	switched-element array antenna
directive gain	8 dBi / 10 dBi
tracking	electronic beamsteering
error correction	$R=1/2$, $K=4$ convolutional coding / 3 bit soft decision Viterbi decording
modulation / demodulation	offset-QPSK coherent detection
CODEC	8 kbps / 16 kbps APC-AB
data bit rate	18 kbps / 34 kbps

Fig. 6 Mobile Terminals for Message Services. (receiving only terminal (left), two-way mode terminal (right).)

Fig.7 Schematic Diagram of the two-way mode terminal.

Table 2. The Major Parameters of Two-way Mode Message Mobile Terminal

antenna type	quadrifilar helix antenna
directive gain	7 dBi / 3 dBi
tracking	
error correction	$R = 1/2$, $K = 4$ convolutional coding / 3 bit soft decision Viterbi decoding with interleave
modulation / demodulation	FSK discriminator detection
data bit rate	150 bps

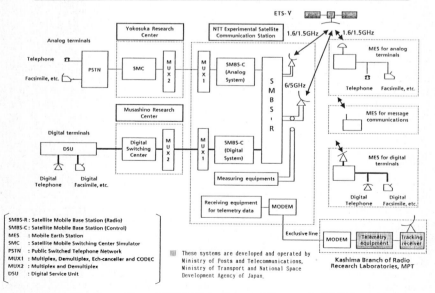

Fig.8 Configuration of NTT's Experimental System in EMSS program

LAND MOBILE SATELLITES USING HIGHLY ELLIPTIC ORBITS

B. G. Evans, L. N. Chung
University of Surrey, Guildford, Surrey, GU2 5XH, U.K.

This paper looks at the current state of land-mobile satellite systems in Europe and advances the advantages of highly elliptic orbits for improved coverage and service provision. A description is then given of the U.K. Technology (T-SAT) satellite which is an advanced on-board processing mobile payload satellite aimed at demonstrating both the use of the Molniya orbit and of the advantages of on-board processing to such missions.

1. INTRODUCTION

Currently all civil mobile-satellite communication systems are based on the use of satellites in the geostationary orbit operating at frequencies in the 1.5/1.6 GHz band using mobile terminals with either low gain omnidirectional (land mobile and aeronautical) or high gain steerable (maritime) antennas. The geostationary orbit from three satellite positions provides virtually global coverage within latitudes ± 75° which is ideal for most maritime and aeronautical applications. However, elevation angles at the extremes of the coverage area for the geostationary orbit are very small and for land-mobile applications, even at moderate elevations, severe multipath propagation problems ensue. Fade margins of the order of 18 dB have been shown necessary [1] to provide adequate service for 90% availability in urban areas at moderate elevations. For a totally European coverage to the northerly limit and accounting also for the mountainous terrain in both northerly and southerly locations an even greater margin may be needed. As land-mobile systems require simple and low cost, non-tracking mobile antennas, the requirements on a geostationary satellite to provide economic LMS services exceed the use of current technology. For example, the eirp requirements would require large deployable antennas to generate a multi-spot beam coverage and possibly on-board processing to provide interbeam connectivity. Such technology is not considered to mature, in space qualified form, until around 2000.

This paper looks at the use of highly elliptic orbits for a European LMS service, which will give improved elevation coverage and reduced fade-margins and possibly allow economic services to be provided by the mid 1990's with conventional technology. The ability of such systems to also cover the polar regions have possible applications for land, maritime and aeronautical systems.

2. ADVANTAGES OF HIGHLY ELLIPTIC ORBITS

In 1981, a University consortium proposed the use of the highly elliptic orbit to overcome problems with geostationary satellites for LMS services. This proposal originally named CERS [2] and now renamed T-SAT, was the forerunner of the ESA Archimedes study and will be described in section 3.

Several alternative satellite orbits are possible which allow the overall eirp of the satellite to be reduced. Constellations of low-earth orbits have been proposed but by far the more interesting are the highly elliptic set of orbits. Satellites in these orbits have relatively low velocity near apogee and when this is matched to the Earth's rotational velocity, the satellite will appear to be almost stationary ('apogee dwell') over a particular geographical area for a substantial part of its orbital period. If the period is synchronous or sub-synchronous, this effect will recur at the same longitude on successive days, but in general the procession of the line of apsides will slowly change the latitude at which apogee occurs.

However, at the particular inclination of 63.4° there is no precession of perigee. Two highly elliptic orbits which exhibit these characteristics are the Molniya and Tundra. Both have inclinations of 63.4° and their main differences are;

	MOLNIYA	TUNDRA
Orbital period	12 hours	24 hours
Apogee altitude	39,500 km	46,300 km
Perigee altitude	1,000 km	25,300 km

The relationship between the above and the geostationary orbit is shown in Figure 1. Both orbits provide a near zenith satellite position, when viewed from the earth at modest latitudes, for ± 4 hours around apogee. The 12 hour Molniya orbit provides a further 8 hours on its next orbit for a region at the same latitude but separated by 180° in longitude as shown in Figure 2, before returning on the third orbit to its original position relative to the earth. For this orbit, a 24 hour coverage for one region requires three satellites in three places at 120° to each other. Clearly the same constellation gives a 24 hour coverage for the other coverage region.

The advantages for high elevation angles experienced in Europe, and for polar regions, are demonstrated in Figure 3 which shows a satellite's view of the earth from an apogee located at 3.5°W longitude as compared with the equivalent geostationary orbit position. Clearly multipath due to buildings, mountains etc. can be almost eliminated at these high elevations and for most parts of Europe the elevation angle is well above 50°. Hence simple high gain non-tracking antennas can be used on the mobile, giving around 15 dB advantage. This added to the 15 dB or so improvement due to multipath yields a total system link budget improvement of around 30 dB, which can be presented as a 1000 times the capacity!

The Tundra orbit [3] is similar, but of 24 hour period and lower eccentricity; it is geosynchronous and its perigee is much higher reducing its radiation exposure; but the apogee dwell is less, requiring more earth-station tracking.

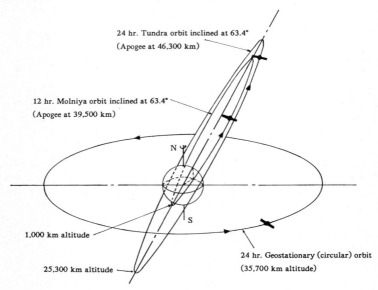

Figure 1. Relationship Between The Geostationary, Molniya and Tundra Orbits

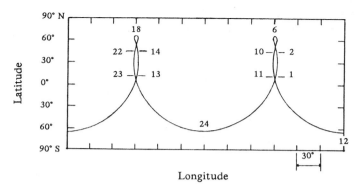

Figure 2. Ground Track Followed By Satellites in 12-hour Highly Elliptical Orbits

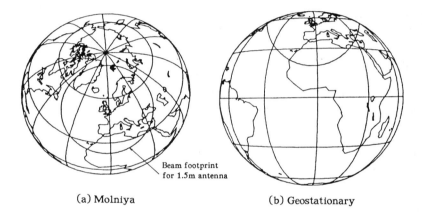

(a) Molniya   (b) Geostationary

Figure 3. Comparison of View From The Satellite of The Earth From The Apogee of A 12 Hour Molniya Orbit (a) and The Equivalent Geostationary (b) Position Both at 3.5°W Centred on The U.K.

The advantage of the Tundra orbit is that it is possible to communicate with both satellites for periods and thus hand-over is improved.

It is possible to use a two satellite system (spaced 12 hours apart) or a three satellite system (spaced 8 hours apart) as per the Molniya. A true comparison between the two options [4] involves a complicated trade-off between the following:

- Spacecraft design and complexity and stability and power
- Satellite lifetimes
- Ease of handover between satellites
- Launcher costs
- Earth coverage
- Earth-station complexity - tracking and Doppler shifts
- Interface to the geostationary orbit

An alternative arrangement called the Loopus system [5] provides a similar constellation of satellites and some investigation of the hand-over procedures has taken place in this study.

Although whichever system is used, it is necessary to provide several satellites in a constellation to provide 24 hour coverage, the launch energy required to place an equivalent satellite into elliptic orbit is approximately one half that of the geostationary satellite. However, satellite lifetimes remain an unknown quantity and this could make the Molniya system in particular, quite costly. Comparisons with geostationary satellites are, however, invidious in that the latter cannot provide a satisfactory service till the 2000's.

3. THE U.K. TECHNOLOGY SATELLITE T-SAT PROGRAMME

For several years now and borne out of the CERS [2] programme, a consortium of U.K. Universities (Table 1) have been researching an advanced on-board processing experimental land mobile payload for a Molniya orbit satellite now known as T-SAT [6]. They have been joined by a second consortium of Universities who are researching an experimental Molniya spacecraft capable of carrying the land-mobile into orbit. The purpose of the programme is to demonstrate the possibilities of LMS service using a highly elliptic orbit, to acquire more data on lifetimes in the Molniya orbit and to demonstrate the all important use of on-board processing which will be essential for second generation LMS satellites.

Table 1. Land Mobile Payload Studies Team

Organisation	Role
University of Surrey	System design, on-board processing
University of Bradford	Mobile terminal design and systems concepts
University of Loughborough	Modulation systems
University of Manchester	Coding systems
Kings College, University of London	RF systems
Portsmouth Polytechnic	Doppler correction and beacon payload
QMC, University of London	Antenna design
RAL	Coordination and overall system design

3.1. T-SAT Spacecraft

The T-SAT satellite is currently baselined on a 600 kg (including injection motor) three-axis stabilised system with dimensions 1.75 m high and 1.6 m in section. Alternative technology flat-plate antennas for the mobile communications payload are mounted on the opposite faces of the spacecraft. In operation, only one antenna is in use at any time with its axis pointing at the centre of the earth. Steerable solar arrays coupled with the ability to manoeuvre the spacecraft about all three axes will enable the solar cells to maintain sun pointing.

3.2. Land Mobile Payload [8]

The land-mobile payload for T-SAT has been designed by a consortium of U.K. Universities, each taking responsibilities as shown in Table 1. The payload is based upon the use of on-board processing techniques and its design incorporates the possibility for flexible experimentation in order to provide experimental evidence on the performance of various completing techniques ahead of the need to commit to an operational payload.

3.2.1. System description

Two possible access systems have been investigated, namely Single Channel Per Carrier (SCPC) on the up link from the mobile with Time Division Multiplexing (TDM) on the down link to the mobile and Time Division Multiple Access (TDMA/TDM) on the up and down link respectively. Both methods have potential advantages for mobile satellite systems. The SCPC approach allows a lower power transmitter on the mobile and better link margin at the expense of greater frequency stability and a complex digital demultiplexing process on the satellite. The TDMA approach, because of its higher bit rate, relaxes the frequency stability requirement but requires a high power mobile transmitter. It also produces a more flexible system in that, short coded data messages, bull voice traffic and other modes of transmission such as facsimile can all make use of the same channel, merely by altering their time allocation requirement within the frame.

On the down link, both systems are identical. The TDM approach, encompasses all the desirable features of the TDMA system, and has the additional virtue of providing continuous synchronization at the mobile, when on-board processing is used.

The details of the signalling procedures for each method have been specified and link budget calculation performed. Formats for signalling messages are also specified. The link budget calculations indicate that, for the SCPC/TDM system, an uncoded $E_b/N_o = 13.0$ dB on the up link and 8.8 dB on the down link could be achieved at data rates of 16 Kbit/s and 256 Kbit/s respectively, provided the mobile and satellite transmitter powers were 5W and 20W respectively. On the TDMA system, at a data rate of 256 Kbit/s the achieved E_b/N_o would fall to 7 dB, even with 20W mobile transmitters. (The down link E_b/N_o is similar to the SCPC/TDM case). As these calculations make certain optimistic assumptions, i.e. high gain mobile antennas (15 dB) and < 1.0 dB fading margin, the performance figures should be considered as goals rather than achievable under all conditions. However, for a demonstration and evaluation exercise, they represent a useful target in order to provide the stimulus to develop the necessary novel technology.

The architecture for the mobile payload design follows logically from the overall system description. A block diagram of the proposed design is shown in Figure 4 where the two different options SCPC/TDM and TDMA/TDM are shown as separate chains in the payload.

3.2.2. Antenna system

The antenna is required to operate simultaneously at 1.655 and 1.555 GHz on the up and down links respectively and provide a maximum gain within an overall physical diameter, which would be determined by the spacecraft and launcher consideration of an actual mission. In this case, it has been assumed that the maximum diameter of dish would be 1.5m. Several antenna concepts were evaluated. However a conventional reflector antenna, using either a crossed dipole or a quadrifilar helix feed, was chosen as the base line for the systems. A minimum gain of 23.5 dB is achieved over the coverage area of the system (UK) by this antenna system, when all considerations such as off-axis beam pattern, spacecraft pointing requirements, etc., were investigated.

3.2.3. RF systems

On the receiver system, an overall noise figure of less than 2.5 dB appears to be achievable. The IF output frequency at 70 MHz can be channelled either to the demodulators or to transparent mode operation. The transmitter subsystem input frequency is again 70 MHz and produces an output power of 12 dBW with an overall efficiency of 26%. The overall mass of the system is expected to be near 10 kg, with total power consumption less than 90W.

Monolithic Microwave Integrated Circuits (MMIC) have also been studied as a possible advanced technology option for the RF sub systems. The complete system could be constructed from a set of custom GaAs chips. Integration on one or two chips does not yet appear feasible, due to the different requirements of the various sub-systems. However, a useful gain in mass and repeatability of performance could be gained by this approach, especially if many channels were required for a commercial equivalent to this prototype system.

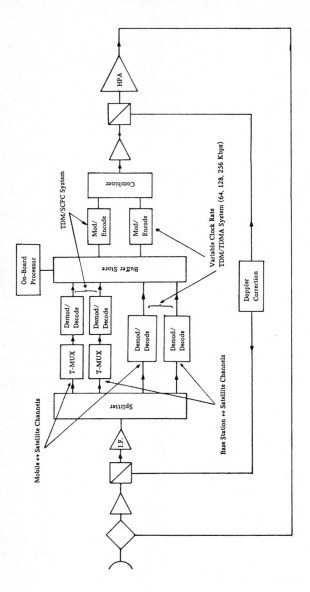

Figure 4. Land Mobile Payload System For T-SAT

3.2.4. Modem design

The modem has been designed to provide flexible operation so that transmission rates from 64 Kbit/s to 512 Kbit/s would be possible. Modulation schemes such as BPSK, QPSK and OQPSK could also be required. Thus a flexible design with an element of programming to realise different functions seemed essential. The study has covered modem requirements for both schemes proposed, i.e. TDMA/TDM at high transmission rates, and a digital demultiplexer/demodulator for the SCPC system at 16 Kbit/s operation. Construction and testing of the modulator had been completed by the end of the initial phase. The digital signal processing has been achieved using systems based on TMS320 architecture.

3.2.5. Coding schemes

A number of coding schemes were studied and have included hard decision minimum weight-decoding, minimum weight decoding for Reed Solomon Codes, reduced set zero neighbour decoding algorithms and soft decision minimum weight decoding. An extended BCH code has been proposed as the best compromise with half rate coding. A basic block length of 128 bits has been chosen given a coding gain of 2-3 dB. Again TMS 320 based architecture is proposed for implementation of the decoding algorithms.

3.2.6. On-board processor

The control of the overall system is performed through the on-board processor which is required to implement the concepts discussed. Two basic approaches to interconnection architecture were considered and consist of either a bus or pipeline approach. The design of on-board processor systems has to be able to operate under a variety of constraints i.e. low power, space environment and interconnection to other items, such as modem, codecs and frequency control of local oscillators. CMOS technology has the required low power consumption and the 8088/8086 processor is undergoing space qualification procedures. Bus architecture tends to have a higher power consumption than a pipeline approach. However, the MIL STD - 1553B bus system seems a desirable approach, if the power constraint is not critical, as it is used already for aerospace applications. However, for implementation of a bread board model, a bus architecture based on the IBM-PC system, which uses 8086 processors, has been chosen.

3.2.7. Doppler correction system

The radial motion of the satellite in an elliptical orbit, such as the Molniya, is considerably more than that experienced in a geostationary orbit. Thus if left uncorrected, the apparent frequencies of signals, received by the satellite and transmitted from it, would exhibit Doppler shifts of ~ 10 kHz at L-Band. Although such Doppler shift could be accommodated by a suitable system design, this would increase complexity at the mobile receivers and cause poor spectrum utilization. One method proposed to overcome this Doppler effect, would be to steer the satellite local oscillator frequencies of both the receiving and transmitting systems. The frequency would be controlled by a digital store with a resolution of ~ 30 Hz per bit. Over the 12 hour operating period, changes in frequency occur approximately once every 20s to maintain the frequency within the resolution of the system. A prototype system has been constructed using a 1.5 GHz SAW oscillator. Tests have demonstrated that the frequency can be maintained within one and a half times the resolution of one bit.

3.2.8. Mobile terminal

The mobile terminal, in effect, has to mirror the operation of the satellite payload. Thus all modulation and coding schemes and signalling protocols need to be compatible. However, the implementation of these concepts can differ, especially as the mobile terminal, in any commercial system, would have to be designed for low cost production. Thus the prototype system, although required to demonstrate both SCPC/TDM and TDMA/TDM operations, would have to be of a suitable design for volume production. The effect of the RF noise environment on the operation of the mobile receiver was also investigated.

The study concludes that a low cost 20W transmitter is entirely feasible at L-Band. Overall receiver noise temperature of $\sim 300°$ K are possible.

4. CONCLUSIONS

The advantages of highly elliptic orbit satellites have been demonstrated for land-mobile services in Europe. The design of an advanced on-board-processing experimental demonstrator payload, together with a small Molniya spacecraft which together form the U.K. T-SAT project have been discussed. The payload is in an engineering demonstrator format and will be fully tested by the end of 1987. The spacecraft has been designed and work is hoped to commence on a construction during 1988.

REFERENCES

[1] Lutz, E. et al, "Land Mobile Satellite Communications-Channel Model, Modulator and Error Control", Proceedings of an ESA Workshop, ESTEC, ESA SP-259, pp. 37–42, June 1986.

[2] Evans, B.G. and El-Amin, M., "On-board Processing for Future Business Satellite Systems", Comms. 84, Birmingham, May 1984. See also IEE Symposium on CERS Satellite, London, April 1984.

[3] Collings, J. et al, "Technical Military Communications by Satellite Relay at High Latitudes", AGARD Conf. CP 344, Oct. 1984.

[4] Lee-Bapty, I.P., "Satellite Coverage of High Latitude Areas", RAE Tech., Report 86011, Feb. 1986.

[5] Dondle, P., "Loopus Opens a New Dimension in Satellite Communications", Int. Journal Satellite Comms., Vol. 2 1984, pp. 241-250.

[6] Gardiner, J.G., "The T-SAT Communications Satellite Payload", IEE Conf. Comms-86, No. 262, pp 109-112.

[7] RAL (SERC), "T-SAT - Report on the Technology Satellite Design Study", Vol. 1 Technical Study Rutherford and Appleton Laboratory, Oct. 86.

[8] Norbury, J.R., "Report on 'The Proof of Concept' Phase of the Mobile Payload Associated with T-SAT", SERC, London, Jan. 1987.

AN ON-BOARD PROCESSING SATELLITE PAYLOAD FOR A EUROPEAN LAND MOBILE SATELLITE SYSTEM

I.E. Casewell	B.G. Evans	A.D. Craig
Racal-Decca Advanced Development Ltd.,	University of Surrey,	British Aerospace Plc.,
Walton-on-Thames, UK.	Guildford, UK.	Stevenage, UK.

The paper considers the application of satellite on-board processing (OBP) technology for land mobile applications for possible introduction in the late 1990s. Initially, a system outline is presented followed by an brief overview of the work being carried out as part of an ESA study, highlighting some enhancements made. Comments concerning the economic feasibility of such a system is also presented. The paper then consider in more detail the functions and implementation of one of the major processing module, the On-board Control Processor & Network Control System Processor module.

1. INTRODUCTION

Current operational mobile satellite systems provide service to a limited section of the community (maritime) via earth-coverage beams and transparent transponders. The key items in serving other sections of the mobile community (aeronautical & land) and in extending the maritime service to meet the growing traffic demands in this sector are Multiple beam antennas and On-board processing.

This paper outlines the results of study work carried out for ESTEC on an on-board processing concept for the land-mobile market. However, the broad conclusions will in fact be similar for an advanced satellite to serve the aeronautical/maritime mobile areas.

The work concentrates on the technical feasibility of an on-board processing payload for use in a land mobile system capable of demonstration during the mid to late 1990's. The study concentrated on the application of digital signal processing techniques applicable to the on-board processor itself, i.e. consideration of the antenna and RF sub-systems beyond that required at the system level is excluded.

The main objective of the system is to provide the land mobile user with a low speed, two way telegraphy and a full duplex telephony connection to the terrestrial network. In order to provide a complementary service to that offered by the envisaged pan-European cellular network a service area covering the whole of Western Europe and preferably the Middle East/North Africa must be considered.

Unlike most existing satellite systems, any future land mobile system must be designed from the outset with a large potential user population in mind. Even the most conservative estimates exceed 10,000. This leads to a shift in emphasis from minimizing space segment cost to minimizing mobile terminal cost, in order to achieve a lower overall system cost.

Simple terminals imply the adoption of a mobile antenna which requires a minimal form of pointing. The G/T of the MES is therefore, at best, around -20dB/K. This implies the use of very high antenna gains on the satellite. A secondary reason for using high gain antennas on the satellite is to facilitate frequency reuse, via spatial diversity, both between land mobile systems and within a given land mobile system.

A major problem with a European Land Mobile system using a geostationary satellite is additional path losses attributable to signal shadowing by obstructions and multipath fading caused by the vehicle's motion. The shadowing loss is the dominant effect in all but the most open of environments even though low directivity mobile antennas are used. The use of advanced ARQ schemes to combat these fading effects makes a high integrity telegraphy service a technical feasibility even in the worst of environments. However, the telephony service, because of its real time nature, is not amenable to such techniques and can only be realistically provided in the intervals between shadowing events. To reiterate, it is not economically feasible to provide sufficient link margin to maintain communication during deep fades, thus it is necessary to exploit the good periods between fades to the full.

2. LAND MOBILE SYSTEM STUDY

The land-mobile satellite system studied provides a circuit switched service for reduced quality telephony and low data rate telegraphy between the Mobile Earth Station (MES) and a Fixed Earth Station (FES). Two classes of MES have been identified, namely a Type I terminal providing either telephony or telegraphy facilities and a simpler Type II terminal providing telegraphy only.

To enable growth, the system design employs a modular configuration. Each module supports a population of 10,000 speech and data terminals, and 10,000 data only terminals. A spacecraft will support one module, and the system can be expanded by adding additional modules and consequently using extra frequency spectrum.

The feeder link is at Ku band and uses a single beam to illuminate the same area covered by the mobile link. The mobile link will operates at L-band and uses twelve 1.1° spot beams to cover Western Europe (see Figure 1). Two hypothetical traffic distributions have been considered: a uniform distribution over all the beams, and a non-uniform distribution (see Table 1). The frequency plan for the non-uniform traffic distribution (see Figure 2) case requires a total bandwidth of 1213kHz in the forward link and 1475kHz in the return link.

Table 1. Hypothetical Traffic Distribution

Beam No.	1	2	3	4	5	6	7	8	9	10	11	12
Traffic (Erlangs)	1.6	7.0	14.8	36.6	46.0	18.8	20.8	24.4	8.0	5.4	7.6	9.0
% of Traffic	0.8	3.5	7.4	18.3	23.0	9.4	10.4	12.2	4.0	2.7	3.8	4.5

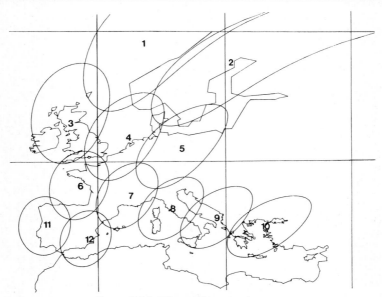

Figure 1. Multibeam Coverage Area

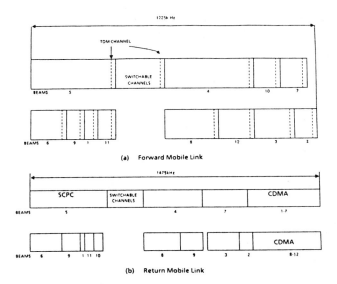

Figure 2. Frequency Plan For The Non-Uniform Traffic Distribution

Table 2 summarises the access schemes, modulation and coding methods used. Details of the system has been reported in earlier publications [1], [2] and [3]. The following sections will discuss some important features of the system and enchancements that have been made since those publications.

Table 2. Access, Modulation and Coding Scheme Summary

Link	Access	Modulation	Coding
Forward Feeder Link	TDMA	DQPSK at 4.096Mb/s	BCH (384,192)
Forward Mobile Link			
i) Telephony	FDM/SCPC	DECQPSK at 9.6kb/s	None
ii) Telegraphy	TDM	DECPSK at 4.8kb/s	RS (64,32), RS(64,4)
Return Mobile Link			
i) Telephony	FDM/SCPC	DECQPSK at 9.6kb/s	None
ii) Telegraphy	CDMA	Spread Spectrum/DPSK	RS (32,24), RS (64,48), RS (16,4)
Return Feeder Link	TDM	QPSK at 4.096Mb/s	BCH (384,192)

2.1 Access Schemes

For the feeder link, where spectral efficiency was the major feature a TDMA/TDM solution is proposed. A frame/superframe structure is employed to take account of the mix of telephony and telegraphy services. There are 25, 40ms frames in a 1sec superframe. The use of the superframe is referred to each FES transmitting a telegraphy channel. Access messages, including channel requests, are transmitted on a slotted-Aloha basis with one slot provided at the beginning of each frame. A Hybrid Network Control System (HNCS) burst is provided once every 4 frames to communicate network control information to the On-Board NCS (OBNCS).

The forward mobile link telegraphy will use a single TDM carrier per beam. Each TDM frame is 1.1866 second long, containing 4 signalling slots (for polling and access messages) and 10 traffic slots (for telegraphy data or ARQ acknowledgement messages).

The return mobile data link uses a code division multiple access (CDMA) scheme in order to provide an efficient random access channel facility at the low data rates necessary to enable a data only terminal to be an attractive option. Data are transmitted at 400 bps using a 403-bit tiered code. The use of a tiered code and a synchronous form of spread spectrum is to minimize the acquisition time, which is essential for an efficient random access channel. All mobiles operating within one system module will transmit on a common nominal centre frequency.

The majority of the access signalling is handled by the OBNCS and are performed out of band. System throughput is enhanced by introducing a "beam logon" system whereby the MES will inform the OBNCS whenever it moves from one beam to another. This will considerablely reduced the amount of polling delay and remove the need to have separate polling slots (hence improve channel utilization). Further, the polling algorithm complexity is reduced and the need for a mobile location database at the GCNCS is removed.

2.2 On-Board Processor Architecture

The overall OBP architecture is in the form of a functionally distributed system, with each subsystem being autonomous in its own right but under the control of the On-Board Control Processor (OBCP). The design optimizes the complexity, flexibility and reliability of the overall system. Aspects of the OBP architecture studied included topology, fault tolerance, system partitioning and software requirements.

Figure 3 & 4 show the block diagram of the forward and return link processor architecture. Signal flow between various modules is pipelined, while control information is carried via a serial control bus. The dual redundant MIL-STD-1553B serial bus is proposed. The main control processor module consists of the OBCP and OBNCS in a tightly coupled configuration. This will be discuss in more detail in section 3.

Enhancements are being made on the Forward Telegraphy subsystem, the Return Telegraphy subsystems and the Transmultiplexers, to increase the degree of integration, by the use of DSP techniques. These result in a significant reduction in the cost, power, mass and volume requirements of the payload. The VLSI technology baseline has also been revised.

On the forward link (Figure 3), the telephony bursts are separated from the telegraphy & NCS bursts, after demodulation. The telegraphy and NCS data are buffered and decoded before being split again. The NCS data will go to the on-board processor, while the telegraphy data are mixed with control information before being encoded and stored in the corresponding downlink TDM buffer. Instead of having a separate modulator per beam, all the downlink TDM data are multiplexed and fed into the Transmultiplexer, together with the telephony traffic. The Transmultiplexer carries out the modulation, frequency multiplexing and routing functions, and produces the complete mobile downlink telegraphy & telephony signals.

On the return link (Figure 4), signals from MESs are separated into one of three paths, telephony traffic, telegraphy traffic and NCS traffic. The telephony traffic from each beam is down converted and filtered to isolate the telephony signals. These are then digitized and fed into the FDM to TDM Transmultiplexer, which will demultiplex it such that after demodulation the slots in the resulting TDM are grouped in terms of originating beam. The telegraphy and NCS signals share one set of downconverters and A/D's per beam. Digital routeing is preferred to an I.F. switching matrix approach for the two banks of telegraphy and NCS CDMA receivers. After demodulation the NCS messages are buffered and decoded before being passed to the OBNCS. The telegraphy traffic, after demodulation & decoding, are combined with the downlink NCS data from the OBNCS before being encoded with a BCH encoder. The encoded traffic is then combined with the telephony traffic at the TDM formatting buffer to form the feeder downlink TDM.

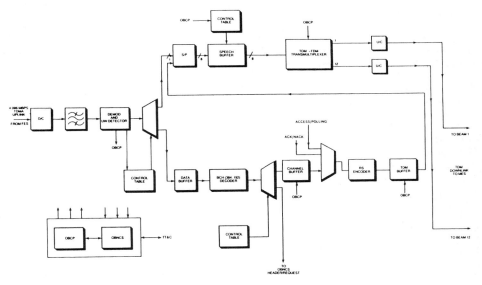

Figure 3. On-Board Processor Forward Link

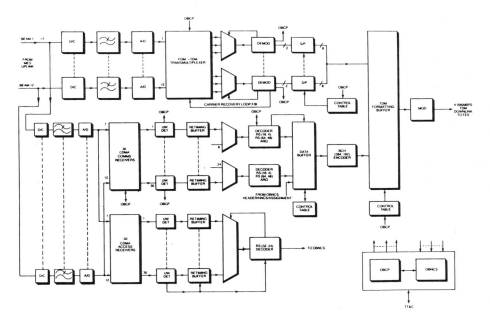

Figure 4. On-Board Processor Return Link

2.3 Digital Signal Processing Technologies

During the course of the study various technologies were considered. For the majority of time critical functions, where semi-custom VLSI circuits are required, a bulk silicon $1.25\mu m$ CMOS technology has been selected. It is assumed that, within the time scale of this project, such technology will become space qualifiable and that the standard cell custom design method will mature.

It should be noted that additional high speed interface circuits are required to buffer interchip connections. Since these consume significant amounts of power, careful consideration of system partitioning is required in order to minimize the number and speed of these interfaces. In general, a horizontal (as opposed to vertical) partitioning is preferred. This also tends to provide greater reliability and flexibility.

Based on an analysis of the selected implementation of each function, estimates of the power consumption, number of PCBs and the mass & volume of the payload have been made. The on-board processor is estimated to have a power consumption of 171W, a mass of 46kg and a volume of $40dm^3$. These parameters are believed to be consistent with the realisation of a 200 channel transponder on an Olympus class spacecraft.

2.4 System Economics

An approximate cost model has been developed for the complete land mobile system incorporating the onboard processor. The variation of overall system cost and other system parameters with various key design parameters are studied. It shows that for the baseline system the cost of the space segment is significantly higher then that of the ground segment. Significant saving can be achieved by using a high gain antenna with azimuth steering on the MES. The study also shows that system cost increases gradually with increased number of beams and so frequency reuse can only be provided at increased cost. An investigation of the scaling of system parameters with number of channels shows that significant saving in terms of cost per channel can be achieved with higher capacity systems. A transponder capacity of at least 800 channels is necessary for a land mobile system to be considered economically viable.

3. THE OBCP/OBNCS SUBSYSTEM

The OBCP/OBNCS subsystem is one of the most important processing modules within the OBP system, handling all the high level protocols and coordinating the operations of other modules. In this section, we will discuss its major functions and implementation.

The main functions of the OBNCS processor are to process reservation requests, allocate channel resources, and implement the high level call setup and clear-down protocols. A database of all the FESs and MESs and their activity status is also maintained. It has to respond to various erroneous network conditions (e.g. loss of FES synchronization, loss of mobile carriers etc.). It also has to perform on-board testing and maintenance functions including a switch-over to completely on-ground NCS mode when necessary.

The OBCP deals directly with the running of all the on-board subsystems. It derives from the channel allocation information, given by the OBNCS, all the necessary control information (e.g. beam configuration table form the Transmultiplexer, control table for the TDM formatting buffer etc.) for the various subsystems. The OBCP also feeds status information obtained from the various subsystems back to the OBNCS to enable it to carry out further allocations. This information, indicates among other things the loss of mobile carrier, on-board buffer overflow, FES synchronization timing etc. It is also responsible for routeing information from one subsystem to another.

The OBNCS processing is dominated by message formating (data movement), database access (table searching and list processing), and message decoding. The main limitation on the speed of execution is the response time expected by the mobiles and FESs after sending requests (i.e. time out before re-try). Hence, the time limitation on the OBNCS processor is generally not very tight and a general purpose 16-bit processor (e.g. 68000 or 8086 family) should be adequate.

The on-board control processor which handles the control function has a much tighter time schedule. This is mainly dominated by the feeder link TDMA and TDM control where all the updates have to be performed within one frame time (40ms). Its processing is dominated by the calculation, formating, and outputing of control information. A fast 16-bit processor (e.g. 12-16MHz 68000 type) is required.

To carry out all its functions, the OBNCS will require an operating environment which will coordinate the running of various tasks simultaneously and satisfy the various timing constraint of each task (i.e. a real time operating system). As for the OBCP, which has to meet strict time limits in deriving various control information, the asynchronous operating environment of the OBNCS is not ideal because of the time overhead during task switching. A superior scheme is to use synchronous scheduling and have the OBCP cycling through a fixed task table. The amount of software required for the OBCP/OBNCS module, including the operating system, application software and data bases, is estimated to be of the order of 260-500Kbytes.

The OBCP and OBNCS form a tightly coupled module as their operations are closely related. The OBCP acts as the execution unit and I/O processor of the OBNCS. The OBNCS manages the OBP resources and the OBCP sets up the other processing elements accordingly. The proposed architecture of the two processors is shown in Figure 5. A shared global memory is used as a communication channel where command and data messages are passed between the two processors. Each processor will also have its own separate program and working memory. The I/O controller, which interfaces the module to the rest of the system through the bus, is connected to the OBCP. This means that messages intended for the OBNCS are read by the OBCP before passing them on to the OBNCS via the global memory. Messages arriving from the OBNCS are passed, via the common memory channel, to the OBCP which is responsible for dispatching the message to its intended destination.

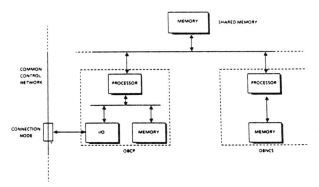

Figure 5. Architecture of OBCP/OBNCS Module

Reliability is essential in an OBP system, especially when the NCS functions are also implemented on-board. Thus fault tolerance, both hardware and software, have to be designed-in. Dual redundancy is being proposed within the OBCP/OBNCS module. Both processors are duplicated. Checking is performed by comparing the output of the processor module and its duplicate which will also be running the same algorithm. If a disagreement occurs, then both processors will attempt to recover with the aid of rollback software. If the recovery procedure fails to remedy the situation, then a switch-over is initiated and normal operation will be interrupted.

Throughout the life-time of the satellite, updates on the on-board software are inevitable. This may be due to discovery of new bugs or reconfiguration of the system to meet changing needs. Thus, a mechanism must be provided to allow for this. For reliability and security reasons, it is proposed to use the TT&C link for this purpose. The updated software is first loaded to the OBCP/OBNCS module and is distributed under the control of the OBCP to the subsystem concerned. In case the software for the OBCP/OBNCS module is to be replaced, a software reset is required to enforce the new changes. For relatively minor updates affecting only a few subroutines, non-interrupting updates can be achieved by employing a vector table mechanism.

Memory devices associated with the processing modules must be adequately protected. Extra memory has to be incorporated such that faulty memory sections can be replaced by spare ones via memory management or chip replacement. The use of byte-wide RAM chips will reduce the total chip count. However, a single error-correcting code would not be very suitable for the byte-wide chips because some of the memory failure modes may cause multi-bit errors within a memory word. It is therefore proposed to use a simple Triple Redundancy technique, so that, by using majority voting, multi-bit errors in a chip can be eliminated.

4. CONCLUSIONS

The study has shown that it will be technically feasible to design and manufacture an on-board processor payload that will satisfy the functional requirements of the system as defined by the study, by the end of the 1990s. The telegraphy service will provide adequate performance in all environments including city areas, whereas in the case of telephony, it is considered only sensible to attempt to provide a service in rural areas. This conclusion is based on the fruition of the technology assumptions made. The study has established that the use of on-board processing reduces spectral requirements and generally increases system efficiency.

The use of digital signal processing techniques has been shown to provide a good solution for most of the major functional blocks, in particular, the merging of routing and transmultiplexing functions into a single element is significant. The inherent attributes of digital signal processing solutions are ideally suited to space applications provided the basic hardware elements are sufficiently robust. Flexibility is considered to be a major requirement, this can be provided by digital signal processing solutions.

The current power, mass and size estimates are believed to be compatible with the payload capacity of one of the large Olympus satellites.

ACKNOWLEDGEMENT

The majority of the work described in this paper was performed by a consortium, led by Racal Decca, under contract to the Telecommunications Directorate of the European Space Agency, with technical assistance from British Aerospace, University of Surrey & DFVLR.

REFERENCES

[1] E. Lutz, W. Papke and E. Plochinger, "Land Mobile Satellite Communications - Channel Model, Modulation and Error Control", 7th Int. Conf. on Digital Satellite Communications, 1986.

[2] I.E. Casewell, "On-Board Processing Payload Technology For An European Land-Mobile Satellite System", Proc. of ESA Workshop on Land-Mobile Services by Satellite, ESA-SP-259, June 1986.

[3] B.G. Evans, I.E. Casewell, A.D. Craig and M.H.M. El Amin, "An On-board Processing Satellite Payload For European Mobile Communications", 7th Int. Conf. on Digital Satellite Communications, 1986.

A HIGHLY EFFICIENT MULTISTAGE APPROACH TO DIGITAL FDM
DEMULTIPLEXING FOR MOBILE SCPC SATELLITE COMMUNICATIONS

Heinz Göckler

ANT Nachrichtentechnik GmbH, Postfach 1120
D-7150 Backnang, F.R.Germany

This paper is focused on the description of the hierarchical multi-stage method (HMM) for digital demultiplexing of an FDM signal being composed of L adjacent SCPC signals. L is (preferably) a power-of-two, here L=32. This HMM approach to FDM demultiplexing applies band-pass sampling and is based on the processing of complex-valued signals by linear-phase FIR filters, where at any stage of processing the respective signals are always oversampled by two.
  The simulation results fully confirm the predicted system performance. Presently, an electrical demonstration model is constructed by cascading six identical specially designed signal processors.

1. INTRODUCTION

In North America, Japan and Europe digital communications with mobile vehicles via satellite is currently systematically investigated [1-7]: A forward link takes messages from an Earth station to the satellite, which retransmits to mobiles. A return link begins at the mobile, goes up to the satellite, and thence to the Earth station.

The satellite will employ spot beams to achieve power gain and to facilitate frequency reuse. It is anticipated that forward links will employ TDM techniques. A mobile will acquire one such TDM signal and extract its own traffic from it.

Each active mobile within a spot beam will be assigned a different operating centre frequency applying a channel frequency spacing of width B. In essence, the mobile-generated signals obtain simultaneous access to the system by frequency multiplexing and by space discrimination afforded by the satellite-antenna pattern.

According to [5], it will subsequently be assumed that the satellite has 19 spot beams, that up to 3600 mobiles are to be served simultaneously, and that up to 800 mobiles may be served in a single beam with a channel frequency spacing of

$$B = 8.4 \text{ kHz} \qquad (1)$$

The last requirement is consistent with appropriately shaped QPSK signals of a data rate of 9.6 kbps to be applied in each mobile transmitter: Square root of 40% cosine roll-off filtering in conjunction with a maximum frequency offset of ±600 Hz due to Doppler shift, oscillator instabilitities etc.

Simple translation of the FDM uplink at L-band to C-band would be an inefficient use of power and spectrum. Furthermore, not all signals in a beam are destined for the same Earth station. Instead, individual TDM streams to each of the Earth stations are required. To accomplish the format change and necessary routing requires extensive signal processing on-board the satellite.

Fig. 1 shows a block diagram of the return link on-board processing. The received signals are separated from one another in a frequency demultiplexer (FDM DEMUX). Each separated signal is passed to a receiving QPSK modem (DEMOD) applying complex signal processing, which produces a digital data stream. These many parallel streams are recombined into serial TDM streams for retransmission to the Earth stations.

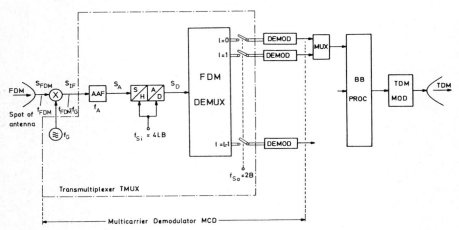

FIGURE 1
Blockdiagram of a multicarrier demodulator for L channels applying digital signal processing (Return link)
AAF: analog anti-aliasing bandpass filter; f_{Si}, f_{So}: input and output sampling frequency of demultiplexer (FDM DEMUX); f_A: centre frequency of AAF.

2. STATEMENT OF THE PROBLEM

This paper concentrates on the description of the hierarchical multistage method (HMM) for digital FDM-demultiplexing of L=32 SCPC signals in conjunction with non-ideal band-limitation of the analog FDM signal in front of the A/D converter. This extended demultiplexer, in Fig. 1 marked by a dashed line, is subsequently called transmultiplexer (TMUX).

It follows a more detailed description of the TMUX design problem (Fig.1). The oscillator frequency f_G of the (analog) down-converter has to be selected such that the desired L real signals from mobiles pass unaffectedly the analog anti-aliasing bandpass filter (AAF) of passband width L B. The final down-conversion to baseband is achieved by sampling the continuous AAF output signal with a sampling rate of at least 2LB consistent with the Sampling Theorem, using a fast sample-and-hold circuit (S/H) cascaded by a relatively slow analog-to-digital converter (A/D). In contrast to the minimum possible sampling rate, however, subsequently oversampling by a factor of two is anticipated at the DEMUX input. With this approach the specifications of the AAF and DEMUX filters are greatly relaxed. This is further supported, if the transition from real to complex (analytic) signal processing is performed as close to the DEMUX input as possible [1]. Therefore, the present TMUX design problem is stated as follows:

Design a highly modular 32-channel HMM DEMUX for real input and complex output sequences. Apply complex signal processing with oversampling by a factor of two throughout the DEMUX. Select the AAF centre frequency f_A and the channel allocation within the AAF passband relative to the input sampling rate

$$f_{Si} = 4LB = 4 \cdot 32 \cdot 8.4 \text{ kHz} = 1075.2 \text{ kHz} \tag{2}$$

such that the most efficient HMM DEMUX implementation results.

3. THE DIGITAL HMM-DEMULTIPLEXER

The 32-channel analog SCPC-FDM signal to be demultiplexed digitally is centred at an IF of about 17.5 MHz. For satisfaction of the Sampling Theorem, the FDM signal is band-limited by a crystal bandpass filter (AAF) such that it can be (over-)sampled with f_{Si} according to (2). With this bandpass sampling scheme, demanding for a fast and accurate yet feasible S/H circuit and a slow A/D converter, the FDM spectrum of the digitized DEMUX input signal $s_D(kT)$ is folded down to the centre frequency of

$$f_D = f_{Si}/4 = 32B = 268.8 \text{ kHz} \tag{3}$$

The tree structure of the most efficient and highly modular hierarchical multistage method adopted for demultiplexing is depicted in Fig.2 [1,8,9]. It is assumed that its 32-channel SCPC-FDM input signal is given by a complex-valued (analytic) sequence $\underline{s}_D(2kT) = \underline{s}^0(2kT)$, which is the result of preprocessing in a digital anti-aliasing filter (DAF). The DAF [10] performs the transition from real to complex signal representation of the digitized FDM signal $s_D(kT)$, and decimates by two. Subsequently, underlining indicates that the associated signals or filter coefficients are complex-valued.

In the HMM of Fig.2 each cell (block) splits its complex input signal into two complex output sequences, each decimated by two. Different stages and their respective input sampling rates are distinguished by a (Roman) superscript

$$\varkappa \in \{0, I, II, III, IV, V\} \tag{4}$$

$$f_{Si}^{\varkappa} = f_{Si} / \prod_{\varrho=I}^{\varkappa} 2 = 1/T^{\varkappa} \tag{5}$$

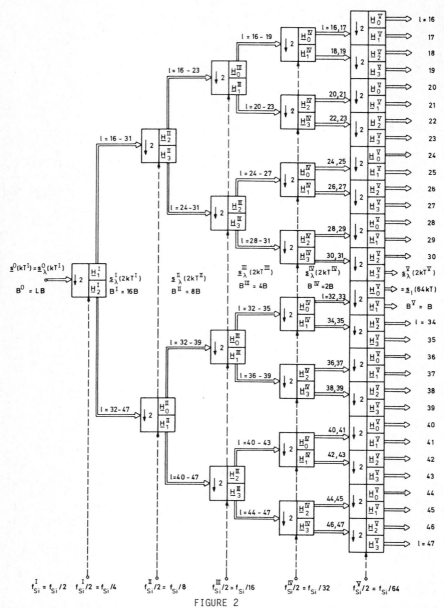

FIGURE 2
Overall block diagram of HMM for 32 usable slots (odd channel allocation scheme), fed by DAF output sequence $\underline{s}^0(kT^I)=\underline{s}_D(2kT)$; stage numbering by Roman numbers: $\varkappa \in \{I,II,III,IV,V\}$, $f_{Si}^\varkappa = 1/T^\varkappa$

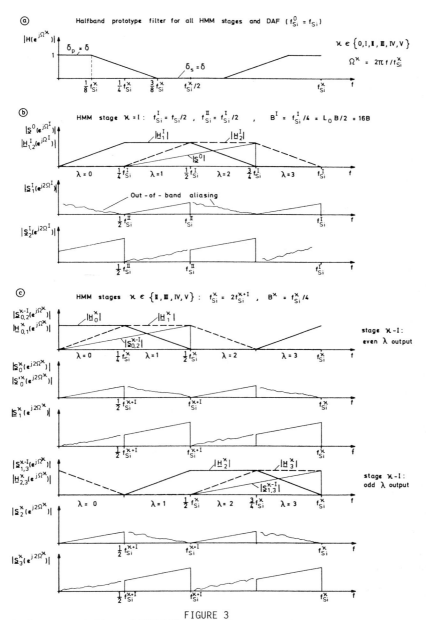

FIGURE 3
Spectral representation of HMM-DEMUX according to Fig.2
(a) Prototype halfband filter for all stages
(b) HMM stage $\varkappa=I$ with $|\underline{S}^0(e^{j\Omega^I})| = |S_D(e^{j\Omega^I})|$
(c) HMM stages $\varkappa \in \{II,III,IV,V\}$; λ: slot number of considered HMM stage, $\varkappa-I$: preceding stage, $\varkappa+I$: following stage.

where $\varkappa=0$ stands for DAF ($f_{Si}^{0} = f_{Si}$). Each HMM cell is able to deliver $\Lambda=4$ different output signals. From these output sequences of each cell always two are discarded. The four passbands of an HMM cell, each of bandwidth $B^{\varkappa}=f_{Si}^{\varkappa}/4$ are distinguished by $\lambda \in \{0,1,2,3\}$. The two out of four slot transfer functions $\underline{H}_{\lambda}^{\varkappa}(e^{j\Omega^{\varkappa}})$, $\Omega^{\varkappa}=2\pi f/f_{Si}^{\varkappa}$ to be realized by the various HMM cells are indicated in Fig.2 by its subscript λ.

A more profound understanding of the HMM can be gained from the associated spectral representation; Fig.3. To this end, first stage I is considered (Fig.3b). Here, the DAF output spectrum, centred at $f_{Si}^{I}/2$, is split into two. This is achieved by filtering the lower 16 of the desired SCPC signals by $\underline{H}_{-1}^{I}(e^{j\Omega^{I}})$ and the upper 16 by $\underline{H}_{-2}^{I}(e^{j\Omega^{I}})$. Decimation by two leaves the output spectrum $\underline{S}_{-1}^{I}(e^{j\Omega^{I}})$ of slot $\lambda=1$ in its original location, whereas $\underline{S}_{-2}^{I}(e^{j\Omega^{I}})$ is shifted to the frequency interval $[0, f_{Si}^{II}/2]$. Similarly, Fig.3c shows the spectral representation of the subsequent stages. Here it must be distinguished, whether the HMM cell under consideration is fed by a signal stemming from a slot filter of odd or even λ (cf. Fig.2).

As a consequence of the bandpass sampling scheme based on oversampling by two at any stage of DEMUX inherent signal processing, the complex-valued coefficients of the DAF and all HMM cell filters can be derived from a common linear-phase finite impulse response (FIR) halfband prototype filter with real coefficients [11], the frequency response of which is shown in Fig.3a. Furthermore, it should be recognized that all HMM cells can be made identical by the introduction of two trivial frequency shift operations.

As it is obvious from Fig.3c, each DEMUX output channel is still loaded by a repeatedly aliased spectral portion adjacent to the spectrum of the usable slot signal. This calls for a final band-limitation, which is accomplished by a non-decimating linear-phase FIR halfband filter with complex coefficients [12]. Note that this halfband filter (HBF) is not included in Fig.2.

The filters of the HMM-DEMUX have been designed such that a minimum signal to distortion ratio of 30 dB is achieved for all 32 DEMUX output signals after final band-limitation. Thereby an ideal FDM signal is anticipated at the input port of the AAF in front of the A/D converter. As a result, the DAF and HMM prototype calls for a symmetric FIR halfband filter of length 11, whereas the length of the symmetric halfband filter for final band-limitation should not be shorter than 19.

The reported results have been verified by extensive simulation of the system taking into account the impact of the analog anti-aliasing filter, as can be read from Table 1. Presently, an electrical demonstration model (EDM) of the DEMUX is being constructed by cascading six identical versatile stage processors (VSP), as outlined in Fig.4. Each VSP performs all operations of the respective stage $\varkappa \in \{0,I,II,III,IV,V\}$ of the HMM-DEMUX.

TABLE 1
Signal to distortion ratio obtained by simulation with QPSK stimulation of the DEMUX

Signal wordlengths	Signal to Distortion Ratio S/D in dB					
	SET 1 w_c = 16 bit		SET 2 w_c = 10 bit		SET 3 w_c = 16 bit	
$\{w, w_F, w_i\}$	l = 18	l = 24	l = 18	l = 24	l = 18	l = 24
$\{\infty, \infty, \infty\}$	31.9	34.5	32.4	35.2	25.7	25.1
$\{11, 15, 32\}$	31.4	33.8	31.8	34.5	25.4	24.9
$\{11, 12, 14\}$	30.3	32.1	30.8	33.1	24.9	24.6

Coefficient SET 1&2: Filter length n=11, coefficient SET 3: n=7
Signal wordlengths: w: A/D conversion, w_F: between HMM cells, w_i: cell inherent.

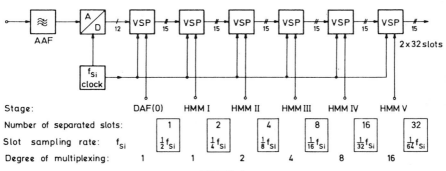

FIGURE 4
Hardware multiplexing scheme of HMM TMUX

4. CONCLUSION

In this paper the hierarchical multistage method (HMM) has been described as very efficient and highly modular approach to SCPC-FDM demultiplexing on-board a satellite. On a bit level, the operation rate of this method is well below that of the filter method described in [13], and even below that of a novel approach to the polyphase method of transmultiplexing [9,14]. Additional key

features of the HMM-DEMUX are:
- Bandpass sampling scheme
- Oversampling by two at any stage of DSP
- Processing of complex-valued (analytic) signals
- Exclusive use of linear-phase FIR filters in order to minimally degrade the phase-sensitive QPSK-SCPC signals to be processed in the DEMUX.

Due to its high degree of modularity (cf. Fig.2), the HMM-DEMUX has the potential of optimizability (short filters), ease of manufacturing and testing, suitability to (dedicated) VLSI realization, high reliability with or without redundant VSPs (cf. Fig.4), and reduced amount of overhead circuitry.

ACKNOWLEDGEMENTS

This work was supported by the European Space Technology Centre (ESA-ESTEC), Noordwijk, The Netherlands under Contract No. 6497/85. However, the opinion expressed in this paper is not necessarily shared by ESTEC. In particular, the author is greatly indebted to G.Björnström, ESTEC and to H.Eyssele, ANT for their support in promoting the reported investigations, and to his colleagues M.Hagen, H.Scheuermann and A.Szillus for various stimulating discussions on topics treated in this paper.

[1] F.M.Gardner: On-board processing for mobile-satellite communications. Final Report: ESTEC Contract 5889/84, Palo Alto, CA: Gardner Research Co., 2 May 1985.
[2] Study of Systems and Repeaters for Future Narrowband Communication Satellites, Phase 2 Final Report, Telespazio, July 1985, ESTEC Contract No. 5484/83/NL/GM(SC).
[3] G.Colombo, W.Heine, K.Jesche, W.Schreitmüller and F.Settimo: System architecture and management of advanced regional satellites for land mobile applications. Globecom 1985, New Orleans, LA, pp.38.2.1-38.2.7.
[4] W.Kriedte and A.Vernucci: Advanced regional mobile satellite system for the nineties. Globecom 1985, New Orleans, LA, pp.38.1.1-38.1.6.
[5] ESA Invitation to tender AO/1-1816/85/NL/MS: Study and development of on-board multicarrier demodulator for mobile satellite communications. Noordwijk, September 1985.
[6] F.Ananasso and E.Saggese: A survey on the technology of multicarrier demodulators for FDMA/TDM user-oriented satellite systems. Globecom 1985, New Orleans, LA, pp.6.1.1-6.1.7.
[7] INMARSAT Request for proposal No.114: Study of a digital on-board multicarrier demultiplexer. London, May 1987.
[8] H.Göckler and H.Scheuermann: A modular approach to a digital 60-channel transmultiplexer using directional filters. IEEE Trans.COM-30 (1982) 7, pp.1598-1613.
[9] H.Göckler: German Patent Application P3,610,195.
[10] H.Göckler: German Patent Application P3,621,737.
[11] R.E.Crochiere and L.R.Rabiner: Multirate digital signal processing. Englewood Cliffs NJ: Prentice Hall Inc., 1983.
[12] H.Göckler: German Patent Application P3,705,206.
[13] E.Del Re and P.L.Emiliani: An analytic signal approach for transmultiplexers: Theory and design.IEEE Trans.Comm.COM-30 (1982)7,pp.1623-1628.
[14] H.Göckler: An analytic signal processing approach to digital polyphase transmultiplexing: Theory and design. In preparation.

MULTICARRIER DEMODULATOR (MCD) USING ANALOG AND
DIGITAL SIGNAL PROCESSING

P.M. Bakken, V. Ringset, A. Rønnekleiv*, E. Olsen

ELAB, The Norwegian Inst. of Technology, N-7034 Trondheim, Norway.

*Department of Electrical Engineering and Computer Science,
The Norwegian Inst. of Technology, N-7034 Trondheim, Norway.

Future generations of communication satellites for mobile services would benefit from on-board demodulation and remodulation of each carrier from the mobile stations. The carriers access the satellite in frequency multiplex division.

This paper demonstrates that analog signal processing based on surface acoustic waves (SAW) can transform 300 received carriers to a time division multiplex format which lends itself to sampling, A/D conversion, further filtering and demodulation by digital processing. This is done in a SAW based multicarrier demodulator (MCD). Analysis and simulations reported in the paper show that the electrical performance of the MCD is similar to other high quality demodulators.

1. INTRODUCTION

The trend in satellite communication systems today is towards reduced cost for the ground segment, in particular for the mobile stations. To this end new and more low cost technology is introduced in ground stations and a larger part of the total system resources are put in the satellite payload. Multiple beam antennas are important to increase the receive sensitivity and the EIRP of the space segment and such antennas are under intense development by ESA and elsewhere. On-board demodulation would be another major step towards lower cost mobile units because it would both improve the transmission budget and give flexibility to reroute the traffic between the beams of a multiple beam payload.

An on-board demodulator must be able to demodulate low-rate carriers within mass and power budgets much stricter than for a conventional demodulator. ELAB has, sponsored by ESA, worked to develop effective multicarrier demodulators (MCD) for on-board use in satellite communication systems since 1984. ELAB's MCD concept is based on innovative application of two signal processing methods; analog signal processing (ASP) by surface acoustic waves (SAW) and digital signal processing (DSP) by application specific integrated circuits (ASIC). This paper explains the principles of the MCD, reports the most important performance analysis carried out and the current status of the MCD project.

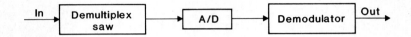

Fig. 1. Principle operations of a SAW-based multicarrier demodulator.

2. PRINCIPLES

An MCD must carry out a set of signal processing operations. The function of these operations are well known, but to minimize the power consumption it is important to use efficient signal processing methods. As indicated in Fig. 1, the operations may be split in two different classes. First, the FDMA signal is demultiplexed into separate channels. We assume a fixed channel plan, and the demultiplexing operation can be carried out independently of the modulated signals occupying the slots of the frequency plan. The demultiplexer does not require flexibility, but a high volume of processing, observations which strongly indicate that ASP should be used for demultiplexing. As demonstrated in Ch.3 SAW ASP is suited.

The demodulation of each carrier must be able to adjust to the symbol timing of the carrier and to the RF phase for coherent detection. A certain degree of flexibility is required, and DSP has been preferred for the demodulation.

3. SAW DEMULTIPLEXER

The demultiplexer should preferably carry out the following functions:

A. *Bandpass filtering.* One filter per carrier is required. The filter bandwidth determines the digital sampling rate and should be kept to approximately the width of an individual channel slot. However, the bandpass filter must also allow for frequency drift and doppler offsets of the carrier without causing significant distortion.

B. *FDM to TDM conversion.* The FDM/TDM conversion makes it possible to use one A/D converter for all carriers and also to time multiplex the DSP hardware between a number of carriers. Function B is a key to overall efficiency of the MCD.

The SAW implemented chirp Fourier transform (CFT) [1, 2] with a suitable processing frame length and window function has the properties A and B and has been the basis for our work [3, 4].

To allow long input processing time frames we use the CMC-realization of
the SAW-based chirp Fourier-transform. It may be described as a convolution,
multiplication and convolution, CMC, of the input signal by chirps. The
transform is performed in a circuit as shown in Fig. 2. Each processed input
time frame has a duration of $2.2\ T_s$. The bandwidth of the input signal is B_s.
Signal and noise outside this range will be removed by the first chirp filter.

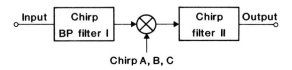

Chirp signals A, B and C have duration $<3T_s$, are periodic with
periods = $3\ T_s$, and relative timedelays of T_s, down chirp, see below.

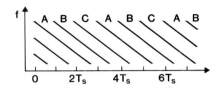

Fig. 2.   Circuit to perform the CMC-transform.

Fig. 3. shows in a frequency-time plot how one input time frame is transformed
in the circuit. As seen from the figure, the input time frame which has a
duration of $2.2\ T_s$ is actually dispersed over a time T_{out} by the first chirp
filter to give a total duration of $2.2\ T_s + T_{out}$ at the multiplier stage.

The transform of one input signal frame appears at the output with duration

$$T_{out} = kT_s \qquad (3.1)$$

which requires the chirp rate μ' to be

$$\mu' = \frac{B_s}{kT_s} \qquad (3.2)$$

and the duration of each multiplying chirp (see below)

$$T_1 \approx 2.2\ T_s + B_s/\mu' = (2.2 + k)T_s \qquad (3.3)$$

With k equal to 0.8 (or less) this requires three multipliers operating in parallel, to perform one CMC- transform every T_s.

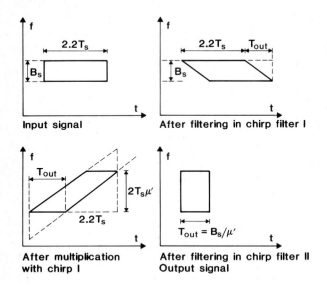

Fig. 3. The CMC-transform shown in a frequency versus time plane.

In this structure the window filtering may be included in the last chirp filter. This will remove signals which fall outside the nominal input time frame (time frequency window), but which without window filtering would be processed in the multiplier due to the dispersion in the first chirp filter of the input time frequency window.

Signals outside the input bandwidth B_s may give responses at the output appearing outside the output time frame T_{out}, and might therefore interfere with neighbour output time frames. This determines the maximum allowable bandwidth of the first chirp.

The second operation in the CMC-transform is a multiplication of the incoming signal with three chirping signals. We have chosen to use only one mixer where the three chirp signals are added together before they enter the mixer. The adverse effects of 3rd order intermodulation are analyzed. Due to the high regularity of the three chirp signals, many of the intermodulation products fall on the desired chirps and may not be distinguished from these. However, they will change when a chirp starts or stops. For these products only the discontinuities which appear when a chirp starts or stops are really harmful. Other products will fall outside the used (critical) frequency range for the chirps if the centre frequency of the chirps is chosen sufficiently high. This will be the case in our system where the centre frequency is 109.3 MHz. In order to cause a degradation of less than 0.01 dB our analysis has shown that the two tone intermodulation products should be suppressed by at least 36.4 dB. According to measurements made on an active mixer this value should be obtainable.

Deviation from the ideal response in the second chirp filter and from ideal amplitude and phase in the chirps will result in adjacent channel leakage. By assuming an average E_b/N_o equal to 17 dB for all carriers we have found that an accuracy of 0.11 dB in amplitude and 0.75^0 in phase are necessary in order to cause a degradation of less than 0.1 dB, when the effects are combined. The first filter could hardly contribute to the degradation of demodulation, since it will only affect the in-band filtering, and even this is a rather moderate way since it is broadband. The chirp lines are made on ST-cut quartz to achieve high temperature stability. The insertion loss in the second line which is the more critical one, is estimated to be in the range 38 to 46 dB, where 10 dB is due to array reflection losses with 8 mm aperture of the reflecting arrays at 39.32 MHz.

The delay change in this chirp line due to temperature variations is ± 4 ns with a deviation from the turnover temperature of 20 K. The turnover temperature of the substrate is the temperature where the 1st order temperature coefficient of delay is zero and may be chosen by changing the cut angle. This delay change can be accepted without compensation. The temperature coefficient of delay in ST quarts is different for the direction along the arrays and perpendicular to it. This means that the optimum angle of the reflecting grooves in the arrays will change versus temperature. This will lead to a temperature dependent loss which is found to be 0.2 dB for a temperature change of ± 20 K.

The SAW CFT will be used to receive a large number of carriers. The window function applied in the CFT will act as a prefilter with an equivalent frequency response W(f). The window function is optimized to serve as an anti aliasing filter for the digital filter. The resulting equivalent low-pass frequency response is shown in Fig. 4.

The signal at the input of the MCD is in FDM format. The signal at the output of the CFT is in TDM format and the wanted channel is selected simply by choosing the correct sampling instant. A timing error in the sampling can

be of two types:
- jitter in the sampling from frame to frame
- constant offset in sampling time.

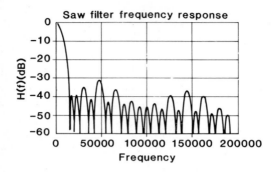

Fig. 4. Frequency response of equivalent anti-aliasing filter.

Analysis have shown that a constant timing offset of ± 12 ns can be accepted without degrading the performance by more than 0.05 dB even for a worst case combination of frequency offsets of carriers (± 600 Hz). Jitter in the sampling instant will cause a more serious degradation. The jitter must be kept smaller than ± 1.5 ns in order to keep the degradation less than 0.01 dB.

The signal is digitized in the A/D converter. The dynamic range of the A/D converter is chosen such that an input level of E_b/N_o equal to 20 dB shall not cause clipping. The number of bits must be chosen such that the desired distance between the quantization noise and the thermal noise is obtained. In our case it is necessary to use 7 bits which will cause a degradation of 0.17 dB in signal to noise ratio. A lower degradation due to quantization can be obtained by using more bits (8 bits gives 0.04 dB at E_b/N_o = 20 dB) or by reducing the dynamic range of the A/D converter such that moderate clipping occur at the highest input signal levels.

The signal can be converted from IF to baseband and sampled in different ways:
- Bandpass sampling, real or complex
- Down conversion by means of analog mixers and quadrature local oscillators. A/D conversion in I and Q arms.

If bandpass sampling is used, sampling and down conversion are done in the same process. We have, however, chosen the latter method which involves mixing down to almost zero frequency by means of analog mixers. This gives the simplest solution and the degradation due to errors in the analog components are found to be 0.07 dB. It is assumed that a phase error of 2 degrees and an amplitude balance of 0.6 dB are obtainable values.

4. DIGITAL DEMODULATOR

Individual digital signal processing is carried out for each carrier. The complexity in the digital part of the receiver is measured in number of real multiplications per symbol per carrier. This figure is found to be $12\,N_s+5$, where N_s is the number of samples per symbol produced by the analog part of the MCD. Increasing the sample rate will reduce the requirements to the relative accuracy in the analog part while the complexity in the digital part goes up. The choice of N_s is therefore a tradeoff between complexity in the digital and the analog part. Our approach is to reduce the complexity in the digital part as much as possible. The limitation is the physical length of the quartz crystal in the SAW line. This results in a sampling rate of four samples per symbol which gives 53 real multiplications per symbol in the digital part of the demodulator. The response length of the SAW line is 114 μs which give a physical crystal length of 21 cm.

The digital filter is a FIR-filter with 19 taps. The filter taps are optimized by using a computer program developed at ELAB [5]. The filter is optimized to give
- small intersymbol interference
- sufficient stopband attenuation.

The first criterion was not difficult to fulfil. In order to fulfil the last criterion a filter length of 19 taps (impulse response length 4.75 symbols) was necessary. The word length in the coefficients is chosen to be seven bits. The internal data word length is eleven bits and the output data is quantized to seven bits. Only one sample per symbol is computed at the output of the filter.

The sampling in the A/D converter is asynchronous to each channel. A signal sample at the optimum timing instant is obtained by means of a digital phase locked loop and the selection of one filter out of a set of eight digital FIR polyphase receiver filters. The timing loop controls the timing of output signal from the digital FIR polyphase filter. The loop has two outputs; one "coarse" output which determines the number of shifts in the FIR filters from one symbol to the next, and one "fine" output which determines which set of polyphase filter coefficients is going to be used. The eight sets of filter coefficients give a resolution in time of 0.03125 symbol intervals. The loop adjusts the timing every second symbol.

The incoming signal has a frequency uncertainty of ± 600 Hz. During acquisition of continous, randomly modulated signals a frequency loop is used to compensate for this rather large frequency offset. When the frequency is corrected to a value sufficiently close to its correct value the phase correction loop takes over and phase lock to the incoming signal is obtained. The frequency loop is of a new design and requires seven real multiplications per symbol during acquisition [6, ch 4]. During acquisition of data bursts, an 8-point FFT is used to limit the frequency offset to ± 300 Hz, which is sufficiently close to avoid false locks.

The carrier phase detector is referred to as a modified Costas loop detector. The output from the numerically controlled oscillator is fed back to the input phase shifter, which shifts the signal down to exactly zero frequency. The following detectors are used to control the demodulator
- Data present detector
- Frequency lock detector
- False frequency lock detector
- Phase lock detector.
- Unique word detector (for burst mode only)

The data present detector detects if data is present at the input of the digital part. The frequency lock detector detects if the frequency estimate provided by the frequency loop is acceptable. If false frequency lock is detected by the false frequency lock detector the frequency is corrected. The phase lock detector detects when the phase of the locally generated carrier is close enough to the optimum value. This will also indicate lock of the timing loop since the phase information of the signal is maximized at the optimum timing instant.

In burst reception mode, the frequency, phase and timing acquisition are carried out in a predetermined sequence, therefore only the data present and the unique word detectors are used.

5. SUMMARY
The total degradations in the demodulator are summarized in the table below at E_b/N_o equal to 8.4 dB.

Sample time jitter (±1.5 ns)	0.01 dB
Down conversion (0.6 dB, 2^0)	0.07 dB
Quantization (7 bits)	0.17 dB
Input phase shifter (7 bits)	0.02 dB
Timing recovery (8 polyphase filters)	0.03 dB
Carrier recovery (Loop BW=50 Hz)	0.10 dB
Distortion in FIR filters	0.01 dB
Carrier to filter frequency offsets (Inband distortion in SAW-filter)	0.02 dB
Total degradation	0.43 dB
Mixer IM (CMC transform)	0.01 dB
Sample time offset (±12.5 ns)	0.05 dB
Adjacent channel leakage $(E_b/N_o)_{adj}$=20 dB	0.20 dB
Additional adjacent channel leakage $(E_b/N_o)_{adj}$=17 dB (Amplitude and phase errors in SAW-line II and chirps)	0.10 dB
Degradation with adjacent channel interf.	0.79 dB

The power consumption for a slightly different scheme was evaluated in [1], and was found to be 15 mW per carrier. In phase two of this project an electrical demonstration model will be built, and a more accurate estimate will be established.

6. CONCLUSIONS

Combinations of SAW ASP and DSP are promizing to achieve efficient on-board demodulation of 300 mobile type digital single channel carriers. Analysis and simulations demonstrate that the performance is similar to other high quality demodulators when the carrier power levels are reasonably balanced as in satellite communications. An electrical demonstration project is currently in progress at ELAB.

ACKNOWLEDGEMENT

The authors are grateful to G. Bjørnstrøm, ESTEC for the valuable comments and discussions during the MCD project.

REFERENCES

[1] Jack M.A. Paige E.G.S.: "Fourier Transforms Processors Based on Surface Acoustic Wave Filters", Wave Electronics $\underline{3}$ 1978, pp 229-247.

[2] Williamson R.C., Dolat V.S., Rhodes R.R., Boroson D.M.: "A Satellite-Born SAW Chirp-Transform System for Uplink Demodulation of FSK Communication Signals", IEEE 1979 Ultrasonics Symp. Proc., pp 741-747.

[3] Rønnekliev A., Bakken P.M., Ingebrigtsen K.A., Lier E.: "Application of Surface Acoustic Wave devices in Communication Satellites". ESTEC/Contract no 5509/83/NL/GM(SC), final report ELAB report no STF44 F85153, Sept. 1985.

[4] Rønnekliev A., Bakken P.M., Lier E.: "Multicarrier Demodulator Based on the SAW Chirp Fourier-transform", Proc. from Int. Seminar on Technology for High Speed Signal Processing, The Norwegian Inst. of Technology, Trondheim, Norway, 21-23 Aug. 1985, pp 209-213.

[5] Bakken P.M., Lier E.: "Documentation of Computer Programs for Optimalization and Simulation of Multicarrier SAW-based Demodulator", ELAB report no STF44 A86005, Jan. 1986.

[6] Ringset V., Rønnekliev A., Bakken P.M., Olsen E.: "Study of On-board Multicarrier Demodulator with Analog and Digital Processing for Mobile Communication". Final report ESTEC contract no 6899/86/NL/JG/(SC), ELAB report no STF44 F87086, June 1987.

A SAW-BASED INTEGRATED QPSK COHERENT DEMODULATOR

P.Tortoli, F.Andreuccetti, G.Manes

Dipartimento di Ingegneria Elettronica, v.S.Marta, 3, FI - Italy

R.Giubilei *, D.Gerli **

* Selenia Spazio SpA, Roma - ** Micrel Srl, Firenze

The implementation and experimentation of a prototype coherent demodulator for QPSK sequences are described. The system makes use of Surface Acoustic Wave (SAW) devices for IF filtering and carrier recovering. This approach was encouraged by some of the key features of SAW devices (e.g. reliability, compact size, low power consumption), which make them ideal candidates for use on-board the satellite in a digital communication link. The possible application of the system for multicarrier demodulation in on-board regenerative repeaters is shortly discussed.

1. INTRODUCTION

Surface Acoustic Wave (SAW) devices have already been successfully applied in satellite communication systems. A typical example is represented by their use as building blocks in many of the subunits involved in the terrestrial equipment of a satellite digital communication link |1|.

In particular, key features such as compact size and low power consumption make SAW devices suitable to be used on board the satellite. For example, a satellite-borne SAW-based system for uplink demodulation of FSK signals from up 100 users is described in |2|.

SAW devices thus appear ideal candidates to be used in on board regenerative repeaters |3|, where weight and power consumption of the demodulator are critical issues. In this context, a FDMA/TDM system has been recently proposed |4| which uses FDMA in the uplink by several carriers digitally modulated at low data rate (up to 2 Mbit/s), demodulation and regeneration on board, conversion in a single TDM high data rate stream that will be sent on the down link.

The heart of the system is a multicarrier demodulator (MCD) |5| on board which demodulates the FDM signal. The MCD consists of a bank of RF filters which drive coherent QPSK demodulators. Each coherent QPSK demodulator then requires a carrier recovery circuit, where the received IF is first multiplied by 4. In this way any modulation is removed, since the phase is changed back to 2π, whatever the original phase shift. After filtering with a high-Q resonator, a digital division by 4 recovers the original frequency.

SAW technology offers ideal solutions for both the above mentioned filters. In particular, SAW transversal filters can guarantee an excellent phase linearity, as requested in the channel filters in order to minimize the intersymbol distorsion.

The critical performances in terms of selectivity as well as stability involved in the carrier recovery circuit, can then be satisfied by means of a voltage controlled stable oscillator used in a classical PLL tracking filter mechanisation. As the IF carrier frequency ranges between 70 and 200 MHz, an oscillator based on a SAW resonator (SAWR) is ideally suited as a VCXO working at its fundamental frequency in the loop. It exhibits the attractive capability of allowing for compensation of frequency shifts originated by the satellite itself, while providing a satisfactory phase noise.

A prototype QPSK coherent demodulator, including a SAW channel filter for 2 Mbit/s sequences, a SAWR based oscillator tracking filter and related circuitry, has been implemented and experimented. In this paper, a technical description of the system is given and the main performances obtained by each constitutive block are reported.

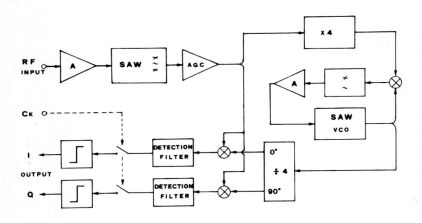

Fig.1   Block diagram of the coherent QPSK demodulator

2. CONFIGURATION OF THE IMPLEMENTED SYSTEM

A block diagram of the implemented coherent QPSK demodulator is shown in fig.1. It basically consists of an IF filter which is followed by a frequency multiplier for the recovery of an unmodulated carrier. Filtering is then achieved by means of a VCXO which is tracked to the carrier thanks to a classic PLL circuit. Division by 4 then allows the original frequency to be restored on two quadrature channels. Finally, the digital bit stream is detected through demodulators consisting of mixers, low pass detection filters and sampling circuits.

The main features of each implemented block are reported in the next sub-sections.

2.1 IF receiver

The IF receiver is constituted by a low noise wideband amplifier implemented by means of monolithic integrated circuits with minimum use of external components. A noise figure of about 2.5 dB was measured. The signal is then passed through a SAW bandpass filter having a nominal 3 dB bandwidth of 2.6 MHz. This value allows an unfiltered QPSK carrier with over than 2 Mega bit/s rate to be accomodated in mild-bandlimiting conditions.

The SAW filter exhibited, in untuned conditions, an insertion loss of about 39 dB. The measure of the input and output impedances, which essentially consist of series RC circuits, allowed simple matching networks to be developed. By properly selecting RF transformers and tuning coils the insertion loss has been reduced to about 20 dB and the effective amplitude and phase responses of fig.2 were measured. They show a passband ripple lower than 0.1 dB, a 1 dB bandwidth of 1.7 MHz and ultimate rejection (+/- 4 MHz) of about 60 dB. In particular, the excellent linearity of the phase vs frequency can be appreciated if fig.2b.

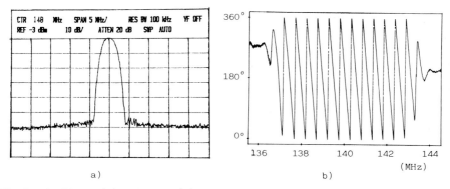

Fig.2   Amplitude (a) and phase (b) response of the SAW bandpass filter

The SAW filter is followed by an other wideband monolithic amplifier with external gain control, inserted in an AGC circuit capable of recovering a fixed power level independent of the input signal level. Variations over a 15 dB range are thus equalized. The output of this gain block is then spitted into two parts, one towards the frequency multiplier, and the other to the properly so-called demodulator, respectively.

2.2 Carrier recovery circuit

Carrier recovery first requires a signal frequency multiplication by 4 in order to eliminate any modulation by changing the phase back to a multiple of 2π, independently of the original phase shift. This is accomplished by means of an analog multiplier composed of a diode working in saturation and a low-Q resonant circuit around 4 times the IF.

After filtering, the output signal is phase compared with the reference oscillation generated by a VCO. This represents the key element of the full system, since its performances can heavily condition the final bit error rate (BER).

The VCO was thus implemented by using a SAWR |6| inserted in a regenerative loop. SAWR's are in fact devices capable of providing high-Q (>10000) and low phase noise characteristics while exhibiting an extremely low size. Aging is typically less than 10 ppm/year, and temperature coefficients as low as 1 ppm/(°C)^2 can be obtained when high quality Q_2 substrate is used.

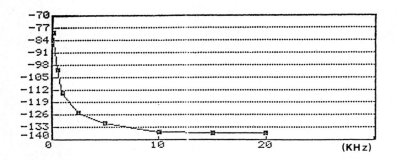

Fig.3 Phase noise of the SAWR oscillator

In our prototype system a very low phase noise (> 100 dBc at 1 KHz) was thus obtained by the combination of the high Q (≈ 5000) resonator with the bipolar transistor as active element. A tuning circuit, based on a voltage controlled diode, was introduced without considerably raising the phase noise, and allowing over than 100 ppm of electronic tuning to be obtained. Armonics resulted lower than 35 dBc, while non-armonic spurious were lower than 90 dBc.

The VCO was inserted in a second order loop obtained by implementing an active RC low pass filter at the phase detector output. The filter was designed by taking into account that an equivalent noise bandwidth of about 0.8 KHz with a 0.7 damping factor had to be reached |7|.

Acquisition has been made easier by means of a proper circuit which is enabled only in unlock conditions. It causes a voltage ramp to appear at the input of the VCO, which is thus forced to sweep until it reaches the input frequency. At this time, the loop locks up and the slewing current is shut off to avoid that it is balanced by a compensating phase error. To guarantee a high probability of acquisition (>90%) the sweep rate is kept lower than 300 KHz/s. By this method the PLL has been demonstrated capable of locking all input frequencies falling within a range of about 60 KHz around 591.420 MHz.

The output of the oscillator is sent to a first emitter coupled logic divider featuring ECL 10K compatible complementary outputs. These latter are sent to a further couple of dividers, so that in phase and quadrature outputs are directly obtained, independently of the signal frequency. As a result, a residual phase error lower than +/- 3° has been measured within the full frequency range.

2.3 Demodulator

The quadrature reference waves obtained by the VCO are first phase compared with the incoming signal (fig.1). The mixer outputs are then sent in to 3-poles low pass Butterworth filters with B = 500 KHz single-sided 3-dB bandwidth. This was designed on the basis of the considerations developed by Austin |8| who indicated a product B·T = 0.5 (with T the symbol duration) as optimum for the detection filter in QPSK signaling. As an example, the output of one detection filter, obtained by demodulating a pseudo random QPSK signal, is shown in fig.4. It exhibits the typical "eye pattern", with a very small amount of intersymbol interference.

The sampling circuit has been conceived by taking into account that it has the specific aim of detecting the sign of the signal emerging from the filter at given instants (i.e., at the center of the eye pattern). A simple comparator has thus been used, followed by a digital latch which is clocked synchronously with the estimated bit arrival time.

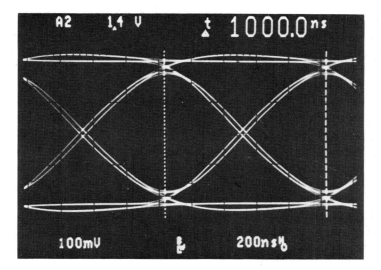

Fig.4 "Eye pattern" of the demodulator filter output

3. CONCLUSION

A prototype QPSK coherent demodulator has been discussed and demonstrated. SAW devices have been used for bandpass filtering and carrier recovering. Good performances have been obtained making this approach attractive in terms of potential integrability and low cost.

The system is currently under test for a quantitative evaluation of the attainable Bit Error Rate. In the next future, a comparison will be made with the performances obtained with a digital demodulator which has been independently developed in Selenia Spazio Spa.

ACKNOWLEDGEMENTS

This work was supported under contract # B3240570 by Selenia Spazio, Rome.
The authors wish to gratefully acknowledge the valid contribution of Dr. A.Taiti, SMA Spa, for his numerous and useful suggestions during the prototype implementation.

REFERENCES

|1| J.Henaff and P.C.Brossard, "Implementation of Satellite Communication Systems Using Surface Acoustic Waves", IEEE Trans. on Microwave Theory and Techniques, Vol.MTT-29, N.5, May 1981, pp.439-450.

|2| R.C.Williamson, V.S.Dolat, R.R.Rhodes, D.M.Boroson, "A Satellite-Borne SAW Chirp-Transform System for Uplink Demodulation of FSK Communication Signals", 1979 IEEE Ultrasonic Symposium Proceedings, pp.741-747.

|3| K.Koga, T.Muratani and A.Ogawa, "On-board Regenerative Repeaters Applied to Digital Satellite Communications", Proc. of the IEEE, Vol.65, N.3, March 1977, pp.401-410.

|4| G.Perrotta, G.Losquadro, R.Giubilei, "Satellite Communication System for Domestic/Business Service", Proc. of ICDSC -7- München, May 1986.

|5| F.Ananasso, E.Del Re, "On board Multicarrier Demodulators", these Proceedings.

|6| T.O'Shea, V.Sullivan, R.Kindell, "Precision L-band SAW Oscillator for Satellite Application", Proc. of the 38-th annual Freq. Symp., 1984.

|7| J.P.Frazier, J.Page, "Phase-locked Loop Frequency Acquisition Study", Trans. IRE, Set-8, Sept. 1962, pp.210-227.

|8| M.C.Austin, M.U.Chang, D.F.Horwood, R.A.Haslov, "QPSK, Staggered QPSK, and MSK. A Comparative Evaluation", IEEE Trans. on Comm., Vol.COM-31, Febr. 1983, pp.171-182.

SATELLITE TERMINALS IN THE ITALIAN NAVY AND USER EXPERIENCE

CV(AN) Roberto PALANDRI
CF(AN) Raffaele AZZARONE
TV(AN) Giovanni Battista DURANDO

Istituto per l'Elettronica e le Telecomunicazioni
"GIANCARLO VALLAURI" - LIVORNO

1. INTRODUCTION

As in all Navies, in Italy we are highly interested in the evolution of communication means to improve their overall reliability; in particular, since several years, some experimental and operational activities have been conducted by means of Telecommunication Satellites.

It is appropriate to mention the international communication tests undertaken with USA satellites in 1970-1975 years (TACSATCOM) and the partecipation to the Italian SIRIO (12-18 GHz band) satellite project (1979-1981) with the realization of a naval terminal and its installation on board of the DDG "ARDITO".

Besides these activities, the Italian Navy has been operatively faced with the problem of satellite communication in two particular emergency circumstances: the tragic earthquake in the Italian central region in November 1980 and the Mission of Peace of Italian Forces in Lebanon (1982-1984).

In both occasions the satellite channels utilized, operating in UHF band, were kindly conceded by USN Authorities.

The experience derived from the design, the realization and the operation of the UHF terminals are the subjects of this paper.

2. OPERATION OF UHF SATELLITE TERMINALS IN TWO EMERGENCY EVENTS

a.- In November 1980 in the Italian central region occurred a severe earthquake which, besides an heavy cost in human lives and the destruction of built-up areas, caused sensible failures to existing communication lines (telephone and radio relais); therefore some difficulties arose for the communications between the Co-ordinating Centers.

Since Italian Navy was carrying on the SIRIO Communication satellite experiment, MOD Italy suggested to move the mobile terminal from the ship "ARDITO" to the earthquaked regions.

This operation appeared very complex, both for the particular type of installation and the dimensions of the terminal.

Besides that, the operation required an amount of time for the disassembling and the reassembling of the whole system, not compatible with the emergency state.

Therefore, consequently to the assurance of the assignement of a UHF satellite channel by the USN Authorities, the realization of two UHF satellite terminals in the MARITELERADAR Institute in Livorno was ordered.

The two terminals, thanks to the experience acquired in satellite communications, were completed in a few days utilizing the new generation of the Italian Navy UHF tactical transceivers and helix antennas designed and built in Mariteleradar Institute.

The first terminal was installed inside a "shelter" (fig.1), equipped for mobile-operations in the earthquaked region, the second was assembled for the utilization in fixed installation.

The terminals were completed at the end of December 1980 and delivered to Italian Army for communication tests between M.O.D. in Roma (fig.2) and the mobile unit (fig.3) with satisfactory results.

b.- On the occasion of the Italian Mission of Peace in Lebanon (Autumn 1982), due to low reliability of civilian telephone lines and HF radio links, MOD Italy ordered again to Italian Navy to provide three satellite terminals, after a renewed assignement of the channel by USN Authorities.

The satellite terminals were built up, realized, installed and operational in two weeks, utilizing again the UHF tactical transceivers and the helix antennas; the first one, sheltered, was installed in the Italian Headquarter at Beiruth (figg. 4-5-6), the other two were located respectively in MOD-Army and MOD-Navy buildings in Roma.

The terminals correctly operated from October 1982 till March 1984.

3. THE SATELLITE TERMINAL OUTLINE

The two above-mentioned realization of UHF satellite terminals, were completed in a very short time and with low costs because the main components

Fig. 1

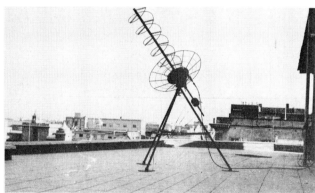

Fig. 2

Fig. 3

Fig. 4

Fig. 5

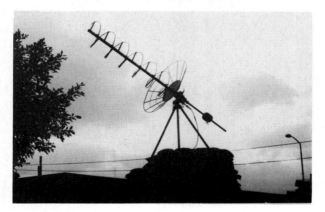

Fig. 6

utilized were the same UHF transceivers used by Italian Navy for tactical communications.

Each terminal was formed by the following components (see block diagram and figg. 7-8):

- no.1 UHF-FM simplex transceiver used for the "up-link" frequency trasmission;
- no.1 UHF-FM simplex transceiver for receiving the "down-link" frequency;
- no.1 RF "up-link" bandpass filter;
- no.1 Diplexer for decoupling the trasmitter from the receiver, firstly realized with resonant cavities and then with automatic selective filters;
- no.1 Helix antenna (right hand circularly polarized)

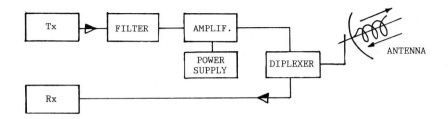

4. GRAPHICAL CALCULATION OF ANTENNA AZ-EL DATA

Setting up operationally the antennas turned out very simple, both for the geostationarity of assigned satellites and for the relatively wide-angle radiation lobe of the antenna used.

Accepting a tolerance of 2-3 degrees in pointing, it was sufficient to know only the latitude and longitude data of the earth-station and the longitude of satellite (since, being the satellite geostationary, its latitude is 0° and its height 36000 Km).

Rather than resolving trigonometric equations, not very practical in emergency situations, resulted very useful the following diagram in which are drawn the lines indicating:

$|\Delta| = |$ longitude of the satellite - longitude of the station $|$

$|\Gamma| = |$ latitude of the station $|$

The coordinates of the interception point of the two lines represent the azimuth and the elevation angles required for the antenna.

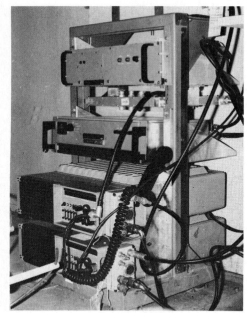

Fig. 7

Fig. 8

GRAPHICAL CALCULATION OF AZ-EL DATA

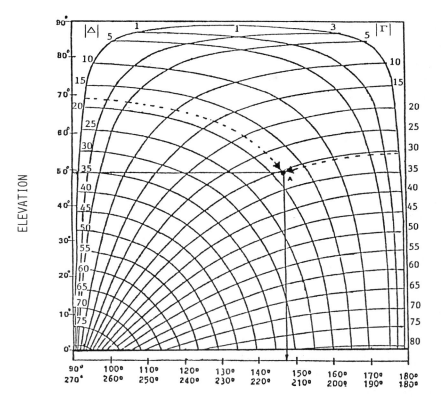

5. CONSIDERATIONS

From above-mentioned experiences the following important considerations have been derived for satellite communications in emergency events:

- The satellites, due to their earth-surface events insensivity, are available for communications both for short and long distances;

- The satellites, utilizing dedicated channels, are free from the frequency congestion and propagation problems, typical of HF band;

- The geostationary satellites, in particular, are easier to be pointed with the antenna, which also reduces cost and complexity of the terminals;

- The UHF satellite band is the most appropriate resource for the "first-voice" link in emergency situation due to the low-cost mobile terminals and simple to point low-directive antennas;

- The realization of satellite terminals with military equipment satisfies the requisite of a high reliability for operations in adverse enviromental conditions; furthermore it offers the considerable advantage that operators and technicians of the Armed Forces are already skilled in their use/maintenance.

The maintenability is also assured by a large provision of military tactical transceivers and relative spare parts and by the existence of military specialized laboratories and facilities;

- A few mobile terminals for emergency communications might be kept ready for use in appropriate Technical Centers; it would be also possible to realize, for special purposes, both "man-pack" terminals, characterized by a low power and sufficient antenna directivity (this latest to be mounted in loco), and terminals with high power trasmitter with omnidirectional and/or low directional antennas, to be installed on ships and aircrafts to eliminate or reducing the problem of the antenna pointing.

SESSION 4

ON BOARD PROCESSING

Chairman: P. De Santis *(INTELSAT, USA)*

PERFORMANCE EVALUATION OF REGENERATIVE DIGITAL SATELLITE LINKS WITH FEC CODECS

Neville A. MATHEWS

INTELSAT
3400 International Drive, N.W.
Washington D.C. 20008-3098, U.S.A.

This paper addresses methodologies for evaluating BER performances of all basic regenerative digital satellite link configurations with FEC codecs. Initially, BER performance analyses pertinent to regenerative links with hard-decision FEC codecs employing random error-correcting codes are considered. Next, regenerative links utilizing soft-decision Viterbi FEC codecs are treated. Some computations of iso-BER curves for both regenerative and non-regenerative links are included to illustrate the link-budget benefits associated with on-board regeneration.

1. INTRODUCTION

Although the concept of on-board signal processing in a digital satellite communication system is not necessarily synonymous with on-board regeneration, the many varied benefits associated with the former can be realized only with regenerative digital satellite repeaters /1/, /2/. Perhaps, the two major benefits accrued by incorporating on-board regeneration are:

(i) Isolation between uplink and downlink BER performances resulting in an improved link budget. Furthermore, there exists the flexibility to allow on-board encoding/decoding, transmission-rate changing and reformatting of digital modulation prior to downlink transmission. Thus, the differing power and bandwidth capabilities of the up- and down-links can be utilized for improved communications efficiency. This has particular relevance with the trend to use smaller earth stations in dispersed networks.

(ii) Due to the availability of baseband signals on board the satellite, dynamic connectivity among all beams and channels is possible through the use of a baseband switch or processor.

In this paper, only methodologies useful in assessing the link-budget benefits associated with on-board regeneration are addressed. In particular, BER performance analyses of all basic regenerative digital satellite link configurations with FEC codecs are considered. The methods described here are based on a judicious combination of analysis and computer simulation. Further, they are applicable to a variety of arbitrary two-dimensional signaling formats, e.g., PSK, Offset-QPSK, MSK, APK etc., relevant to future regenerative digital satellite links.

2. BASIC CONFIGURATIONS OF NON-REGENERATIVE AND REGENERATIVE DIGITAL SATELLITE LINKS WITH AND WITHOUT FEC CODECS

Fig. 1 schematically illustrates all basic configurations of non-regenerative and regenerative digital satellite links with and without FEC codecs*. Note that the configurations are enumerated in increasing order of complexity of implementation.

Link #1, which has neither on-board regeneration nor FEC codecs, is admittedly the simplest and also has the poorest BER performance. Although this configuration is seldom used in digital satellite communications, it is included here for completeness. Virtually all present-day digital satellite links correspond to the non-regenerative configuration of Link #2, since encoders and decoders are easily implemented at earth stations. Links #3 to #7 all include

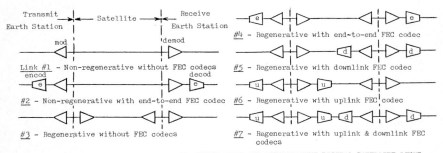

Fig. 1 - BASIC CONFIGURATIONS OF NON-REGENERATIVE AND REGENERATIVE DIGITAL SATELLITE LINKS WITH AND WITHOUT FEC CODECS

on-board regeneration, although Link #3 does not employ any FEC codecs. Link #4 is the only regenerative link configuration where the encoding/decoding and regeneration processes are concatenated. This feature presents no problems from the viewpoint of BER analysis so long as hard-decision detection prior to decoding is used. However, with soft-decision detection prior to decoding, careful consideration must be given to BER analysis since regeneration is essentially a hard-decision process. This point will be taken up in Section 5.

It is perhaps relevant to note here that only Links #5, #6 and #7 incorporate encoders and/or decoders on board the satellite. From a hardware implementations viewpoint, Link #5 is considered to be less complex than Link #6 since the former employs an on-board encoder while the latter requires an on-board decoder. On the other hand, Link #7 clearly involves the greatest degree

---

*All link configurations employing FEC codecs in Fig. 1 can be further extended to include code concatenation schemes by inclusion of an outer FEC codec.

of complexity of implementation since it incorporates both encoding and decoding functions on board the satellite.

3. GENERAL APPROACH FOR EVALUATING BER PERFORMANCES OF REGENERATIVE DIGITAL SATELLITE LINKS WITH FEC CODECS

From the standpoint of BER analysis, all regenerative links incorporating FEC codecs in Fig. 1, with the exception of Link #4*, may be regarded as a pair of BSCs (<u>B</u>inary <u>S</u>ymmetric <u>C</u>hannels) in cascade. Fig. 2a illustrates this schematically, where P_{bu} and P_{bd} denote the information-BERs of the uplink and downlink BSCs, respectively. It is now relatively easy to show that the cascaded BSCs reduce to an equivalent BSC (Fig. 2b) whose information-BER P_b is given by:

$$P_b = (1-P_{bu})P_{bd} + P_{bu}(1-P_{bd})$$
$$= (P_{bu} + P_{bd})\left[1 - 2(1/P_{bu} + 1/P_{bd})^{-1}\right]$$

For most practical cases of interest, both P_{bu} and P_{bd} will be less than 10^{-2}, and the above equation is very well approximated by

$$P_b = P_{bu} + P_{bd} \qquad \qquad \dots (1)$$

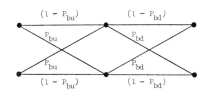

Fig. 2a - UPLINK BSC AND DOWNLINK BSC IN CASCADE

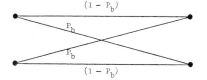

Fig 2b - EQUIVALENT BSC OF REGENERATIVE LINK

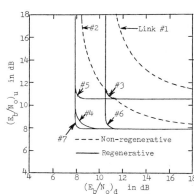

Fig. 3 - ISO-BER CURVES (BER=10^{-6}) FOR LINKS WITH AND WITHOUT HARD-DECISION FEC CODECS

4. METHODOLOGIES FOR EVALUATING BER PERFORMANCES OF REGENERATIVE DIGITAL SATELLITE LINKS WITH HARD-DECISION FEC CODECS EMPLOYING RANDOM ERROR-CORRECTING CODES

A practical approach to evaluating BER performances of regenerative satellite links is to consider, separately, (i) Characterization of the hard-decision FEC
─────────
*This configuration will be treated later on in Sections 4 and 5.

codecs, and (ii) Evaluation of channel-BERs associated with the uplink and downlink paths of the regenerative link. BER performances associated with any of the regenerative link configurations with FEC codecs in Fig. 1 can then be evaluated from equation (1). This procedure will now be discussed in some detail.

The error-correcting performance of a hard-decision FEC codec employing a random error-correcting code is characterized by its information-BER (P) versus channel-BER (P*) relationship:

$$P = F(P^*) \qquad \ldots \quad (2)$$

where $F(.)$ denotes the functional describing the relationship.

Now, making use of (2) in (1), it is evident that the following expressions will be obtained for the information-BERs of the various regenerative link configurations with hard-decision FEC codecs (with the exception of Link #4) in Fig. 1 :

$$P_b(\text{Link } \#5) = P^*_{bu} + F_d(P^*_{bd}) \qquad \ldots \quad (3a)$$

$$P_b(\text{Link } \#6) = F_u(P^*_{bu}) + P^*_{bd} \qquad \ldots \quad (3b)$$

$$P_b(\text{Link } \#7) = F_u(P^*_{bu}) + F_d(P^*_{bd}) \qquad \ldots \quad (3c)$$

where the subscripts associated with the functionals depend on whether there are uplink and/or downlink FEC codecs. Note that the quantities with asterisks denote channel-BERs whereas those without are information-BERs.

So far, Link #4 has been excluded from the analysis. Clearly, for this link configuration the effective channel-BER of the regeneratively-cascaded up- and downlinks is very nearly equal to $P^*_{bu} + P^*_{bd}$ (following the approach taken in Section 3), and the information-BER of the overall link is:

$$P_b(\text{Link } \#4) = F_e(P^*_{bu} + P^*_{bd}) \qquad \ldots \quad (3d)$$

where the subscript associated with the functional now indicates an end-to-end codec arrangement.

In general, the functional in (3a)-(3d) for a hard-decision FEC codec employing a random error-correcting code with a t-error-correcting capability is given by:

$$F(.) = \sum_{e=t+1}^{n} \frac{N_e}{n} \binom{n}{e} (.)^e \left[1 - (.)\right]^{n-e} \qquad \ldots \quad (4)$$

where n is the block length of the code and N_e is the average number of errors in the decoded code block when the number of input errors in the code block is e.

The quantity N_e can be determined from computer simulation of the decoding process by considering all possible e-bit error-patterns in a block. If the number of bits in a block is large, resulting in an excessive number of possible e-bit error-patterns, a Monte Carlo approach is adopted to randomly select a sufficiently large number, say 1000, of the possible e-bit error-patterns.

Based on the above procedures, information-BER versus channel-BER relationships for several hard-decision FEC codecs used in the INTELSAT TDMA/DSI and SCPC systems have been deduced /3/. For information-BERs of interest ($< 10^{-3}$), these relationships display a fairly log-log characteristic between information- and channel-BER. For example, based on this latter feature, the following simple codec functional is deduced for the rate 7/8 (128,112) BCH code used in the INTELSAT TDMA/DSI system:

$$F(.) = 3373(.)^{2.91} \qquad \ldots (5)$$

So far, we have tacitly assumed knowledge of the channel-BERs P^*_{bu} and P^*_{bd} associated with the uplink and downlink paths, respectively, of the regenerative satellite link. Analytical expressions for these channel-BERs are well known for a variety of two-dimensional signaling formats, viz., PSK, Offset-QPSK, MSK, APK etc., when the uplink and downlink paths are ideal Nyquist-filtered linear channels each corrupted with AWGN (Additive White Gaussian Noise). Real-life satellite uplink and downlink paths are, however, bandlimited nonlinear channels each corrupted with Gaussian noise and interferences such as co-channel and adjacent-channel. In such situations, uplink and downlink channel-BERs can be obtained by Channel Simulation Computer Programs, based on FFT (Fast Fourier Transform) and inverse FFT techniques /4/. Channel simulation is first performed in the absence of noise and the effect of the latter is introduced analytically by considering each received signal point in two-dimensional signal space taking into account relevant decision boundaries.

Some computations of iso-BER curves for all the non-regenerative and regenerative link configurations in Fig. 1 are included here to illustrate the link-budget benefits associated with on-board regeneration and hard-decision FEC codecs. Fig. 3 shows the results of these computations assuming: (i) hard-decision FEC codecs, described by the functional in (5), are used, and (ii) coherent BPSK or QPSK signaling ensues through ideal Nyquist-filtered linear uplink and downlink paths each corrupted with AWGN alone. Note that iso-BER curves depict the required uplink and downlink Bit-energy/Noise-density ratios, $(E_b/N_o)_u$ and $(E_b/N_o)_d$, necessary to maintain an objective information-BER*.

As expected, the regenerative link configuration with FEC codecs on both

___
*The bit-energy E_b is implicitly taken to be the energy associated with an information bit, and N_o is the one-sided power spectral density of AWGN.

uplink and downlink (Link #7) and therefore having the greatest degree of implementation complexity, displays the best performance from a link-budget viewpoint. On the other hand, the regenerative link with and end-to-end FEC codec arrangement (Link #4) which entails the least degree of implementation complexity in that both encoder and decoder are located at earth stations, demonstrates a better link-budget performance than either Link #5 or Link #6, which have either on-board encoder or decoder. Furthermore, the performance of Link #4 compares fairly well with that of Link #7. Consequently, it appears that the regenerative link with an end-to-end FEC codec arrangement provides the best tradeoff between link-budget performance and complexity of implementation.

5. METHODOLOGIES FOR EVALUATING BER PERFORMANCES OF REGENERATIVE DIGITAL SATELLITE LINKS WITH VITERBI FEC CODECS EMPLOYING CONVOLUTIONAL ENCODING AND SOFT-DECISION VITERBI DECODING

Here, we initially exclude the regenerative link with an end-to-end soft-decision Viterbi FEC codec (Link #4) from our discussion.

BER performances of regenerative links with uplink and/or downlink soft-decision Viterbi FEC codecs (Links #5, #6 and #7) can be deduced by separately considering the uplink and downlink information-BERs and summing the result as in (1).

With soft-decision Viterbi decoding there is strictly no unique functional relationship between the information-BER and channel-BER which holds for typical satellite interference environments, as is the case with hard-decision FEC codecs. This is due to the fact that the maximum likelihood metric in practical Viterbi decoders is based on optimum decoding in an AWGN environment. Consequently, two differing channel environments, which give rise to identical channel-BERs, will not necessarily result in the same decoded- or information-BER. In such situations, the information-BER can be deduced from the Viterbi BER Upper Bound for a (n_o, k_o) convolutional code (rate k_o/n_o) given by /5/ :

$$P_b \leq \frac{1}{k_o} \sum_{k=d_{min}}^{\infty} c_k P_k \qquad \ldots \quad (6)$$

where: P_b is the decoded- or information-BER; d_{min} is the minimum free distance of the code; c_k is the total number of error bits included in all incorrect paths whose free distance from the correct path (usually the 'all-zero' path) equals k ; P_k is the pairwise error probability, which corresponds to the probability that any of the incorrect paths is selected instead of the correct path.

The c_k's and d_{min} in (6) depend solely on the trellis structure of the particular code, whereas the P_k's are influenced by the coding channel environment including the degree of quantization prior to Viterbi decoding.

The convolutional codes relevant to INTELSAT IBS and IDR digital carriers are the rate 1/2, constraint length K = 7, code /6/ and the rate 3/4 (punctured) code /7/. Appropriate values of d_{min} and the first few non-zero c_k's for these codes are: (i) rate 1/2; $d_{min} = 10$, $c_{10} = 36$, $c_{12} = 211$, $c_{14} = 1404$, $c_{16} = 11633$, $c_{18} = 77433$ (ii) rate 3/4; $d_{min} = 5$, $c_5 = 42$, $c_6 = 201$, $c_7 = 1492$, $c_8 = 10469$, $c_9 = 62901$, $c_{10} = 377561$.

The P_k's can be evaluated once the coding channel's discrete transition p.d.f. (probability density function) is known. In particular, the P_k's are deduced by obtaining the k-fold convolution of the latter and then integrating the result over the region which incorrectly decides the wrong path whose distance from the correct path (the 'all-zero' path) equals k /5/.

To proceed with the methodology for evaluating the P_k's, consider the coding channel associated with the uplink path (or downlink path) of the regenerative satellite link. Assuming BPSK or QPSK transmission and ideal interleaving, the coding channel with N-level quantization at its output may be viewed as a binary-input/N-level-output memoryless and symmetric channel. Fig. 4a illustrates such a channel model when the output has 4 levels of quantization.

When the uplink path (or downlink path) is assumed to be an ideal Nyquist-filtered linear channel contaminated with only AWGN, it is relatively easy to deduce an analytical expression for the channel discrete transition p.d.f., given the number of quantization levels and the spacing between the soft-decision levels at the input to the quantizer. For a real-life satellite uplink path (or downlink path), the channel discrete transition p.d.f can be deduced by first using a Channel Simulation Computer Program /4/ to obtain the statistical moments associated with the scatter of the received signal in signal space due to bandlimited, nonlinear and interference-corrupted transmission. The effect of AWGN can then be introduced analytically to give an expression, based on a Gram-Charlier series expansion /8/, for the channel continuous transition p.d.f. Finally, the channel discrete transition p.d.f. is determined by analytical integration taking into account both the number of quantization levels and the spacing between the soft-decision levels at the input to the quantizer.

As previously noted, the P_k's can be obtained once the coding channel's discrete transition p.d.f. is known. In particular, the method due to Yasuda, Hirata and Ogawa /9/ can be used. Alternatively, the P_k's may be evaluated via an algorithm for repeated convolutions of a discrete p.d.f. with itself /10/.

Some computations of iso-BER curves for the non-regenerative and regenerative link configurations (with the exception of Link #4) in Fig. 1 are included here to illustrate the link-budget benefits associated with on-board regeneration and soft-decision Viterbi FEC codecs. Fig. 4b shows the results of these computations assuming: (i) soft-decision Viterbi FEC codecs employing the rate 1/2, constraint length $K=7$, code /6/ with 8-level soft-decision Viterbi decoding, and (ii) coherent BPSK or QPSK signaling through ideal Nyquist-filtered linear uplink and downlink paths each corrupted with only AWGN. It should be noted that the spacing between the soft-decision levels is optimized for every iso-BER curve.

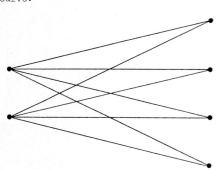

Fig. 4a - MODEL OF CODING CHANNEL REPRESENTING UPLINK (OR DOWNLINK) PATH WITH SOFT-DECISION VITERBI FEC CODEC

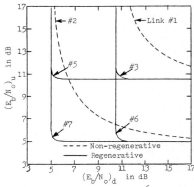

Fig. 4b - ISO-BER CURVES (BER=10^{-6}) FOR LINKS WITH AND WITHOUT SOFT-DECISION VITERBI FEC CODECS. LINK #4 IS OMITTED.

As with the iso-BER curves in Fig. 3, where hard-decision FEC codecs are employed, the iso-BER curves in Fig. 4b clearly indicate that the most favorable link-budget performance under all uplink and downlink path conditions results when FEC codecs are present on both uplink and downlink paths (Link #7). However, only a downlink FEC codec (Link #5) is sufficient if the regenerative link is power-limited on the downlink. Conversely, an uplink FEC codec (Link #6) is only needed if the regenerative link is power-limited on the uplink.

So far we have excluded the regenerative link with an end-to-end FEC codec (Link #4) from our discussion. As noted in Section 2, this is the only regenerative link configuration where the encoding/decoding and regeneration processes are concatenated. In the following, we assume ideal interleaving prior to transmission.

Due to regeneration being a hard-decision process, the uplink path of the link is viewed as a memoryless BSC with crossover probability p_u equal to the channel-BER of the uplink path. On the other hand, with finite-level soft-decision Viterbi decoding on the downlink path, the latter can be viewed as a binary-input/N-level-output memoryless symmetric channel. The overall coding channel may then be viewed as a cascade of the uplink BSC and the downlink binary-input/N-level-output channel. Fig. 5a illustrates such a cascaded channel when the final output has 4 levels of quantization.

When both uplink and downlink paths are assumed to be ideal Nyquist-filtered linear channels each corrupted with AWGN, it is relatively easy to deduce an analytical expression for the discrete transition p.d.f. of the overall coding channel, given both the number of quantization levels and the spacing between the soft-decision levels at the input to the quantizer. For real-life satellite uplink and downlink paths, the transition probabilities associated with the component coding channels in Fig. 5a can each be determined by a hybrid approach involving channel computer simulation and analytical techniques previously described. The discrete transition p.d.f. of the overall coding channel is then easily determined from the transition probabilities of these component channels. Subsequently, the pairwise error probabilities, i.e. the P_k's , are evaluated and used in (6) to determine the information-BER of the regenerative link with an end-to-end soft-decision Viterbi FEC codec.

Some computations of iso-BER curves for both non-regenerative and regenerative links with end-to-end soft-decision Viterbi FEC codecs are included here to illustrate the link-budget benefits accrued from employing on-board regeneration. The 'dashed' and 'solid' iso-BER curves in Fig. 5b illustrate results of these computations assuming: (i) soft-decision Viterbi codecs employing the rate 1/2, constraint length $K=7$, convolutional code /6/ with 8-level soft-decision Viterbi decoding, and (ii) coherent BPSK or QPSK signaling through ideal Nyquist-filtered linear uplink and downlink paths each corrupted with only AWGN. Further, it should be noted that the spacing between the soft-decision levels is optimized for each combination of uplink and downlink AWGN.

The 'hatched' region in Fig. 5b illustrates link-budget benefits due to on-board regeneration. This indicates that benefits become greater as the regene-

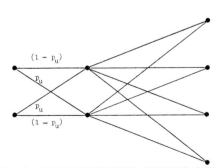

Fig. 5a - MODEL OF CODING CHANNEL REPRESENTING UPLINK AND DOWNLINK PATHS OF REGENERATIVE LINK WITH END-TO-END SOFT-DECISION VITERBI FEC CODEC

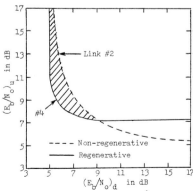

Fig. 5b - ISO-BER CURVES (BER = 10^{-6}) FOR NON-REGENERATIVE AND REGENERATIVE LINKS WITH END-TO-END SOFT-DECISION VITERBI FEC CODECS

rative link is increasingly power-limited on the downlink path. On the other hand, the iso-BER curves indicate a crossover point, beyond which the non-regenerative link is more advantageous from a link-budget viewpoint. This latter feature can be explained by noting that hard-decision Viterbi decoding effectively ensues in the regenerative link as the downlink AWGN becomes vanishingly small, even though there is still 8-level quantization prior to decoding on the downlink. The non-regenerative link, on the other hand, always experiences 8-level soft-decision Viterbi decoding. It is well known that hard-decision Viterbi decoding requires about a 2 dB increase in (E_b/N_o), relative to 8-level soft-decision Viterbi decoding, to maintain the same information-BER.

On comparing the 'solid' iso-BER curve in Fig. 5b with those of Links #5, #6 and #7 in Fig. 4b, we note that the regenerative link with an end-to-end soft-decision Viterbi FEC codec (Link #4) exhibits: (i) a better link-budget performance than the regenerative link with a soft-decision Viterbi FEC codec on the downlink path only (Link #5); (ii) a better link-budget performance than the regenerative link with a soft-decision Viterbi FEC codec on the uplink path only (Link #6), when the downlink path is power-limited; (iii) a poorer link-budget performance than the regenerative link with soft-decision Viterbi FEC codecs on both uplink and downlink paths.

6. CONCLUSIONS

This paper has addressed methodologies useful in evaluating BER performances of regenerative digital satellite links with FEC codecs. Both, hard-decision FEC codecs, employing random error-correcting codes, and soft-decision Viterbi FEC codecs have been considered.

Some computations of iso-BER curves for both non-regenerative and regenerative satellite links were included to illustrate the link-budget benefits accrued by employing on-board regeneration. These computations assume ideal BPSK or QPSK transmission through Nyquist-filtered linear uplink and downlink paths, each corrupted with AWGN. For regenerative links employing hard-decision FEC codecs, the results indicate that the link configuration with an end-to-end codec arrangement provides the best tradeoff between link-budget performance and complexity of implementation. With soft-decision Viterbi FEC codecs, however, the regenerative link with codecs on both uplink and downlink paths displays the best link-budget performance. Nevertheless, the regenerative link with an end-to-end soft-decision Viterbi FEC codec provides a reasonable tradeoff between link-budget performance and complexity of implementation when the downlink path becomes increasingly power-limited.

REFERENCES

/1/ K. Koga, T. Muratani and A. Ogawa : "On-board regenerative repeaters applied to digital satellite communications", Proc. IEEE, vol. 65, No. 3, pp. 401-410, March 1977.

/2/ P. Nuspl, R. Peters, T. Abdel-Nabi and N. Mathews : "On-board processing for communications satellites : systems and benefits", Int. J. Satell. Commun., vol. 5, pp. 65-76, 1987.

/3/ Final Study Report - IS-838 : High speed forward error correction codec, KDD Co. Ltd., Japan, Section 3, January 1978.

/4/ L.C. Palmer : "Computer modeling and simulation of communications satellite channels", IEEE J. Selected Areas in Commun., vol. SAC-2, No. 1, pp. 89-102, January 1984.

/5/ A.J. Viterbi : "Convolutional codes and their performance in communications systems", IEEE Tran. on Commun., vol. COM-19, No. 5, pp. 751-771, October 1971.

/6/ J.P. Odenwalder : "Optimum decoding of convolutional codes", Doctoral dissertation, School Eng. Appl. Sci., Univ. California, p. 64, 1970.

/7/ Y. Yasuda, K. Kashiki and Y. Hirata : "High-rate punctured convolutional codes for soft-decision Viterbi decoding", IEEE Trans. on Commun., vol. COM-32, pp. 315-319, March 1984.

/8/ H. Cramer, "Mathematical Methods of Statistics', Princeton, NJ : Princeton Univ. Press, pp. 227-230, 1946.

/9/ Y. Yasuda, Y. Hirata and A. Ogawa : "Optimum soft-decision for Viterbi decoding", Proc. ICDSC-5, Genoa, Italy, pp. 251-258, March 1981.

/10/ N.A. Mathews, unpublished work.

Satellite Integrated Communications Networks
E. Del Re, P. Barthelomé and P.P. Nuspl (eds.)
© Elsevier Science Publishers B.V., 1988

ACTS: THE FIRST STEP TOWARD A SWITCHBOARD IN THE SKY

F. Michael Naderi

National Aeronautics and Space Administration (NASA),
Headquarters, Code EC, Washington, D.C., 20546, U.S.A.

The Advanced Communications Technology Satellite (ACTS), now under development and scheduled for launch in the first quarter of 1991, is presently the main focus of NASA's communications program. Key technologies for ACTS include electronically hopping spot beams antennas, on-board processing and circuit switching, Ka-band transmission, and laser intersatellite links.

1. INTRODUCTION

The National Aeronautics and Space Administration (NASA) is developing the Advanced Communications Technology Satellite (ACTS) to pave the way for future communications satellites. After the launch, it will be made available to interested U.S. experimenters, free of charge, as a national facility to test the utility of the ACTS technologies for various applications. This paper provides an overview of the satellite system.

1.1. Coverage and Frequency Plan

To experiment with high-gain multiple beam technology, the ACTS antenna system is designed to provide three types of spot beams: stationary, electronically hopping, and mechanically steered. The footprints for the first two are shown in Figure 1.

The ACTS antenna system has three stationary and two hopping beams. One of the two hopping beams can hop to six discrete locations and anywhere within a contiguous area called the west sector. The second hopping beam can hop to seven discrete locations and anywhere within a second contiguous area called the east sector. Altogether, these beams cover 20% of the continental United States, which is deemed sufficient for an experimental system.

On ACTS, the half-power beamwidth for both the stationary and the hopping beams is approximately 0.3 degree. To extend the coverage beyond that shown in Figure 1, a mechanically steerable antenna with a 1.0-degree beamwidth has also been incorporated into ACTS. This beam, not shown in Figure 1, can be pointed anywhere within the disk of the Earth as seen from the 100-degree longitude location of ACTS.

The uplink and downlink transmissions take place within two 1-GHz Ka bands, 29.0-30.0 GHz and 19.2-20.2 GHz, respectively. The ACTS system uses both spatial and polarization diversity to achieve frequency reuse. The three stationary beams use the same frequency, but the Cleveland polarization is orthogonal to the

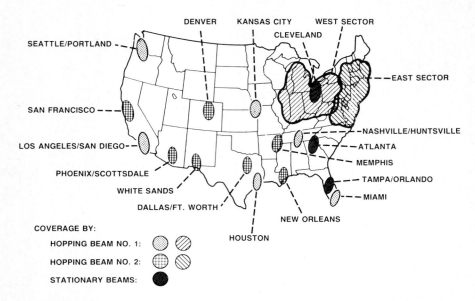

FIGURE 1. The ACTS Coverage by Three Stationary and Two Hopping Beams.

other two beams. The two hopping beams also use the same frequency but employ opposite polarizations.

1.2. Modes of Operation

Two types of switching, IF and baseband, have been incorporated into ACTS. The IF switch interconnects the three stationary beams in a satellite switched time division multiple access (SS/TDMA) mode of operation. The baseband switch interconnects the two hopping beams using baseband circuit switched TDMA. Figure 2 is a simplified block diagram of the satellite payload that supports these two modes of operation.

2. SPACE SEGMENT DESCRIPTION

The ACTS contains two cross-connected payloads: one is an optical payload using a heterodyne transmitter and a direct detection transceiver; the other employs Ka-band multiple beam antennas and on-board switching. Major elements of the microwave payload, the subject of this paper, are the multiple beam antennas, the Ka-band transmitters and receivers, and the two types of switches, i.e., the baseband processor and the microwave switch matrix.

2.1. Multiple Beam Antenna

The ACTS Multiple Beam Antenna (MBA) configuration is shown in Figure 3. It consists of two electrically identical MBAs, one at 30 GHz for receiving and the other at 20 GHz for transmitting.

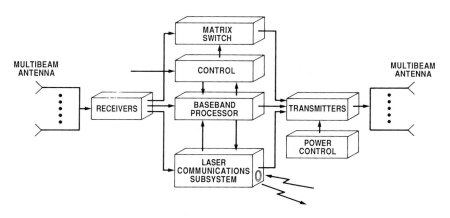

FIGURE 2. Simplified Block Diagram of the ACTS Payload.

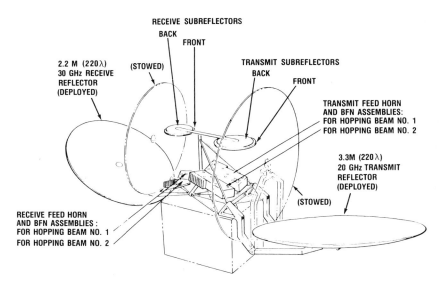

FIGURE 3. The ACTS Multibeam Antenna Configuration.

Both are offset Cassegrainian antennas, each with two polarization-selective hyperbolic subreflectors arranged in a piggyback configuration. The front subreflectors are gridded to pass one sense of linear polarization and to reflect the orthogonal sense. The back subreflectors are solid and reflect the polarization passed through the front subreflectors [1].

The ACTS has two uplink and two downlink beams; each pair of uplink and downlink beams corresponds to one of the hopping beams shown in Figure 1. To produce these four beams, there are two transmit and two receive feed assemblies on ACTS (Figure 3). Each transmit feed assembly uses one of the two transmit subreflectors to provide one of the two downlink hopping beams. In a similar fashion, the receive feed assemblies produce the two uplink hopping beams.

2.2. The Beam Forming Network

Figure 4 shows the schematic of a typical feed assembly consisting of feed horns and a beam forming network. At the heart of the Beam Forming Network (BFN) are the ferrite circulating switches with extremely low insertion loss (approximately 0.1 dB). The switches are interconnected by electroformed copper waveguides. Switch states are controlled by the hopping beam controller under instruction from a central processor that is commanded from the ACTS master control station on the ground. To reverse switching states, brief latching pulses are sent over flexible transmission cables from the controlling electronics to the switches. By controlling these switches, the hopping beam controller directs a signal through the BFN tree to a particular feed horn.

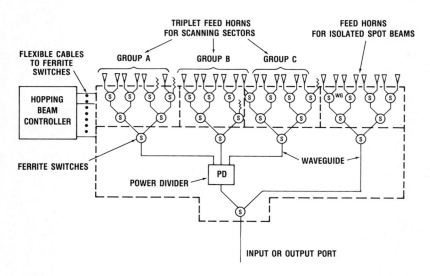

FIGURE 4. A Typical ACTS Beam Forming Network Arrangement.

The isolated spot beams in Figure 1 are formed from individual multiflare feed horns. However, to move the hopping beams within the scan sectors, a triplet of horns is used, one each from groups A, B, and C, shown in Figure 4. Different sets of triplets direct the beam to different locations within the scan sector.

The controller also controls the hopping sequence of the beams and their dwell duration by rapidly changing the states of the switches. The switches in the BFN must be fast enough to allow the hopping beams to hop to many locations during the 1-ms TDMA frame period. Accordingly, their specification calls for a 1-μs switching time.

2.3. Ka-Band Components

The low-noise figure at the receiver is achieved through the use of high-electron-mobility transistors (HEMT). It is expected that a 5-dB noise figure across the 1-GHz receive band can be achieved. Taking into account other losses, this results in a nominal system G/T of 17 to 19 dB in various spot beams.

The 20-GHz traveling wave tube amplifiers (TWTAs) on the ACTS have two ground-commandable power levels of 11 and 46 watts. Once commanded, the transition from low-power to high-power mode takes less than 1 s. The TWTAs have 1-GHz bandwidth and use solid-state drivers. Using the higher power level, the nominal achievable EIRP at various spot beams ranges from 57 to 60 dB.

2.4. Ka-Band Fade Detection and Compensation

While the Ka-band transmission offers the opportunity to tap an unused portion of the frequency spectrum, it also presents a considerable challenge because of its severe fade degradation. For the Ka band to be useful, a reasonable solution to the fade problem must be found. The brute-force approach of providing a large link margin to combat fading leads to an extravagant and costly design. So, in ACTS an adaptive technique is used whereby only those terminals experiencing fade are provided additional protection.

In the baseband processor mode, the link is designed for a 5-dB clear weather margin, but terminals experiencing fade can be dynamically provided a further 10 dB fade protection. Beacon signals transmitted by ACTS at approximately 20 and 27 GHz are continually received and processed at each ground terminal for fade detection. Once the received beacon power drops below a certain threshold, the affected terminal sends a request over ACTS to the network's master control station (MCS) for further protection.

In response, the MCS takes two actions. First, it sends a command, over ACTS, back to the terminal instructing it to invoke coding and also sends a command to the ACTS instructing the baseband processor to switch on an on-board decoder and encoder during processing of the signal to and from the affected terminal. The rate 1/2 code provides a 4-dB gain. Second, the MCS instructs the affected terminal to reduce its uplink burst rate by a factor of 4 and also commands the satellite to reduce its downlink burst by a factor of 4 when transmitting to that specific ground terminal. The burst rate reduction provides an additional saving of 6 dB, which when coupled with the 4-dB coding gain provides an additional 10-dB margin.

The fade compensation technique used in the SS/TDMA mode is unlike the baseband processor mode where the traveling wave tube amplifiers (TWTAs) are always operated at the high-power level. In the SS/TDMA operation the ground terminals are so designed that in clear weather the TWTAs can operate at the low-power level. Upon request by a terminal for fade protection, the MCS instructs the affected terminal and the ACTS to increase their transmission power. On ACTS this is done by switching the dual power TWTAs from the low to high level, thus providing a 6-dB gain.

2.5. Baseband Processor

The baseband processor block diagram is shown in Figure 5. Its major elements are modems, codecs, input/output memories, a routing switch, and a central processor [2].

The modulators and the demodulators in the baseband processor are based on the serial implementation of MSK, referred to as SMSK. The theoretical performance of SMSK is identical to MSK; however, at high data rates, SMSK has a simpler implementation. SMSK has a null-to-null bandwidth of 1.5 times the symbol rate. Additionally, it exhibits minimal sidelobe regrowth when passed through system nonlinearities. During any instant in time, on each of the two uplink hopping beams, the baseband processor can receive data on either one 110-Mbps channel or two 27.5-Mbps channels. Accordingly, there are three demodulators per uplink beam on ACTS.

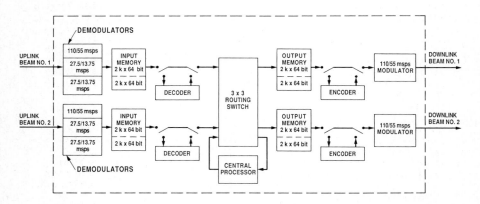

FIGURE 5. Simplified ACTS Baseband Processor Block Diagram.

As was discussed earlier, ground terminals experiencing fade reduce their information rate by a factor of 4 and also invoke rate 1/2 coding. To accommodate reduced burst rates during fade intervals, the baseband processor modems are dual rate. The 110-Mbps demodulator can also accept 55-Msps bursts, and the 27.5-Mbps demodulator can also process 13.75-Msps bursts. The maximum baseband processor throughput is 110 Mbps per beam, which results in a total maximum throughput of 220 Mbps for the two beams.

The IF input to the demodulators is at approximately 3 GHz. Once demodulated, the baseband output is routed to the input memory described next. In the baseband processor mode, capacity is allocated to requesting ground terminals in increments of 64-Kbps channels. Therefore, for each 64-Kbps channel assigned, there must be a time slot allocation, within the 1-ms TDMA frame interval, sufficient for transmission of 64 bits. Since the highest transmission rate in the baseband processor mode is 110 Mbps (actually, 110.592), a maximum of 1728 64-bit words can be transmitted in one frame. Some of these words are reserved for preambles and control channels while the rest are the actual

information data. To be able to store data received during each
frame interval, the input and output memories are sized for
storage of 2000 64-bit words.

Up to three frame durations may be needed to move data through
the baseband processor. During the first frame period, the
content of the frame is written into the input memory. In the
second period, this data is switched, 64 bits at a time, to
appropriate locations in the output memory. Finally, in the third
frame interval, the data is read out of the output memory and
transmitted on the downlink beams. To allow this to be
accomplished in the time span of no more than three frames, both
the input and the output memories consist of two 2-k x 64-bit
storage areas. During a given frame interval, the data is read
into one storage area while at the same time data received during
the previous frame is read out of the other. These memories
alternate the read and write functions in successive frames.

Adaptive forward error correction (FEC) can be provided by the
baseband processor under the direction of the ground master
control station. The ACTS uses a constraint length 5 rate 1/2
convolutional encoder and a maximum likelihood convolutional
decoder (MCD) with 2-bit soft decision and a path memory length of
28. The MCD implementation is a single chip CMOS design. When
decoding is necessary, the output of the input memory is routed
through a parallel-to-serial converter to the MCD (Figure 5).

2.6. Microwave Switch Matrix

The Microwave Switch Matrix (MSM) interconnects the three
stationary beams. Unlike the baseband processor which regenerates
and switches data at the baseband, the MSM switches at the IF
level (3 GHZ). The MSM is a solid-state (dual gate GaAs FET),
programmable crossbar switch with a switching time of less than
20 ns; it is a 4 x 4 switch, but only 3 input and 3 output ports
are used at any given time.

In this mode of operation, the modulation technique is not
restricted to SMSK, and the data rate can vary up to 220 Mbps per
input/output port for a total throughput of 660 Mbps. (The
bandwidth of the MSM as well as that of the ACTS receivers and
transmitters is 1 GHz. Thus, a higher data rate than 220 Mbps
through the ACTS can be accommodated subject to the availability
of proper ground terminals.)

2.7. The Spacecraft Bus

The ACTS bus is based on the RCA series 4000 which has been
used on previous commercial communication satellites. However,
some modifications have been made to accommodate ACTS-specific
requirements [3]. Figure 6 shows the exploded view of the
spacecraft. The rectangular bus structure has a cylindrical
center structure to house the apogee kick motor (AKM). The
payload electronics are mounted on the north and south panels of
the bus, and the MBA structure is supported by the antenna panel
facing the Earth.

The ACTS employs momentum wheels to achieve three-axis
stabilization. Using an Earth sensor for attitude reference, a
roll, pitch, and yaw nominal accuracy of 0.04, 0.06, and 0.15

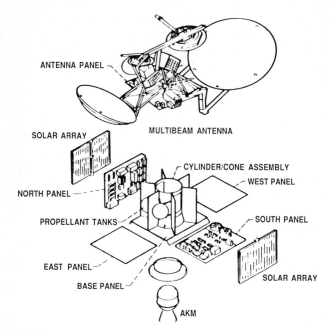

FIGURE 6. Exploded View of the ACTS Spacecraft.

degrees, respectively, can be achieved. To improve on this, the receive MBA will have a monopulse tracking capability using the command uplink from the master control station located in Cleveland. The detected elevation and azimuth error signals generated by an autotrack receiver are provided to the spacecraft attitude control subsystem. Using the autotrack, rather than the Earth sensor, for the attitude reference improves the pointing accuracy in roll and pitch to 0.02 and 0.03 degrees, respectively.

From four solar array panels (135 sq. ft.) ACTS will have a beginning life power of 1770 watts. Given the experimental nature of the spacecraft, batteries are provided to supply power only to essential bus loads during eclipse periods.

3. LAUNCH CONSIDERATIONS

The ACTS is currently on the manifest for a launch aboard the space transportation system (STS) in January 1991. After deployment in the STS parking orbit, a perigee kick motor (PKM) will be fired to place the spacecraft in an elliptical transfer orbit; the PKM selected for ACTS is the Transfer Orbit Stage (TOS).

To achieve drift orbit, a solid-propellant apogee kick motor (AKM) of the STAR-37FM class will be fired. Then, ACTS will use its thrusters and some of its hydrozine to make the final adjustments necessary to attain geostationary orbit.

ACKNOWLEDGEMENT

This paper was prepared by the Jet Propulsion Laboratory, California Institute of Technology, under a contract with the National Aeronautics and Space Administration.

REFERENCES

[1] Choung, Y.H., and Wong, W., Multibeam Antenna Design and Development for NASA Advanced Communications Technology Satellite (ACTS), pp. 568-573, IEEE GLOBCOM 1986, Houston, Texas.
[2] Moat, R., ACTS Baseband Processing, pp. 578-583, IEEE GLOBCOM 1986, Houston, Texas.
[3] Grabner, J.C., and Cashman, W.F., Advanced Communications Technology Satellite: Systems Description, pp. 559-567, IEEE GLOBCOM, 1986, Houston Texas.

RECENT DEVELOPMENTS OF ON-BOARD PROCESSING TECHNOLOGIES IN JAPAN

Fumio TAKAHATA, Hideyuki SHINONAGA and Michihisa OHKAWA

KDD R a D Laboratories
1-23, Nakameguro 2-chome, Meguro-ku, Tokyo 153, Japan

Japan's activities in on-board signal processing technologies both for domestic and international satellite communications are surveyed, focusing on on-board baseband signal processing applicable to regenerative satellites. Japan's satcom development programs are also presented in order to show the application of these technologies to satellites in terms of calendar year.

1. INTRODUCTION

All satellites presently operating all over the world are analog repeaters with simple functions such as frequency conversion, power amplification and static interbeam connection. On the other hand, it is generally recognized that an on-board signal processing gives potential benefits to satellites for improving transmission quality, connectivity, satellite resource utilization and operational flexibility. On-board signal processing technologies are basically classified into two categories. One category is an advanced on-board analog signal processing technology employed on-board the multi-beam satellites including dynamic switching functions at RF (Radio Frequency) or IF (Intermediate Frequency) bands. The INTELSAT VI satellites with MSM's (Microwave Switching Matrices) are now being manufactured for international satellite communication use. The other is an on-board baseband signal processing applied to regenerative satellites.

In this paper, activities in on-board signal processing technologies in Japan are divided into two application fields; domestic and international satellite communication systems. Section 3 describes the on-board baseband signal processing technologies which have been developed mainly by NTT (Nippon Telegraph and Telephone Corporation) for domestic satellites. NTT has been intensively carrying out the research and development of on-board baseband signal processing based on TDMA (Time Division Multiple Access) and SCPC (Single Channel Per Carrier) applicable to future regenerative satellites together with the development of a large-scale multi-beam satellite with the IF-band switching function. The on-board baseband signal processing includes TDMA equipment (burst combiner/divider, compression/expansion buffers, Viterbi decoder, etc.), burst modem's, baseband switches and multi-channel demodulators for SCPC signals. Japan's space development programs for domestic satellite communications are also introduced in Section 2, in order to show the application of these technologies to the satellites in terms of calendar year.

In Section 4, the on-board signal processing technologies for international satellites are presented. KDD (Kokusai Denshin Denwa Co., Ltd.) has been researching and developing on-board signal processing technologies for possible use in future INTELSAT satellites, including an SS/FDMA (Subchannel Switched/Frequency Division Multiple Access) equipment. Mitsubishi Electric Corporation and NEC Corporation have conducted, under the INTELSAT study contracts, the research and development of an on-board burst modem and a baseband switching matrix, respectively.

2. JAPAN'S SPACE DEVELOPMENT PROGRAMS FOR DOMESTIC SATELLITE COMMUNICATIONS

Fig. 1 shows chronologically the Japan's satcom development scenario which has been reviewed by the Space Communications Policy Advisory Council of MPT (Ministry of Posts and Telecommunications). Japan now operates the communications satellite CS-2. A test communications satellite, CS-1, was launched in 1977, and CS-3 will be launched in 1988 as a successor to CS-2. On the other hand, in the area of mobile satellite communications development, Japan has been proceeding with an experiment program, ETS-V/EMSS (Engineering Test Satellite-V/Experimental Mobile Satellite System), for which the satellite was successfully launched in August 27, 1987. An experiment ETS-VI program is also planned as a follow-up and a two-ton class satellite for this program will be launched in 1992. Japan has begun an experimental platform study as a step toward the geostationary communications platform.[1], [2]

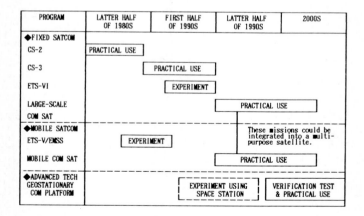

FIGURE 1
Communication satellite development scenario in Japan.

NTT has been developing advanced technologies associated with multi-beam systems both in the fixed and mobile satellite communications fields. A Ka-band (30/20 GHz) multi-beam fixed satellite system including IF-band dynamic switches and an S-band (2.6/2.5 GHz) multi-beam maritime satellite system are planned to be applied to the large-scale communication satellite which will be launched in the latter half of 1990's.[3], [4]

The ETS-VI program includes the above-mentioned fixed and mobile multi-beam satellite communication experiments. Furthermore, NASDA (National Space Development Agency of Japan) and RRL (Radio Research Laboratories) plan to conduct an intersatellite link (ISL) experiment using the ETS-VI for future development of a data relay satellite which establishes communications links between geostationary satellites and low earth orbit satellites or between geostationary satellites.

Regenerative satellites equipping the on-board baseband signal processing such as modem's and baseband switches are not included in the present Japan's satcom development scenario. This type of satellites will be launched after the year 2000 in Japan.

3. ON-BOARD SIGNAL PROCESSING FOR DOMESTIC SATELLITE COMMUNICATIONS

NTT has been extensively researching on-board baseband signal processing technologies related to TDMA and SCPC, including device technologies, toward the realization of future advanced regenerative satellites.[5] In this section, the on-board signal processing technologies developed by NTT are shown in detail.

3.1. TDMA Equipment

Six kinds of CMOS masterslice LSIC's with maximum operation speed of 25 MHz have been fabricated for TDMA equipment in earth stations.[6] For the earth station application, parallel processing of certain speed LSIC's will be the best in terms of LSIC development costs.

With respect to the on-board application, a selection of devices and processing methods has been carried out so as to minimize power consumption, weight and foot print, taking account of radiation hardness against heavy ions in transfer and geostationary orbits.[7] Based on the device selection guideline for the on-board TDMA equipment, a configuration shown in Fig. 2 is proposed as a 100 MHz on-board LSIC-implemented TDMA equipment around the year 2000, which is realized by two 100 kgate class CMOS masterslice LSIC's and microprocessors in addition to CMOS SRAM's and ROM's.

3.2. Burst Modem

NTT developed a coherent QPSK on-board burst modem with a single tuned tank clock recovery circuit and an analog baseband Costas-type carrier recovery

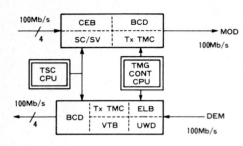

FIGURE 2
100 Mbit/s on-board LSIC-implemented TDMA configuration
with 100 kgate class masterslice LSIC's.

circuit.[8] After that, several key elements for the coherent demodulation have been fabricated, based on MMIC (Monolithic Microwave Integrated Circuit) approach which can reduce hardware size and power consumption especially in relatively low-frequency bands. A coherent (O)QPSK reverse modulation modem with narrow-band filter and limiter has been implemented in terms of fast acquisition and low cycle slip rate even at low C/N conditions.[9], [10] MMIC technologies using bipolar devices and using SST (Super Self-Aligned Process Technology) process for double-balanced mixers, hybrids, burst switches, etc., and digital LSIC technologies for baseband components such as AFC (Automatic Frequency Control) circuits have been employed in the implementation of the modem. A total power consumption of 8 W per modem is estimated.

3.3. Baseband Switch

Two different architectures for digital baseband switches have been studied in terms of a required throughput, bit rate and TDMA frame length along with trends of the device technology development in the years from 1990 to 2000.[8], [11] One architecture is a single T (Time)-switch and the other is a T-S-T (Time-Space-Time) or S-T-S switch, in which memory LSIC's and logic gate LSIC's are employed as a T-switch and an S-switch, respectively. Taking account of a high component density in CMOS devices and a high operation speed in bipolar or GaAs devices, it is concluded that the single T-switch architecture with CMOS memories and the T-S-T (or S-T-S) switch architecture with the S-switch of bipolar and GaAs gate array LSIC's will be employed in long and short TDMA frame lengths, respectively. According to an estimation of required number of LSIC's under the condition of CMOS LSIC employment and a frame length of 12 msec, a duplex 12,500 channel baseband switch with full connectivity will be realized by about 300 CMOS LSIC's in the year 1990 for the single T-switch architecture (including peripheral and control circuits for telemetry command systems).

3.4. Multi-channel Demodulator for SCPC Signals

Regenerative satellite systems transmitted in an SCPC-FDMA mode in up-links and a TDM (Time Division Multiplex) mode in down-links have also been researched and developed. Four kinds of configurations (Conventional, COMMON D/C a SEPARATE DEM, DIGITAL TMPX a DIGITAL C-DEM and ANALOG TMPX a DIGITAL C-DEM) have been implemented for the up-link on-board demodulation. In "COMMON D/C a SEPARATE DEM" scheme, a down converter is commonly used to several SCPC signals by demodulating these signals simultaneously with a single DEM-LSI.[12] "DIGITAL and ANALOG TMPX a DIGITAL C-DEM" schemes make advantage of an FDM (Frequency Division Multiplex)/TDM conversion in the digital transmultiplexer and the chirp transform, respectively.[13], [14] Based on the implementation results, an estimation of weight and power required in each configuration is made for the years from 1990 to 2000, as a function of number of channels to be regenerated. It is concluded that "DIGITAL TMPX a DIGITAL C-DEM" scheme will be suitable in the range of number of channels (each 16 kbit/s) up to about 200 in the year 2000.

4. ON-BOARD SIGNAL PROCESSING FOR INTERNATIONAL SATELLITE COMMUNICATIONS

KDD, Mitsubishi and NEC have been researching and developing on-board signal processing technologies for possible use in future INTELSAT satellites.

4.1. TDMA Equipment

KDD has developed a proof-of-concept (POC) model of the on-board baseband processor for a regenerative SS (Satellite Switched)/TDMA system operating with digital ISL's.[15] This system improves the link performance and geographical connectivity, while maintaining compatibility with the current INTELSAT TDMA system. The satellites employ on-board modem's for regeneration, baseband switch matrices (BSM's) with distribution control unit (DCU) for interbeam connections, retiming circuits to align the timing of P and Q channel data regenerated by the on-board demodulators, and ISL synchronizers. Based upon the POC model circuit design, an estimation of power consumption, mass and volume of the on-board baseband processor is conducted for the LSI implementation assuming current technologies. The results show that the total power consumption, mass and volume of the on-board processor for a six beam satellite are approximately to be 15 W, 4,000 g, and 260 x 260 x 70 mm^3, respectively, including a DC/DC converter with efficiency of 70 percent.

4.2. Burst Modem

KDD has developed an engineering model of 120 Mbit/s on-board CQPSK burst modem for possible use in the INTELSAT TDMA system.[16] In order to realize on-board hardwares with light weight, small size and low power consumption, particular considerations are paid on the filter configuration and a carrier recovery circuit. Roll-off filterings are carried out at the RF-band in input and output satellite multiplexers along with the rejection of adjacent channel interference. A phase comparator with hysteresis characteristics is employed in the PLL (Phase Locked Loop)-based carrier recovery circuit to avoid the hangup phenomenon and to achieve fast lock-in time. Volume, weight and power consumption of the developed modem are 100 x 140 x 20 mm^3, 390 g and 0.62 W for the modulator and 100 x 140 x 60 mm, 750 g and 2.48 W for the demodulator, respectively.

Mitsubishi has also developed an 120 Mbit/s on-board CQPSK burst modem under the INTELSAT study contract.[17] The same design philosophy as that of earth station modem's is basically adopted for most circuits. The receive (transmit) filter is realized by 10th-order filter sections and 3rd- (5th-) order group delay equalizer sections using discrete components. A times-4 multiplier/tank filter/divided-by-4 method and an IF squaring extraction method are employed in the carrier and clock recovery circuits, respectively. Volume, weight and power consumption in total of the modulator and demodulator are 164.6 x 100.4 x 99.0 mm^3, 1.6 kg and 2.94 W, respectively.

4.3. Baseband Switch

NEC has implemented a 16 x 4 on-board BSM based on a GaAs digital LSI technology for applying to 120 Mbit/s TDMA signals, under the INTELSAT study contract.[18] A buffered FET logic (BFL) constructed with the depletion type FET's is adopted to realize the BSM. An assembled BSM IC chip designed in the combination of BFL NOR gates contains 1,292 FET's and 212 shifting diodes. The rise/fall times at the BSM outputs were 0.6 ns and 1.5 ns. Size, weight and power consumption of the chip are 2.8 x 3.32 mm, 1.2 g and 150 mW, respectively.

4.4. Multi-channel Modem for SCPC Signals

KDD in cooperation with NEC has developed an experimental CQPSK group modem, which simultaneously modulates and demodulates 12 PSK carriers, each with a data rate of 64 kbit/s complying with the INTELSAT SCPC system specification.[19] This group modem is realized by making advantage of digital signal processing consisting of filtering in digital transmultiplexers and baseband modulation/demodulation on a per-channel basis.

4.5. SS/FDMA Equipment

KDD has proposed a new concept named "subchannel switched FDMA (SS/FDMA) system" in order to enhance the interbeam connectivity of a multi-beam satellite under FDMA operation.[20] Each transponder bandwidth is subdivided into several narrow bands named subchannels occupying a bandwidth of approximately 10 MHz or below, and frequency multiplexed intermediate or narrow band carriers are interbeam-connected on a subchannel by subchannel basis. A multi-beam satellite employed in the SS/FDMA system is schematically illustrated in Fig. 3. Filters used to subdivide each transponder bandwidth into several subchannels are required to have flat passbands, narrow transition bandwidths, low sidelobe levels, and linear phase responses. Hence, SAW (Surface Acoustic Wave) filter is selected as a single solution among various filters used in the RF frequencies, being basically a transversal filter which enables independent control of the amplitude and phase responses. Up to now, 8 channel SAW filter bank was fabricated in the 700 MHz frequency bands using ST cut quartz as a substrate material.

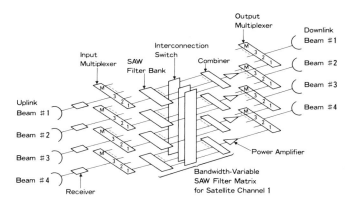

FIGURE 3
Multi-beam satellite employed in the SS/FDMA system.

5. CONCLUSION

Japan's activities in on-board signal processing technologies both for domestic and international satellite communications were surveyed, focusing on baseband signal processing. The intensive research and development for the on-board baseband signal processing have been carried out on the design and implementation of engineering models. From now on, the development will be progressed with a particular emphasis on device technologies toward the realization of operational-use regenerative satellites.

ACKNOWLEDGEMENTS

The authors are grateful to Dr. Y. Ito of KDD and Dr. S. Kato of NTT for useful comments and advices.

REFERENCES

[1] T. Mori and T. Iida,"Japan's space development programs for

communications: An overview," IEEE J. Selected Areas Commun., Vol. SAC-5, No. 4, pp. 624-629, May 1987.
[2] S. Saito,"Space development in Japan and the international cooperation," in Proc. Asia-Pacific Satellite Commun. Symp., Tokyo, Japan, pp. 251-262, April 1987.
[3] K. Miyauchi, H. Yamamoto, K. Morita, K. Kondo and T. Yasaka,"Domestic multi-beam satellite communications system design and performance", in Proc. 15th Int. Symp. on Space Technology and Science, Tokyo, Japan, pp. 939-947, May 1986.
[4] H. Fuketa,"Present satellite communication technologies and future trends," in Proc. Asia-Pacific Satellite Commun. Symp., Tokyo, Japan, pp. 283-294, April 1987.
[5] S. Kato, T. Arita and K. Morita,"Onboard signal processing technologies for present and future TDMA and SCPC systems," IEEE J. Selected Areas Commun., Vol. SAC-5, No. 4, pp. 685-700, May 1987.
[6] S. Kato, M. Morikawa and M. Umehira,"General purpose TDMA LSI development for low cost earth station," in Proc. IEEE Int. Conf. Commun., Toronto, Canada, pp. 16.6.1-16.6.6, June 1986.
[7] K. Enomoto, K. Otani, S. Kato, Y. Sakagawa and N. Shiono,"A study on radiation-hardness of onboard LSIs," (in Japanese), Tech. Group SAT, IECE of Japan, SAT84-51, pp. 1-7, Feb. 1985.
[8] S. Kato, S. Samejima and H. Yamamoto,"An SS-TDMA system using onboard regenerative repeaters and baseband switch," in Proc. Int. Conf. Commun., Amsterdam, North-Holland, pp. 807-812, May 1984.
[9] M. Umehira and S. Kato,"Trade-off study on carrier recovery circuit for on-board burst demodulator," (in Japanese), Tech. Group SAT, IECE of Japan, SAT85-11, pp. 17-24, July 1985.
[10] K. Enomoto, M. Umehira and S. Kato,"Low Eb/No carrier recovery circuit with precise digital Costas APC for TDMA application," (in Japanese), Tech. Group SAT, IECE of Japan, SAT86-35, pp. 49-54, Nov. 1986.
[11] S. Suzuki, T. Mizuno and M. Yabusaki,"A study on on-board baseband switch in multi-beam satellite system," (in Japanese), Tech. Group SE, IECE of Japan, SE85-18, pp. 19-24, May 1985.
[12] K. Ohtani and S. Kato,"An onboard digital demodulator for regenerative SCPC satellite communication systems," in Proc. IEEE Int. Conf. Commun., Toronto, Canada, pp. 56.6.1-56.6.6, June 1986.
[13] T. Izumisawa, S. Kato and T. Kohri,"Regenerative SCPC satellite communication systems," in Proc. AIAA 10th Commun. Satellite Syst. Conf., Orlando, Florida, AIAA-84-0708, pp. 269-275, March 1984.
[14] T. Kohri, M. Morikura and S. Kato,"A 400ch SCPC signal demodulator using chirp transform and correlation detection scheme," to be presented at IEEE Global Telecommun. Conf., Tokyo, Japan, 8.4, Nov. 1987.
[15] H. Shinonaga, G. Satoh and M. Ohkawa,"On-board baseband processor for regenerative SS/TDMA system operating with digital intersatellite links," to be presented at IEEE Global Telecommun. Conf., Tokyo, Japan, pp. 24.4.1-24.4.7, Nov. 1987.
[16] H. Kobayashi, G. Satoh, T. Honda, T. Kimura and M. Ohkawa,"Development of on-board burst mode 120 Mbit/s CQPSK modem," (in Japanese), Tech. Group SAT, IECE of Japan, SAT87-12, pp. 7-14, July 1987.
[17] T. Ishizu, H. Sawada, Y. Kazekami, T. Imatani and K. Betaharon,"On-board modem for regenerative satellite," to be presented at IEEE Global Telecommun. Conf., Tokyo, Japan, 24.2, Nov. 1987.
[18] M. Kudoh, I. Eguchi and K. Kinuhata,"GaAs baseband switching matrix for on-board signal processing," in Proc. IEEE Global Telecommun. Conf., New Orleans, Louisiana, pp. 6.4.1-6.4.4, Dec. 1985.
[19] F. Takahata, M. Yasunaga, Y. Hirata, T. Ohsawa and J. Namiki,"A PSK group modem for satellite communications," IEEE J. Selected Areas Commun., Vol. SAC-5, No. 4, pp. 648-661, May 1987.
[20] H. Shinonaga and Y. Ito,"SS/FDMA system for digital transmission", in Proc. 7th Int. Conf. Digital Satellite Commun., Munich, West Germany, pp. 163-170, May 1986.

TECHNIQUES AND TECHNOLOGIES FOR MULTICARRIER DEMODULATION IN FDMA/TDM SATELLITE SYSTEMS

Fulvio Ananasso
Dip.to Ingegneria Elettronica - University of Rome/ TorVergata
Via Orazio Raimondo - 00173 Roma (ITALY)
(formerly with TELESPAZIO - Via A. Bergamini, 50 - 00159 Roma) (ITALY)

Enrico Del Re
Dip.to Ingegneria Elettronica - University of Florence
Via Santa Marta 3 - 50139 Firenze (ITALY)

ABSTRACT
Regenerative, On-board Processing FDMA/TDM payloads have been recently proposed as valid candidates for user-oriented satellite systems.
Both Business Traffic for Fixed Services (e.g. INTELSAT Business Services, IBS) and Mobile Satellite Systems can potentially take advantages of the peculiarities of such payloads, which substantially require MultiCarrier Demodulation (MCD) of the uplink FDMA carriers to recover the individual modulating streams, which are in turn TDM-formatted to modulate a unique downlink carrier.
The present paper expands on the most suitable techniques/technologies to implement on-board MCD, also providing some recent data on present hardware performance and time-scale for technologies maturity.

The present paper derives by investigations carried out on behalf of Telespazio under Intelsat Contract INTEL-479 by the authors in close co-operation with a number of Telespazio collegues; a special thank is for L. Cellai, G. Chiassarini and E. Ligato.

1. INTRODUCTION

One of the most attractive architectures for business-services satellite systems recently proposed envisages different access methods in the two links, i.e. Frequency Division Multiple Access in the uplink and Time Division Multiplexing in the downlink (Refs. 1-to-3).
In this way, the user uplink RF power requirements are proportional to the individual bandwidths (differently from TDMA which requires power levels proportional to the transponder bandwidth), and, to some extent, network synchronization procedures are not requested.
In the downlink, TDM permits the on-board High Power Amplifier (HPA) to be saturated or slightly backed-off, due to absence of multicarrier intermodulation.
Such an architecture, therefore, attempts to optimize the system RF power resources, however requiring non-trivial access format conversion on-board from FDMA to TDM.
The heart of the (regenerative) payload is then composed by a "MultiCarrier" Demodulator (MCD), which recovers the individual modulating sequences of the (N) FDMA uplink carriers; a subsequent TDM formatter builds-up a higher data rate information stream to modulate a unique downlink carrier (see fig. 1).
Some kind of network synchronization is anyway requested to keep reasonable the MCD complexity.
This requirements is even more stringent in the presence of a high number of low data rate carriers, whose (clock) misalignment at the satellite input may drastically complicate the processing payload, and in limit cases even prevent any reasonable implementation.
A typical INTELSAT-EUTELSAT 120-Mbps transponder (83.3 MHz gros bandwidth) may support several hundred carriers, and the situation is even heavier in Mobile Services, where the transmission rates are of the order of a few Kbps.
Hence, synchronization problems may arise, and should be solved by using proper techniques (Refs. 4 and 6).
In the following sections the most attractive solutions to the MCD problems (with emphasis on the Demultiplexing section) will be reviewed relatively to IBS data rates, which, in the presence of 6.7%-data overhead and rate-1/2 convolutional encoding (3-bit soft-decision Viterbi decoding) and QPSK modulation, range from 137 Kbps to 4.369 Mbps.
In particular, the paper will expand on the following approaches to the MCD implementation problem:

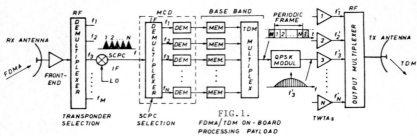

FIG. 1. FDMA/TDM ON-BOARD PROCESSING PAYLOAD

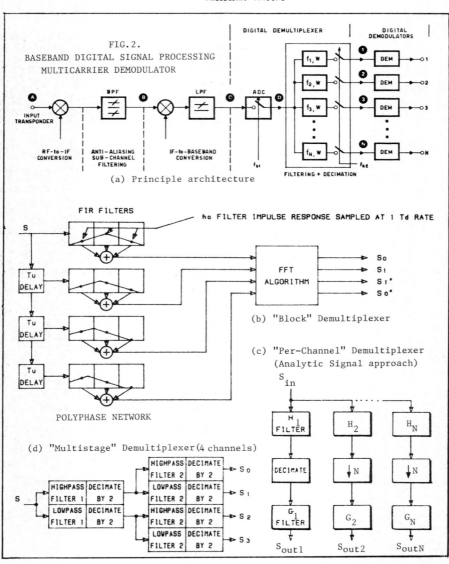

FIG. 2. BASEBAND DIGITAL SIGNAL PROCESSING MULTICARRIER DEMODULATOR

(a) Principle architecture

(b) "Block" Demultiplexer

(c) "Per-Channel" Demultiplexer (Analytic Signal approach)

(d) "Multistage" Demultiplexer (4 channels)

- Baseband Processing
- Chirp Fourier Transform (CFT) SAW Processing
- Integrated Optics

2. BASEBAND PROCESSING

2.1. Digital Demultiplexer

From a general point of view, the demultiplexing of uplink SCPCs (to be subsequently demodulated) can be implemented using three different approaches: (i) per-channel, (ii) block and (iii) multistage, as shown in Fig. 2.

The per-channel method performs the demultiplexing operation essentially by means of a bank of bandpass filters.

Selection of each input signal and its translation to a low-frequency band are achieved by a compound operation of digital filtering and decimation (i.e. decrease of the signal sampling rate).

The block method implements the demultiplexing by using a set of digital filters ("Polyphase" Network) followed by a "block" processor - usually of the FFT type - that processes the output signals from the digital filters altogether.

This procedure of processing the input FDMA signal to obtain an output TDM one hystorically derives from studies on "Transmultiplexers" (Refs. 8, 9), which were originally conceived for the inverse transformation TDM-FDM, to transmit PCM (time-division) signals through analog (FDM) channels.

The multistage method can be considered as a binary tree of two-channel demultiplexers. Each demultiplexing stage performs a lowpass and a highpass filtering with subsequent decimation by a factor of two.

In the following we will briefly summarize the main characteristics and performance of the three approaches.

(i) Per-channel methods

Within the class of per-channel methods, an effective solution from the implementation complexity point of view is represented by the Analytic Signal (AS) approach (Ref. 7).

A specific feature of the AS method is to greatly relax the filter specifications in terms of transition bandwidths, thus achieving a lower implementation complexity with respect to other per-channel approaches.

The implementation structure of the Analytic Signal method envisages a cascade of "H_i" and "G_i" digital filters separated by a decimator (by a factor N).

H_i are bandpass filters performing a rough selection of i-th channel (out of N input channels to the Demultiplexer), having passband and transition bandwidths approximately equal to the channel spacing. They operate at the input (high) sampling rate and supply the output at the low rate after decimation of the signal by a factor N.

The G_i filters, operating at the low sampling frequency, carry out the required additional filtering to select the channels and translate them to the proper low-frequency band after the output adder. The sign inversion of every other output sample for the odd channels (i odd) is necessary to get a non-inverted spectrum for those channels.

It should be noted that the AS method requires an excellent (ideally perfect) matching of the H and G filters frequency responses. Such a matching is obtained by designing complex (quadrature) filters whose "in-phase" branch is made up by the filters corresponding to the real part of the complex impulse response, whereas the "quadrature" branch includes the filters corresponding to the imaginary part.

Linear phase FIR filters are conveniently utilized in this kind of approach to avoid phase distorsions and obtain high implementation efficiency of digital multirate structures: as a matter of fact, decimation is achieved with FIR filters merely not computing the samples to be discarded (IIR filters, on the contrary, are computationally less effective, as they in any case compute all the samples included those to be discarded subsequently upon decimation).

The required number of multiplications-per-second and per-channel required by the Analytic Signal method has been derived in the frame of ESA/ESTEC Contract 6096/84/NL/GM ("Multicarrier Demodulator Design") and is given by:

$$M_{AS} = K_{AS} \times W^2 \left[W(N+4)-2B(N+2)\right]/\left[(W-2B)(W-B)\right] \qquad (1)$$

where:

- K_{AS} = (2/3) log $\left[2/(10\,\delta_1\,\delta_2)\right]$
- N = number of demultiplexer input channels
- W = channel spacing

- B = one-side bandwidth of the signal spectrum
- δ_1 = pass band amplitude ripple
- δ_2 = shop band amplitude ripple

(ii) Block methods

As shown in fig. 2, a block processor consists of a bank of N FIR polyphase filters followed by a N-point FFT processor.
Both the filters and the FFT processor operate at the lower (output) sampling frequency, thus reducing the overall computational complexity.
The N channels to be demultiplexed should be a power of two, in order to fully exploit the advantages of FFT processor. In this case the overall number of real multiplications-per-second is given by (same reference):

$$M_{FFT} = 2W \left[L_{FT} + M_{FT} \right] \quad (2)$$

where:
$$L_{FT} = (4/3) \ N \ W \log \left[1/(10\delta_1\delta_2) \right] / (W - 2B) \quad (2')$$
$$M_{FT} = 8 \ W \ N\log_2 N \quad (2'')$$

(iii) Multistage method

As in the block method, the multistage method is rather attractive whenever the number N of the demultiplexer channels is a power of two. The implementation structure for N = 4 is shown in fig. 2. At each stage the signal spectrum is split down into a couple of sub-bands by half-band filters and decimated by a factor of two. After L stages, 2^L = N channel are obtained.
A peculiarity of the method is the possibility to use the same design for all the half-band filters. Assuming to employ complex filters, much relaxed specifications can be required, decreasing the computational complexity.
With the multistage method, the overall number of real multiplications per second has been derived in the frame of INTELSAT Contract INTEL-479 ("Low-bit Rate on-board Demultiplexing and Demodulation"), and is given by:

$$M_{MS} = (64/3) \ N \times W \log \left[1/(10\delta_1\delta_2) \right] \left[\log_2 N + 2B/(W - 2B) \right] \quad (3)$$

Equations (1), (2) and (3) allow to compare the complexities of the different demultiplexing methods.
For small values of N the difference in complexity - function of channel spacing, W, once the other parameters have been fixed - tends to vanish, while it may be relevant for large values of N.
An optimum value of W may be selected to minimize the computional complexity; however, a nearly optimum value of W is determined by trading-off such different factors as spectrum utilization and convenience of selecting an output (low) sampling frequency corresponding to an integer number of samples-per-symbol. In many cases a convenient choice turns out to be 3 samples/symbol, corresponding to a channel spacing W equal to 3/4 times the transmission bit rate.
From an implementation point of view, some considerations and conclusions can be drawn about the three approaches, based upon both theoretical features and results from the mentioned studies for ESA and INTELSAT:

(i) per-channel methods generally have higher computational complexity, smaller finite-precision arithmetic sensivity, greater flexibility, smaller control circuit complexity;

(ii) block methods have lower computational complexity, higher finite-precision arithmetic sensitivity, smaller flexibility, greater control circuit complexity;

(iii) multistage methods have computational complexity comparable with that of block methods, finite-precision arithmetic sensitivity and control circuit complexity comparable to per-channel methods, intermediate degree of flexibility.

On the basis of some obtained results, the finite-precision arithmetic implementation of the block methods generally require 2-to-3 bits more than the other methods, due to the greater finite-precision arithmetic sensitivity of the FFT block processor.
The control circuit complexity for the block method may be comparable with (or in some limit cases even greater than) the complexity of the computational part, although a custom VLSI implementation of the control part might remove this drawback.

Flexibility may be a critical issue for the demultiplexer.
In the IBS applications different transmission rates are foreseen from 137 Kbps to 4.369 Mbps and likely the Multicarrier Demodulator will have to operate on many of them during the satellite lifetime, to allow for reconfigurability of traffic pattern.
The transmission rates are in general multiple of the smallest one by factors as 2, 3, 5 and combinations of them. If we require the demultiplexer to operate at different rates for a fixed processed bandwidth the number of channels N is inversely proportional to the transmission rate. Thus we can observe that:

(i) block methods are able to operate only at a fixed value of N (i.e. number of points of the block processor), therefore a specialised demultiplexer is required for each individual transmission rate.

(ii) per-channel structures are well suited to variations of transmission rates and number of processed channels N: it is only required to vary the filter characteristics (i.e. coefficients) and decimation factor, using only the necessary (N) branches of the structure designed for the highest possible value of N (lowest data rate). Moreover, the per-channel methods allow to process channels with different transmission rates within the same demultiplexer, as the N paths are substantially independent.

(iii) multistage structures have an intermediate degree of flexibility, as it allows variations of the transmission data rates, although limited to powers of two.

2.2. Digital Demodulator

The most demanding parts of the demodulator are the carrier and timing recovery circuits. Their choice depends on whether the MCD works in continuous or burst-mode.
For burst-mode operation (the most demanding), a suitable structure for the carrier recovery circuit could be the digital version of that proposed in Ref. 12 (Non-linear Phase estimation by Viterbi), that guarantees a fixed acquisition time.
Simulations have shown that 6 bit arithmetic is sufficient to obtain a carrier phase estimate very close to the floating-point implementation.
Ref. 13 proposes a suitable and simple digital timing recovery circuit, where only two samples-per-symbol are required. The algorithm is suitable for both acquisition and tracking modes of operation. For a QPSK modulated signal the error e(n) used to adjust a timing error corrector is evaluated for the n-th symbol according to:

$$e(n) = X_I(n - \frac{1}{2})\left[X_I(n) - X_I(n-1)\right] + X_Q(n - \frac{1}{2})\left[X_Q(n) - X_Q(n-1)\right] \qquad (4)$$

where X_I, X_Q are the in-phase and quadrature components of the signal.
Results obtained from theoretical analysis and computer simulation have shown that the degradations introduced by these two recovery circuits are of the order of a few tenths of dB.

3. CHIRP FOURIER TRANSFORM (CFT) IF PROCESSING

Having in mind the IBS transmission rates, the filter bank of fig. 1 might be in principle implemented via analog demultiplexers at a convenient IF, exploiting Surface Acoustic Wave (SAW) devices. Such devices exhibit rather peculiar compactness and lightweight; however, the narrower the (useful and/or transition) bandwidth, the longer the filters are, being substantially composed by InterDigital Transducers (IDTs) etched on a piezoelectric crystal (Quartz, Lithium-Niobate,...), reproducing a sampled version of the filter impulse response (Ref. 14). Hence, practical implementation constraints prevent from envisaging large, narrowband SAW filter banks, as the previously mentioned compactness of the overall Demultiplexer would no longer hold true. This fact requires the user data rate to be around 1 Mpbs or more to use SAW demultiplexers, which in this configuration implement a kind of "per-channel" approach.
However, SAW devices can be effectively be employed in an alternative way to implement a "block" demultiplexer, as depicted in fig. 3 which exploits SAW spectrum analyzers implementing "Chirp" Fourier Transform (CFT). To build-up such a kind of Fourier Transformers, Linear Frequency Modulators (LFM or "Chirp" filters) may be used, having group delay which increases ("up-chirp") or decreases ("down-chirp") linearly with frequency.
A chirp filter or LFM can be suitably realized using SAW Reflective Array Compressor (RAC, Refs. 14 and 15).
Several RAC devices can be combined in a couple of configurations known as "Multiply-Convolve-Multiply" (M-C-M) and "Convolve-Multiply-Convolve" (C-M-C) chirp transformer. The frequencies within the useful bandwidth B - which is a requirement for the design of the Fourier Transformer, as it only can operate over a finite and limited frequency range - are displayed subsequently on a time division basis, the overall band B being converted into a time interval T.

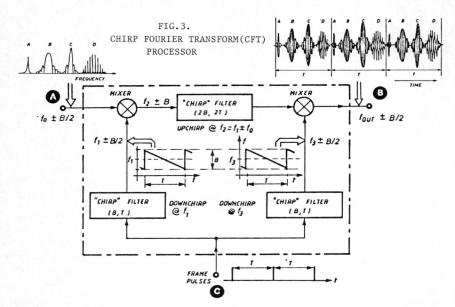

FIG.3. CHIRP FOURIER TRANSFORM(CFT) PROCESSOR

If trains of T-spaced periodical pulses are provided at the input ports of the LFM(s) feeding the multiplier(s), such T-duration time intervals repeat periodically. Inside each time interval the time position of each frequency of the useful band is shifted proportionally to the absolute value of frequency.

Due to the necessary time-windowing to perform the Fourier transform, the displayed spectrum in the time position relative to the particular center frequency is actually the convolution between the true spectrum and a (sinx/x)-like function (Fourier transform of the time-window).

Thus, such an arrangement can be used as Multicarrier Demodulator with the addition of suitable baseband processing (tailored to deal with the selected modulation format) for data demodulation (including carrier and clock recovery).

It is, however, worthwhile to point out that the T frame duration should be chosen equal to (theoretically one but in practice) some multiple of symbol interval, and, in principle, in that interval all the SCPC waveforms should be time-aligned (i.e. the timing of the different uplink carriers made synchronized) in order to allow a short-term spectrum estimation consistent for each carrier in the T frame (one or more symbols) period.

However, some kind of arrangement (i.e. bank of filters at the CFT output to perform a kind of matched receiver by selecting the filter with highest output power) can relax or even remove the mentioned constraint; the subsequent carrier and clock recovery circuit, tailored to the particular CFT structure and modulation format, does not exhibit peculiar criticality, and is not addressed here.

However, the "block" demultiplexing operation does not easily allow for flexibility, as the SCPC data rate should be the same and kept fixed for each CFT demultipexer.

Moreover, a "ping-pong" configuration in generally envisaged to assure a 100%-duty cycle, as the convolution in the (2T, 2B) C_2 chirp filter causes a processing time within the CFT longer than T, not allowing in principle a continuous FDMA input signals demultiplexing.

Lastly, it has to be mentioned that practical BxT limitations constrain the T duration (i.e. one or more symbol periods) not to be higher than 50-100 usecs; the data rates able to be processed by the CFT demultiplexers, therefore, should not be lower than some (2 or 3) hundred Kbps. The upper bound, on the contrary, is totally compatible with user-oriented systems data rates (e.g. 4369 Kbps in the IBS).

4. INTEGRATED OPTICS (BRAGG CELLS)

Integrated Optics Signal Processing is probably one of the most promising techniques for 1990's implementation within a number of spacecraft electronic subsystems like mo-dems, demultiplexers, switching matrices, combiners,....

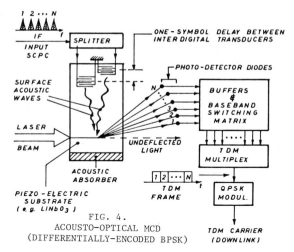

FIG. 4.
ACOUSTO-OPTICAL MCD
(DIFFERENTIALLY-ENCODED BPSK)

The level of integration can be really significant, although a number of technological issues have still to be solved, particularly in the field of temperature stability and frequency resolution.

Concerning our MCD application, one particularly simple method for demultiplexing a SCPC signal (where carrier recovery is not needed) is sketched in fig. 10 of Refs. 3 and 5, valid for differentially-encoded BPSK carriers(reproduced here in fig.4).

Owing to that scheme, demultiplexing and demodulation can be carried out on a single chip using a (modified) Integrated Acousto-Optical Spectrum Analyser (IAO-SA), which serves as a demultiplexing filter for the IF carrier. One optical detector per SCPC carrier would be used. To carry out the demodulation two SAW launchers are used, with a one bit or one symbol delay between the two launchers.

With no phase change between adjacent bits, the acoustic waves from each launcher will add and result in the light being deflected to the appropriate detector. If there is a 180-degree phase change, the two acoustic waves will destructively interfere (i.e. cancel) and there will be no deflection of the incident light to the appropriate detector. Thus, each detector will produce a signal only when there is no change of phase in its carrier (i.e. between adjacent bits). In this way the modified acousto-optical spectrum analizer can demultiplex (and demodulate) many carriers.

It has to be mentioned that, using this arrangements, the photo-detector diodes are actually facing a situation whereby the input signal is substantially ON-OFF keyed, avoiding some non-trivial problems of continuous light-to-electric voltage transduction if a coherent demodulation were envisaged.

Furthermore, the mentioned configuration, although handling differentially-encoded BPSK signals, is not exactly a differential (multicarrier) demodulator, as the recovered signal out-of-photodiodes does depend upon a number of different geometrical and electrical factors rather than merely phase difference between adjacent symbols.

Thus, the degradation figure with respect to coherent detection (about 0.5 dB and 2.3 dB in a binary and quaternary modulation, respectively) is not strictly valid here - probably it is too optimistic -, although a QPSK arrangement - implemented by doubling the fig. 7 circuit in such a way to build-up a quadrature receiver, each path processing a binary sequence at symbol rate - would be anyway preferred to save 3 dB of RF bandwidth.

The timing recovery is performed as usual, whilst the carrier recovery, as already pointed out, is not needed with the mentioned scheme.

Table 1 depicts some key parameters of Integrated Acousto-Optical MCD, as achievable with today technology.

Photodetector diodes are commonly available on the market in chip form with time-multiplexed serial output primarily in powers of two for FFT processors.

512-1024 diode chips are usual, the diode size being rather small such as the frequency resolution (i.e. angular spacing between any two adjacent detectors) is not limited by the diode dimension. However, it is difficult at the present status of technology to go below 100-200 KHz resolution (depending on temperature range and IF bandwidth), so that for the present application only data rates close to the Mbps range sound viable for IAO-SA realization.

TABLE 1 - TECHNICAL CHARACTERISTICS OF SOME BRAGG CELL DEVICES
(as desumed by present production)

CENTER FREQUENCY	0.5 - 3 GHz
BANDWIDTH	100 MHz - 2 GHz
FREQUENCY RESOLUTION	100 KHz - 3 MHz
DYNAMIC RANGE	40 - 70 dB
CROSSTALK	30 - 60 dB
LASER DIODE POWER	1 - 20 mW
DIODE SIZE	10 - 50 μm
DIODE ARRAY	256 - 1024 diodes
SUBSTRATE	$LiNbO_3$, Quartz

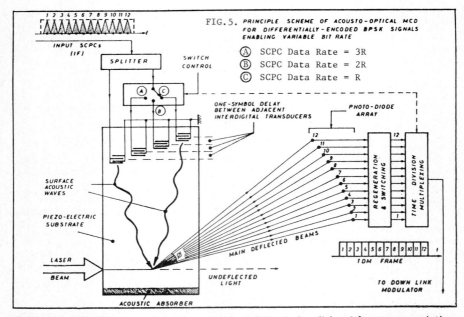

FIG. 5. PRINCIPLE SCHEME OF ACOUSTO-OPTICAL MCD FOR DIFFERENTIALLY-ENCODED BPSK SIGNALS ENABLING VARIABLE BIT RATE
(A) SCPC Data Rate = 3R
(B) SCPC Data Rate = 2R
(C) SCPC Data Rate = R

Open points with this approach are temperature stability (primarily) and frequency resolution, whilst the instantaneous bandwidth is not a problem.
Reconfigurability might be achieved as outlined in fig. 5. An input Single Pole-Triple Throw (SP3T) switch selects one out of three (or more in principle) possible delays between pairs of Interdigital Transducers, allowing three possible data rates (R, 2R, 3R) to be detected on different diodes (minimum data rate: all diodes interested; 2R data rate: alternate diodes; 3R data rate: one diode interested every three ones). More data rates can in principle be envisaged with similar arrangements.

5. COMPARISON AMONG TECHNOLOGIES - CONCLUSIONS

Table 2 resumes some relevant data on the discussed MCD technologies.
Baseband processing MCD: are capable of operate across the whole range of IBS data rates, whilst Acousto-Optics only seem applicable to the highest portion (several hundred Kbps); CFTs only exclude the lowest data rate.
Concerning hardware complexity IAO-MCD and semicustom CMOS assure the lowest weight and power consumption, although Integrated Optics permit the demultiplexing - and associate demodulation - of a higher number of channels within the same chip.
Development times are reasonable for each configuration.
In conclusion, MultiCarrier Demodulators have been shown to be implemented in several technologies, the relative complexities being such as to permit reasonable spacecraft implementation in mid 90's FDMA/TDM satellite systems.
More R+D activity is requested in this area, relevant to the development of user-oriented satellite systems.

TABLE 2. COMPARISON OF MCD TECHNOLOGIES.

Parameter / Technology	Availability	Transmission Rate		Maximum number of demultiplexed channels at rate		Physical parameters for a 4.3 Mbit/s MCD			
		Min (bit/s)	Max (bit/s)	Min	Max	Number of channels	powers consumption (W)	weight (kg)	Development time (months)
Digital Standard Chips	now	137K	4.3M	64	2	2	110	4.5	8
Advanced Digital Semicustom (CMOS)	within 10 years		20M		2	8	20	1	4
IAO-MCD	now	3.2M	4.3M	36	24	24	15	1.5	12 (*)
SAW CFT	now	270K	4.3M	60	22	22	40	7	12 (*)

(*) Extrapolated data from manufacturers

This work was sponsored in part by the International Telecommunication Satellite Organization (INTELSAT) under Contract INTEL-479. The opinions expressed are not necessarily those of INTELSAT.

REFERENCES

1. G. Perrotta: " Advantage of Satellite Regenerative and Processing Repeaters for Low Bit Rate Links Between Small Earth Terminals", Int. Conf. on Space Telecommunications and Radio Broadcasting, Toulouse, March 1979.
2. G.J.P. Lo, R.A. Peters: "Emerging Technologies for Future Communications Satellite Payloads", MSAT'83, March 1983, Addendum.
3. F. Ananasso, E. Saggese: "User-Oriented Satellite Systems for the 1990'S", AIAA 11th CSSC, S. Diego (CA, USA), March 1986.
4. T. Izumisawa, S. Kato, T. Kohri: " Regenerative SCPC Satellite Communications Systems", AIAA 10th CSSC, Orlando (FL, USA), March 1984.
5. F. Ananasso, E. Saggese: "A Survey on the Technology of Multicarrier Demodulators for FDMA/TDM User-Oriented Satellite Systems", IEEE/GLOBECOM-85, New Orleans (LO, USA), December 1985.
6. F. Ananasso, E. Del Re: "Clock and Carrier Synchronization in FDMA/TDM Satellite Systems", IEEE/ICC-87, Seattle (WA, USA), June 1987.
7. E. Del Re, P.L. Emiliani: "An Analytic Signal Approach for Transmultiplexers: Theory and Design", IEEE Trans. on Comm., vol. COM-30, N. 7, July 1982.
8. H. Scheuermann, H. Göckler: "A Comprehensive Survey of Digital Transmultiplexing Methods", IEEE Proceedings, Vol. 69, N. 11, November 1981.
9. M.G. Bellanger, J.L. Daguet: "TDM-FDM Transmultiplexer: Digital Polyphase and FFT", IEEE Trans. on Comm., vol. COM-22, N. 9, September 1974.
10. T. Ohsawa, J. Namiki: "Digital Group Demodulation System for Multiple PSK Carriers", AIAA 11th CSSC, S. Diego (CA,USA), March 1986.
11. S. Bellini, G. Tartara: "On-Board Multicarrier Digital Demodulation in Regenerative Satellites", AIAA 11th CSSC, S. Diego (CA,USA), March 1986.
12. A.J. Viterbi, A.M. Viterbi: "Non Linear Estimator of PSK Modulated Carrier Phase with Application to Burst Digital Transmission", IEEE Trans. on Inf. Theory, vol. IT-29, N. 4, July 1983.
13. F.M. Gardner: "A BPSK/QPSK Timing-Error Detector for Sampled Receivers" IEEE Trans. on Comm., vol. COM-34, N. 5, May 1986.
14. D.P. Morgan: "Surface-Wave Devices for Signal Processing", Elsevier Science Publishers, 1985.
15. R.M. Hays, C.S. Hartmann: "Surface Acoustic Wave Devices for Communications", IEEE Proceedings, vol. 64, N. 5, May 1976.
16. E.H. Young Jr., Shi Kay Yao: "Design Considerations for Acousto-Optical Devices", IEEE Proceedings, vol. 69, N. 1, January 1981.
17. A.E. Spezio, J. Lee, G.W. Anderson: "Acousto-Optics for Systems Applications", MICROWAVE JOURNAL, February 1985.

THE ITALSAT QPSK BURST MODE COHERENT DEMODULATOR

A. D'AMBROSIO - G. ALLETTO

GTE Telecomunicazioni S.p.A. 20060 Cassina De' Pecchi, Milan, Italy

1. GENERAL

ITALSAT Satellite is scheduled for launching end 1989 beginning 1990. The system configuration has already been described elsewhere [1]; it is basically a preoperational, network oriented, high capacity, regenerative SS-TDMA communication satellite.

Its operation in Ka band (30-20 GHz), joined to Time Division Multiple Access imposed stringent specifications to on board equipment, and mainly to the on board demodulator, asking for an equipment able to operate in burst mode with a large dynamic range of the input signal.

The specified dynamic range is in fact 37 dB, which is not compatible with proper operation of conventional coherent demodulators.

The problem can be overcome either by using a hard limiter or a fast ALC loop.

The use of a limiter is anyway affected by some important drawbacks, as it will be shown later.

A second problem is represented by burst-mode operation, implying fast carrier recovery and freedom from hang-up phenomena.

The possibility of using a Costas-loop-like approach was evaluated and experimentally tested.

It needed nevertheless a Start-Of-Burst pulse, to be provided by the on board processor; but no such possibility was made available, so, finally, a tracking filter solution was selected [2].

Unluckyly said approach implies a down conversion. It needs in fact a carrier filter at four times the input frequency, which is kind of a problem in our case, having an IF at 12 GHz.

Being clearly impossible to try a filter with a bandwidth of 0.4 MHz at 48 GHz, a down conversion was foreseen at an IF frequency of 700 MHz, which is the highest frequency still allowing filter implementation using temperature stabilized dielectric materials.

The need of a conversion gave rise to a further problem concerning the best suited reference for L.O.

Conventional quartz stabilized references do not appear to be convenient in this case, since recovered carrier is only demanded to be coherent with the data carrier, which is automatically accomplished by the particular arrangement of the circuit (X4 - filter - :4).

Rather, a phase reference appears to be better suited, since proper demodulation requires a well defined relationship between reference carrier and incoming data constellation.

Lastly, the need of delivering flight standard units in a relatively short time imposed non optimal solutions in some areas, in terms of weight and size, in order to meet with the stringent customer P.A. requirements.

In next paragraphs some problem areas will be discussed in some detail. Lastly a summary of the results will be given.

2. GENERAL BLOCK DIAGRAM

The general layout of the demodulator is given in fig. 1.

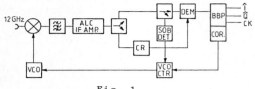

Fig. 1

The incoming QPSK modulated signal is firstly passed through a down converter and then shaped by a dedicated filter, so to have an overall Raised Cosine Spectrum with a roll-off factor of 0.63. Moreover, the total shaping is unevenly split between transmitter and receiver (respectively 64 and 36%) to optimize the performance in presence of channel non linearity (ground TWTA, refer to [3]).

The input large dynamic range (-55 dBm to -18 dBm, partly due to adverse weather conditions and partly to unfavourable ground station position with respect to beam center) is then compressed by a fast ALC loop in a total output variation ranging from +3 dBm to +6 dBm.

The output signal is split in two paths: Demodulator and Carrier Recovery. The way carrier is recovered (X4; filter; :4) guarantees in every case its frequency coherence, not its phase.

Accordingly, a correlator (actually a Costas-like detector) is included in the Baseband Processing Unit, so to give information on the actual carrier position with respect to its theoretically optimum value.
Said information is thus used to drive the VCO Control Unit only in selected time slots, that is only when a burst is present and when the initial acquisition transients have died out. The enabling pulse is provided by the SOB detector.

It is worth to remark that VCO synchronization takes place with a long time constant (several frames); as a consequence there is no possibility to track fast frequency variations (burst-to-burst), which must consequently be kept small (12 KHz p.t.p.).

3. DYNAMIC RANGE

In section 1 it was remarked that, in order to cope with the large dynamic range foreseen for the system, either a hard limiter or a fast ALC loop must be included.

However, a limiter is affected by a number of drawbacks, which are outlined hereunder:
- If the limiter is placed before the receive filter, then, in order to avoid intersymbol interference, a complete Raised Cosine filter with compensation for x/sinx must be used at receive side.
  This arrangement, however, is responsible of a twofold degradation, due to both signal power spreading and out of band noise conversion.
- If the limiter is placed after the filter the noise problem practically disappears, but strong pulse amplitude distortion will arise, (see fig. 2), needing an accurate post demodulation filter.

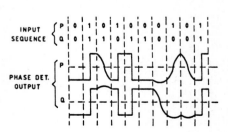

Fig. 2

- A limiter is usually affected by AM to PM conversion, which should be kept below 1°/dB on the whole dynamic range.

A fast ALC loop, conversely, exhibits AM to PM conversion only during transients, that is at the beginning of the preamble, behaving linearly for the whole remainder of the burst.

A problem arising with this approach is represented by the loop bandwidth. If it is too small, large portions of the burst are likely to be cancelled. Conversely, if it is too large, spectrum distortions will occur, since its behaviour gets closer to a real limiter.

A figure of 1.3 MHz was found experimentally: acquisition shows to be very fast and spectrum distortions are still negligible.

A block diagram of the amplifier is given in fig. 3.

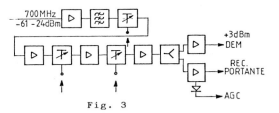

Fig. 3

Three PIN attenuators are used, able to cover a dynamic range larger than 40 dB. The unit features two separate outputs, respectively to the Quadrature Phase Detector and to the Carrier Recovery Circuit.

The shaping filter consists of three sections: a 4 pole Butterworth unit, a couple of poles, 73 MHz apart on each side of the center frequency and a group delay equalizer.

The overall amplifier response is shown in fig. 4, compared with the theoretical curve.

The two sidelobes have negligible effects since channel spacing is sufficiently wide (330 MHz).

The amplifier third order intercept point is higher than +20 dBm.

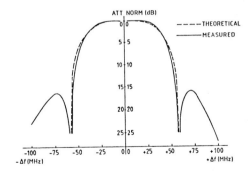

Fig. 4

The loop amplifier consists of a wideband operational amplifier (HA 2539) and a voltage repeater (LH 0002). Its diagram is given in fig. 5.

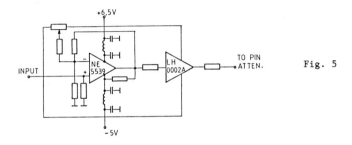

Fig. 5

4. CARRIER RECOVERY

The carrier recovery circuit is likely to represent the most delicate subassembly of the demodulator.

Its basic principle is indeed very simple: a chain consisting of a quadrupler, a narrow filter, and a divider by 4, so to destroy the phase information contained in the modulated signal.

In practice, conversely, its operation is a careful compromise among contrasting requirements and needed, consequently, extended experimental work.

In theory only unfiltered rectangular (NRZ) pulse envelopes do provide a constant output on a single line at quadrupler output.

In presence of filtered (Nyquist) inputs the output pattern is more complex, consisting of a number of discrete lines spaced by the symbol frequency (fs).

The output shows also a background continuous spectrum (Pattern Noise). This noise has two causes. The former is due to the fact that the carrier content is different for different transitions; it is mainly amplitude noise and is approximately Gaussian in a narrow filter centered on any of the discrete lines.

The latter is due to the sidelobes of the elementary pulse and, after quadruplication is basically a phase noise. It tends to approximate a Gaussian distribution only for very low values of the roll-off factor (rectangular channel spectrum). For higher roll-off figures it tends to be a discrete process with almost uniform distribution.

Fig. 6 shows the theoretical output of an ideal quadrupler, represented by an amplitude characteristics of the type x.sgn x and a zonal filter at the 4th harmonic, computed for a roll-off factor of .63.

Fig. 7, conversely gives the actual output of a real quadrupler, implemented with a Step Recovery Diode, fed by a spectrum featuring a roll-off factor of 0.4.

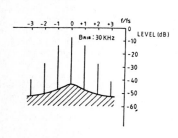
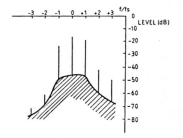

Fig. 6   Fig. 7

Apart some band limitation, clearly visible in fig. 7, the two outputs look relatively close to each other.

As dividers are concerned it should be realized that all frequency dividers are regenerative devices, showing consequently an operational threshold; a limiting amplifier is thus needed, to keep level as constant as possible at its input.

The overall block diagram, then, is shown in fig. 8.

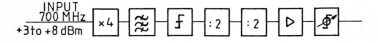

Fig. 8

The limiter is actually implemented with a chain of 5 stage FET amplifier. The useful limitation range is of the order of 15 dB with a negligible AM to PM conversion factor.
When thermal noise is present, however, even the presence of a limiter does not prevent the divider input signal to decrease below the divider threshold or to change drammatically its phase, performing a turn around the origin.
In this case cycle skips will occur, which can be very dangerous if, as in present case, absolute encoding is specified.
It is practically impossible to evaluate cycle skip frequency by computer simulation unless for very low S/N at quadrupler input.
In order to try to evaluate that figure theoretically, the following considerations can be made:
- Being the noise narrowband, its phasor rotates very slowly with respect to carrier phasor, so a quasi-static analysis is justified.
- If the overall phasor falls below divider-threshold, than the probability of having skips is 3/4 for a single divide-by-4 device and something between 1/2 and 3/4 when two cascaded dividers by 2 are used.
- If the overall phasor is higher than the threshold then the skipping probability equals the probability of its phase to pass through the value π in either direction [4].
Assuming a divider threshold = 0, then the cycle skip frequency comes out to be:

$$F_s \cong \frac{BN}{\sqrt{12}} \, \text{erfc} \left[\frac{K}{4} \sqrt{\frac{E_b}{N_o} \frac{f_b}{BN}} \right]$$

where K is the mean value of the carrier amplitude, No is the <u>total</u> noise density and BN is the carrier filter noise band.

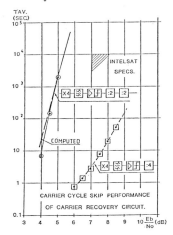

Fig. 9

Fig. 9 shows the measured cycle skip performance of the carrier recovery circuit, obtained with two cascaded dividers by 2.
The performance is also shown of a direct divide-by-4 device. Three different such devices, implemented on the basis of different principles (a Miller Divider, a Parametric Divider and an ILO) gave similarly bad results. The reason of said behaviour is still unknown.
Actual dividers are Miller units implemented with conventional bipolar transistors.
A last remark concerning phase stability: due to the particular VCO control used in the unit (see next paragraph), the overall AM-PM of C.R.C. must be confined in less than 0.5°/dB.
Moreover, the total phase shift of the unit across temperature (0 to 50°) is to be within ±5° w.r.t. the value it has at 25°C.

5. BASEBAND PROCESSING AND VCO CONTROL UNIT

The Baseband Processing Unit (clock extraction and regeneration) looks rather conventional apart two specific features:
- it is implemented with a semicustom ECL Gate Array.
- It includes a digitally processed Costas detector (Correlator).

The purpose of the Correlator is to provide an error input to the VCO control circuit. There is in fact a direct relationship between VCO frequency and carrier phase, via the carrier filter phase characteristics.

Said information (sign [sin 4 ϕ]), must however be forwarded to the VCO only when a burst is present, hence the necessity of a Start of Burst detector.

The diagram of the circuit is given in fig. 10. The SOB detector is a delay line detector, able to recognize the preamble format by providing a suitable pulse, which is then properly delayed before enabling. The delay is necessary to smooth out all transients due to ALC loop and carrier recovery, and provides, consequently, a lower bound to burst length (250 to 300 symbols).

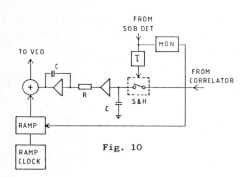

Fig. 10

The Sample and Hold circuit is implemented digitally with a semicustom Gate Array.

This solution was adopted because of the long frame specified for the system (32 mS), which would have rendered difficult the implementation of an analogic S & H circuit.

Loop shaping is obtained with an accumulator, clocked by SOB signal or by ramp clock.

The ramp, necessary during initial acquisition, is automatically generated forcing the accumulator input with a fixed level.

The accumulator output, after D/A conversion, is used for VCO synchronization. The VCO driving voltage quantization implies a frequency jitter:

$$\Delta F = K_V \cdot V_{max} / (2^N - 1),$$

where: K_V = VCO sensitivity (MHz/V),

V_{max} = Maximum output voltage excursion,

N = Number of bits at DAC input.

Said jitter is responsible of a carrier phase error [$\Delta\phi$] and S/N degradation [$\Delta(S/N)$] at carrier filter output, which are respectively:

$\Delta\phi$ = 1/4 arctg (8 $\Delta F/B$),
$\Delta(S/N)$ = 10 log [1 + (8 $\Delta f/B$)^2].

where B is the carrier extraction filter 3 dB bandwidth.

Considering K_V = 0.3 MHz/V, Vmax = 10 V, B = 400 KHz and tolerating a maximum carrier phase error of 1° a minimum frequency step comes out of 3.5 KHz, corresponding to a number of bits $\geq$10. The corresponding degradation of carrier S/N is: $\Delta(S/N)$ = 0.02 dB.

A further degradation is due to the digital correlator, which provides only the sign of the frequency error and not its amplitude. Accordingly, the actual VCO frequency will lie somewhere in between maximum and minimum frequencies inside the frame. The maximum phase error for the specified frequency limits (12 KHz p.t.p.) is then of the order of 4°.

The Gate Array was realized in 1.5 μ HCM technology, with a number of utilized gates equal to 721, and implying a consumption around 50 mW.

6. EXPERIMENTAL RESULTS

The BER performance of the unit is shown in fig. 11 in the following conditions:
Burst under measure: 2218 symbols; level: 30 dB down w.r.t. nominal level.
Interfering burst: Nominal level.
Guard time between burst: Zero
The degradation w.r.t. theory is ≤ 1.7 dB at BER = 10^{-6}.

Fig. 12 shows the degradation due to a frequency error for a BER = 10^{-4}.

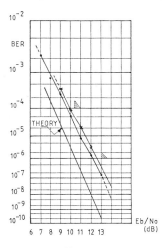

Fig. 11

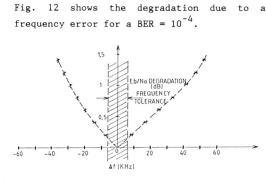

Fig. 12

Lastly, fig. 13 gives the BER performance for the complete 30 GHz receiver, versus input level. The overall receiver noise figure is 4.5 dB. The whole unit is likely to represent one of the most sensitive 30 GHz receivers ever conceived for high capacity satellite communications.

A computer simulation of the ITALSAT demodulator is given in fig. 14, where an assembly of 3 demodulators is shown.

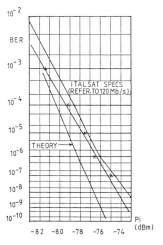

Fig. 13

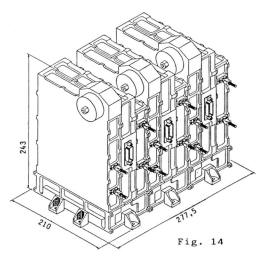

Fig. 14

ACKNOWLEDGEMENT

The authors wish to thank the European Space Agency (ESA) and the Italian National Council of Research (CNR) for their substantial support in the development of the unit.

REFERENCES

[1] F. Valdoni, F. Marconicchio, S. Tirrò: "The ITALSAT Programme", Telecommun. J., Vol 51, pp 70-75.

[2] B. Baccetti, A. D'Ambrosio: "On Board Burst Mode QPSK Demodulation, an experimental investigation for optimal configuration", GLOBECOM '85, New Orleans, pp. 192-195.

[3] R.A. Harris: "Transmission Aspects of the European Communication Satellite (ECS) System", ESA Journal Vol. 2 N. 4, pp. 259-279.

[4] S.O. Rice: "Statistical Properties of a Sine Wave Plus Random Noise", BSTJ 27, January 1948, pp. 109-157.

T-S-T SS/TDMA SYSTEM FOR SERVICES OF DIFFERENT BANDWIDTHS

G.Battista ALARIA (°), Giovanni COLOMBO (°), Giovanni PENNONI (°°)

(°) CSELT - Centro Studi e Laboratori Telecomunicazioni S.p.A. -
    Via G. Reiss Romoli, 274 - 10148 Torino (ITALY)
(°°) ESA: European Space Agency, Noordwijk - THE HETHERLANDS

After a presentation of the main features of the TST/SS-TDMA prototype, a list is given of the control topics that have been solved for carrying wide band services and for accomplishing a dynamic channel assignment to the ground terminals.

1. TRENDS IN SERVICES AND SYSTEMS

Telephone networks are now expanding and offering a large number of telematic services with different bandwidth requirements (B-ISDN).

Moreover, cellular radio systems are offering telephone facilities for cars and trucks to a potentially very large pool of users.

This steady evolution will take place in around thirty years to complete and will cost to each large European country almost 50,000 MECU. According to the current planning of the European PTT's it will essentially involve upgrading the ground network leaving the satellite a negligible role.

True, today's satellite has little to offer to cope with future needs, but the same is true of today's ground network. We will summarize here a number of new satellite technologies that will meet these future services requirements. First there is the large aperture multi-beam antenna which can link a very large number of small earth terminals whilst at the same time increasing the available bandwidth by frequency reuse. This fact reverses the present situation, in which the traffic to the satellite system (e.g. EUTELSAT TDMA/DSI system) is collected at few large stations linked at high hierarchical levels of the national network. In fact the traffic could be collected at local exchanges or even user's premises. From a system point of view, this traffic can be properly handled by real-time on-board routing processors able to interconnect earth terminals of various traffic volume operating optimally at the appropriate data rate (e.g. 131 Mbit/s, 33 Mbit/s, 2 Mbit/s [1], [2]). Furthermore the intrinsic flexibility of the on-board processor allows us to accommodate the requirements of telematic services by having the ability to allocate bandwidth dynamically according to user demand.

To cope with the requirement of a large number of on-board low-rate demodulators, the Agency is currently investigating the feasibility of a 2 Mbit/s Multi-Carrier Demodulator operating according to a Multicarrier-TDMA scheme [3].

Finally, in the front-end area, a number of new developments are in progress that will reduce the weight and power consumption of each component (e.g. low-noise receivers, modems, MMIC or hybrid technology solid state amplifiers, triple mode filters, optical beam-forming network, advanced solid-state antennas, RF and optical ISL).

The above overview primarily addresses satellites for fixed services, but substantial developments are also underway for mobile users.

2. SYSTEM DESCRIPTION

A satellite with onboard TST switching stages can provide some features that could not be efficiently performed by only space stage structure. In particular, every Traffic Terminal (TT), can collect and emit its traffic towards all the other stations by means of a single burst with a clear improvement in frame efficiency. A unique preamble, - i.e. Guard Time, Bit Carrier and Timing Recovery - is present for each terminal. Since this system is able to route each single call on demand, every TT can consider all stations of all transponders as a unique destination station. Furthermore, the onboard frame storage permits a simple approach to the bit-rate conversion, i.e. the possibility of choosing different bit rates, selectable in a prefixed range.

This system feature allows the stations to choose their channel capacity in the expected range consistent with the system performances. In the bread-board implementation only two bit rates are present (133 Mbit/s and 33 Mbit/s), but other bit rates, e.g. 8 Mbit/s and 2 Mbit/s, could easily be introduced into the system. The satellite thus becomes a switching centre (star) of the telecommunication network, integrated into an ISDN environment. This system is very suitable for scenarios with several terminals (hundreds) characterized by different traffic requirements and bandwidths.

The main system characteristics are depicted in Fig. 2.1.

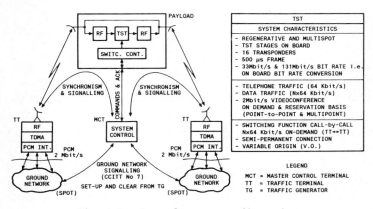

Fig. 2.1 - General system configuration

The Master Control Terminal (MCT) can set the satellite network configuration in terms of:
- bit rate for the 16 transponders (131 Mbit/s or 4 x 33 Mbit/s)
- number of stations per transponder (up to 16)
- number of switching exchanges per station (up to 12)
- number of circuits and burst position per station.

The system features depend strictly on the control mechanisms utilized for sharing the transmission resources and the switching capabilities between different traffic sources and service requirements.

The following section is, in fact, devoted to this question.

3. SYSTEM CONTROL FUNCTIONS

The basic needs for enlarging the convenience region of a TST-SS/TDMA system involve both varying the emission burst of each station and maintaining the system efficiency as high as possible in a wideband service environment. A Variable Origin (VO) procedure has therefore been introduced, together with a feasible onboard routing algorithm; they are described below.

3.1 VARIABLE ORIGIN PROCEDURE

The distribution of the system resources and switching capabilities is summarized in Fig. 3.1.

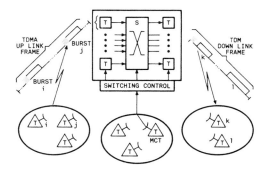

Fig. 3.1 - TST/SS-TDMA switching configuration

Every terminal is assigned a Carrier, supplying a TDMA frame. Due to the presence of the onboard TST network, offering complete connectivity among the up and down links, every terminal emits only once along the frame, hence a TDMA frame appears as a sequence of emission bursts, each one associated with a given terminal. This sequence is called the "emission plan". It is important to allow emission-plan modification, particularly when a large number of small stations share the same frame. This need can arise due either to a variation in the traffic offered (recovered through a statistical estimate) or a significant channel occupancy (wideband connection request).

The possibility of modifying the emission plan on each TDMA frame is provided by the MCT, which manages the switching reconfigurations needed both onboard and on the ground.

Given an initial emission plan, the VO rearrangement algorithm must be capable of accomplishing a new emission plan, consistent with the updating dimensioning needs, and this for each frame.

Obviously, the emission and routing reconfigurations must respect the continuity of the calls in progress. This is achieved by the MCT, by temporarily duplicating the calls in progress when they have to be shifted. In addition, it is necessary to move the preambles to meet the requirements of the new emission plan; this is done by the relevant terminal, under the co-ordination of the MCT, which sends the appropriate messages to the satellite.

Fig. 3.2 can help in understanding the VO procedure.

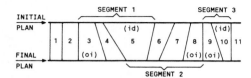

Fig. 3.2 - V.O. plan and segment identification

The station bursts are labelled with a progressive integer; no distinction is made between preambles, guard times and traffic channels.

Here, just a rough outline will be given of the algorithm adopted which is detailed in [4].

A basic choice consists in moving the preamble only when it is possible to give it its final position (univocally defined by the final emission plan). The preamble is just moved when the connections belonging to the contiguous bursts are few enough to allow such shifting, evenutally after an herding action. To reach such conditions, the MCT imposes proper occupancy constraints on the decreasing bursts and then waits until the ending of a suitable number of conversations is registered.

Clearly, increasing and decreasing bursts play an important role in the VO action. They imply the possibility of subdividing the VO modification into a number of "segments" which can be treated independently of each other. The segments are bounded by the so-called "inner decreasing" (**id**), and "outer increasing" (**oi**) bursts, defined as follows:
- inner decreasing bursts: the set of final channels belongs entirely to the set of initial channels (bursts 5 and 10 in the Figure)
- outer increasing bursts: the set of initial channels belongs entirely to the set of final ones (bursts 3, 8, 9 in the Figure).

A segment is thus a set of contiguous bursts, starting with an (id) burst and ending with an (oi) burst.

It can be proved that any VO modification can be partitioned into a number of independent VO modifications (segments) which can take place in parallel, i.e.: the times needed for accomplishing the segment modifications do not affect each other.

The possibility of processing any VO action with a certain degree of parallelism shortering its accomplishment delay, produces a high efficiency even in a non-stationary traffic environment. Moreover, such delays are affected by the residual conversation times, and might assume quite high values (minutes). Fig. 3.3 should clarify the VO delay mechanism; the arrow r_i represents the outcome of the random variable "residual conversation time of the burst i" (it is only applied to decreasing bursts); the segment P_j represents the shifting time of the preamble attacked to the j-th burst.

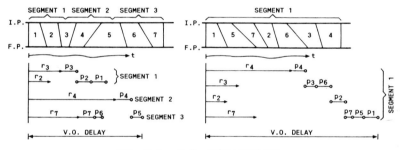

Fig. 3.3 - V.O. delay machine

The cases are equivalent from the dimensioning modification point of view; they differ only in the burst order along the frame. Different orders give rise to a different number of segments, thus implying different degrees of parallelism and consequent distinct VO delays.

The evaluation of the VO delay is at present under study since it can play an important role when a decision has to be taken about the suitability of the VO action itself.

3.2 ONBOARD ROUTING ALGORITHM FOR WIDEBAND SERVICES

The control problems that arise when carrying a multipoint videoconference (VC) will now be discussed since they are representative of the wideband switching environment.

The system can provide a VC service to be supplied on a booking basis; moreover, complete resource sharing is permitted between voice and wideband traffic, in order to increase system efficiency and flexibility. The necessary channels must, of course, be activated with sufficient time anticipation with respect to the VC starting point.

Point-to-point and multipoint VC connections are provided for video codecs working at 2 Mbit/s (a single connection corresponds to 32 telephone connections).

The service characteristics are as follows:
- each VC partner receives the image of the Current Speaker (this one receives the image of the Previous Speaker);
- while a multipoint VC is in progress, the system must assume, in real time, the switching configuration compatible with the actual speech activity of the partners (either a speech-detector mechanism or a "chairman management" can drive this process).

Because of the TST architecture, the problem of accomplishing a VC connection is divided into four (almost) independent tasks; two of them relate to the releasing process of the up and down channels for the wideband connections; the other two directly concern the onboard switching algorithm. A brief description will be given below of the solutions adopted for each task.

TASK 1: Releasing Procedure.

Once booked for the desired time, the VC service must be supplied with absolute certainty, so the channels needed have to be made available promptly, with an associated resource reduction for the voice traffic. This is achieved via the introduction of suitable occupancy constraints on the up and down links.

The number of wideband channels to be released depends on the onboard switching plan (TASK 4); here it is assumed that **b** telephone channels are necessary on both the up and down link for each VC connection.

In order to optimize the link occupancy it is mandatory to have only one wideband channel on the up and down links involved in a VC connection; this will be proved to be possible when TASK 4 is discussed.

Looking at a generic frame, let a_k be the number of VC's involving the **k-th** terminal among the S sharing the frame; let n_1, n_2, ... n_S be the burst dimensionings before the VC's accomplishment and let A_1, A_2, ... A_S be the related offered telephone traffics. It has been decided to modify the burst dimensions n_k in n_k+p_k; $p_k \gtreqless 0$; k = 1,2, ... S where p_k is a solution of the minimum problem:

1) $$\begin{cases} \min_{\{p_k\}} [\max_k \{B_k\}] \\ B_k = B(A_k; n_k+p_k-ba_k) \end{cases}$$

In 1) B_k is the Erlang loss probability suffered by the **k-th** terminal user with $n_k+p_k-ba_k$ channels available for the telephone traffic A_k. The $\{p_k\}$ S-tuple which is solution of 1) uniformly distributes the loss over the terminals for each VC configuration and at the same time completely defines the related VO actions eventually needed (see Fig. 3.4).

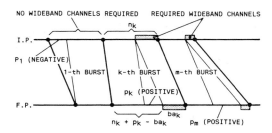

Fig. 3.4 - Emission plan redefinition for creating wideband channels

TASK 2: Safety margin procedure.

This task can be stated as follows:
"Determine a time interval before the VC accomplishment which is enough to guarantee, under the new occupancy constraints (TASK 1), and with a preassigned success probability, that the needed amount of channels is released when the VC starts".

The problem requires, in principle, that the system is studied globally, as a unique set of resources: the up and down link channels of all the transponders. Nevertheless it can be shown that it is conservative to calculate the releasing time of each terminal burst and to hold it as a safety margin, working independently with respect to the other ...es. Since the safety margin equals zero when a burst does not decrease in the new emission plan ($p_k \geqslant ba_k$: bursts k and m, Fig. 3.4), the problem can be reduced to:

"Given a burst whose telephone channel capacity must be reduced from n_k to m_k ($m_k < n_k$), find the distribution of the random variable W: the time necessary for reaching a burst state (number of conversations in progress) not greater than m_k".

With exponential conversation times and and Erlang distribution for the state probabilities, the distribution function $F_W(w)$ has been derived analitically. Now, given the probability α of releasing the necessary number of channels in time, the safety margin τ is calculated in such a way that:

2) $$F_W(\tau) \geqslant \alpha$$

Note that $F_W(w)$ is utilized in the booked VC service as an important operation mechanism but, at the same time, it can measure the service performance (waiting time) when the wideband request is allowed to directly access the system.

The next two tasks concern the onboard switching algorithm.

TASK 3: Rearrangement Procedure.

The availability of the up and down VC channels (assured by TASKS 1 and 2) does not guarantee, by itself, the feasibility of the complete VC connection (internal TST blocking). Finding a rearrangement algorithm capable of creating the onboard wideband paths without interrupting the voice connections in progress, is the object of the present task.

Reference is made to Fig. 3.5 which shows both the spatial representation of the TST network and the matrix representation of a generic switching state. T represents the number of system transponders and N the number of internal slots. Each slot is given a label (a lowercase letter in the figure) univocally associated with its time position along the internal link connecting Time and Space stages.

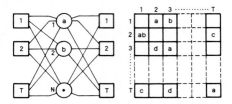

Fig. 3.5 - Spatial representation and equivalent switching configuration of a TST connecting network

The matrix completely defines the onboard switching state (for instance two channels of the uplink 2 are switched towards downlink 1 through the internal slots a and b).

Note that at TST level only the origin and destination frames are significant for a connection routing (the origin and destination terminals have no impact).

A rearrangement algorithm has been defined capable of freeing an internal wideband path whilst minimizing the loss probability suffered onboard by the telephone traffic while the VC connection is in progress. The call movements are performed asynchronously in the sense that call shiftings are not constrained to be accomplished during the frame period. This approach guarantees the connection continuity by temporarily doubling the internal path of the connection to be moved. Moreover the time needed for the operation is not a significant system constraint.

The choice of the internal path is optimized by imposing the constraint that, if possible, any multipoint wideband connection must be performed through the same set of internal TST slots, in order to approach the wide sense nonblocking networks conditions [5].

To give an idea of the convenience of such a rearrangement, Fig. 3.6 shows a case study. The loss probability suffered onboard by a telephone call in the presence of a wideband multipoint connection has been measured (simulation) under two rearrangement criteria:

i. the internal path is made free for the VC through a random rearrangement
ii. the internal path is constructed by means of the proposed algorithm.

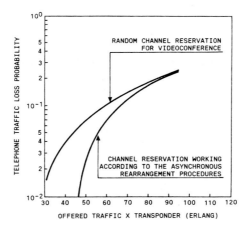

Fig. 3.6 - Network performance in the presence of wideband multipoint connections

TASK 4: Reconfiguration task.

It again refers to the onboard switching rules and is applied only when a multipoint VC connection arises:
"An onboard reconfiguration algorithm must be found such that the number of internal TST time slots is minimized for each multipoint connections. At the same time the algorithm must correctly follow the evolution of the Current and Previous speakers during the VC"

Obviously, knowledge of the amount of internal time slots needed is the starting point for the preceding rearrangement procedure and, moreover, can condition the number of wideband channels on the up and down links (TASK 1).

A unique internal wideband time slot is sufficient for supplying the connectivity required by the service. This has been achieved by introducing a broadcast connection onboard: the image emanating from a VC partner is sent towards all the other partners. The broadcasting is carried out at the S stage of Fig. 3.1.

CONCLUSIONS

An overview has been presented of the control mechanisms inserted in the TST/SS-TDMA prototype developed by CSELT Research Laboratories under ESA contract. The control topics have been studied in order to allow both a suitable handling of different bandwidth services and a modification of the resources allocated per terminal according to the traffic loss variations.

Further work is needed for quantifying the consequent performance levels characterizing the system.

REFERENCES

[1] Alaria G.B., Pennoni G., An SS-TDMA satellite system incorporating an onboard Time/Space/Time switching facility: overall system characteristics and equipment description, Proc. ICC'84 Amsterdam, 14-17 May 1984

[2] Alaria G.B., Pennoni G., SS/TDMA satellite system with on board TST switching stage, CSELT Technical Reports - Vol. XII, Nr 3 - June 1984

[3] Colombo G., Pennoni G., Advanced on board processing for user oriented communication systems, 7th International Conference on Digital Satellite Communications, München, May 12-16, 1986

[4] Annoni M., Settimo F., Ventimiglia G., Problems and algorithms for the introduction of the variable origin in a TST/SS-TDMA system. Submitted to the ICC 1988 - Philadelphia, USA

[5] Benes V.E., Blocking states in Connecting Networks made of Square Switches arranged in Stages. The Bell System Technical Journal, Vo. 60 Nr 4, April 1981

AN ON-BOARD PROCESSOR FOR ISDN AND ISDN-COMPATIBLE APPLICATIONS

H.P. Kuhlen *

MBB- Space Systems Group
P.O.Box 801169, D-8000 Munich 80, West-Germany
Tel. 49 (89) 6000-3542

The new role of satellites in future satellite integrated communications networks is yet to be defined. Among other tasks it is expected that the satellite has to act as an intelligent node of a meshed network such as the ISDN.
In addition, satellite based system designs for the 90´s have to consider new types of services and new categories of users requiring direct access to the satellite. In actual fact the availability of satellite terminals located directly on customer premises individually tailored to their effective needs, will dominate their decision to go for satellite communication in particular where high rate video or file transfer services are required (value-added services). Furthermore, ease of operation and the inherent capability to address a large number of subscriber by using the same ground equipment will also support that decision.
These requirements can only be satisfied, if and when the satellite transponder is designed to provide features in addition to its traditional role of frequency conversion and amplification. A potential system scenario will have several orthogonal spotbeams interconnected through a baseband processor acting as an intelligent node controller of the associated network.

INTRODUCTION

This presentation provides some selected highlights of the on-board processor with ISDN exchange capabilities which is presently under design and development at MBB´s Space Systems Group. The design has been

---

* The work reported here has been funded with MBB R&D budget

based on the system architecture currently in the process of research and development in Europe under the auspieces of ESA/ESTEC. This concept calls for a system where the transponder on-board the satellite plays the role of a "switchboard-in-the-sky" (ESA).

The design and development of the space-borne transponder for telecommunications applications is a typical example of a new way of creative modern system design where it is a challenge to bring together many independent engineering disciplines.

Among these disciplines are the telephone switching and networking techniques, digital signal processing, data handling and computer networking techniques as well as LSI and VLSI technologies and last but not least the peculiarities of the platform, the satellite. All these disciplines contribute to the system design of future satellite integrated communications network.

THE NETWORK INTERFACES

Out of the many concepts for a SS/TDMA transponder architecture with regenerative on-board processing (TST) which have been published in many publications, the ESA switchboard-in-the-sky concept was selected as the baseline for a breadboard realization of the MBB Baseband Processor.

This concept is an example for a network oriented system application. Strictly speaking the entire satellite communications system consisting of the baseband processor (BBP), the associated master control terminal (MCT) and several traffic terminals (TT) represents an exchange facility. In contrary to the usual terrrestrial toll center it is spreading over a wide service area e.g. Europe.

In the case of a user oriented scenario it would provide access points located directly on an end-users site (customer premises station) providing access to the exchange facility "on site" thus avoiding additional microwave or other feeder links often quoted as the most expensive last mile.

In accordance with the ISDN philosophy each system access point is represented by a variable number of transparent data channel of 64kbps with one common signalling channel at 64kbps for each TT.The signalling is not interpreted in the TTs but will be transfered to the MCT. Thus, only the MCT is the central intelligence by coordinating all signalling information of all connected TTs in the spot beams.

The MCT decides if a new call request can be granted, in other words if the required capacity can be assigned or not. Then depending on this decision, the originating and the destination terminal will be connected through the on-board processor. This will be achieved by MCT command using the on-line demand assignment updating message (DAUM) channel.

The structure of the DAUM has been generally adopted from the existing definition (ESTEC) to be compatible with the ESA equipment. However, minor modifications have been introduced which do not affect the com patibility to allow additional features in terms of processor monitoring and trouble shooting.

While the signalling information for any connection must go through the MCT, the user information is routed transparently through the on-board processor. The on-board processor performs two main tasks: first, the fast switching of the incoming data channel and second, the control process where the routing requirements of the MCT are executed.

THE ON-BOARD PROCESSOR

The BBP consists of the TST-stages and the digital control unit which provides the main communication interface to the MCT. The whole BBP is actually slaved to the MCT. This approach unloads the on-board processor considerably from tasks which can easily be performed on ground resulting in less complexity in the space segment. The experimental test set-up with the major digital interfaces is shown in figure 1.

The bread board model comprises only of the digital parts, i.e. no RF-stages or MODEMs are included. The interface between the "space" and the "ground" parts is composed of a digital link carrying four signals: I, Q, symbol clock and a data valid-bit for the simulation of the lock condition of the on-board burst demodulator.

The received frames are loaded into the memory of the input time stages (ITS) while simultaneously a frame is being transmitted (TDM) from the output time stages (OTS) on the downlink. All received frame are loaded into one of two alternating buffer memories consecutively. Hence, at any time there is one memory being loaded while the second is being processed. The effective exchange process between the input -and output time stage occurs on the stable buffer during a frame period (ping-pong principle).

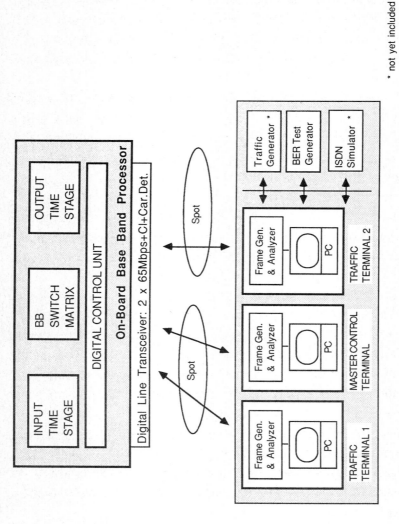

Figure 1: BBP Test & Experimental Set-Up
Ground Station Simulation <==> BBP Interface

* not yet included

The central process in the BBP transfers the information contained in the input memories through the space stage into the output memory. The downlink which is then composed of the information out of the output memory has to be arranged in a defined sequence where the messages for each traffic terminal have to placed in an organized order.

This process performes the routing of 2000 64kbps channel, each channel is represented by 32bit-words, from the input stage of one spot to the output stage of the other spot. At present two "spots" at 132Mbps are available. Due to the modular design of the hardware, extensions for other bitrates e.g. 32Mbps or less are possible.

The design of the BBP is based on a modular concept. Each module has a clearly defined interface. Extensions and/or different configurations of the processor are possible without re-design of the other components. The modules are shown in figure 2 together with the main signal flows. The BBSM has a 4x4 switch capability with options for two more input/output time stages.

The interaction between the BBP and the MCT is one of the key features of the processor design. Therefore it became necessary to at least simulate the main functions of the MCT. Unfortunately, at present, no standard computer can handle bitrates of 132Mbps. This conclusion led to the development of a dedicated high bitrate software controlled interface - the so-called frame generator/ analyzer. In conjunction with a standard computer for the man/machine interface this equipment can act as an MCT or as an TT depending on the loaded software.

A so-called sequencer performs the construction of the frames. For test purposes it is sufficient to control on-line only the very few overhead bits while most of the "user"channel can be filled with random number information from a built-in hardware generator. However, two "user" channel (i.e. one circuit) have been included which can be traced through the on-board processor.

For a given sequence, the generator creates the required 500μsec frames with defined silent periods (simulating guard times). Hence it is possible to simulate any configuration of ground stations by variation of number and/or capacity of TTs in the spot. An example of the generated seqence is shown in figure 3.

Optional interfaces of these equipments provide additional capabilities ranging from simple bit-error ratio tests to sophisticated protocol (ISDN) investigations. This provides a valuable testsystem in particular in view of the anticipated field

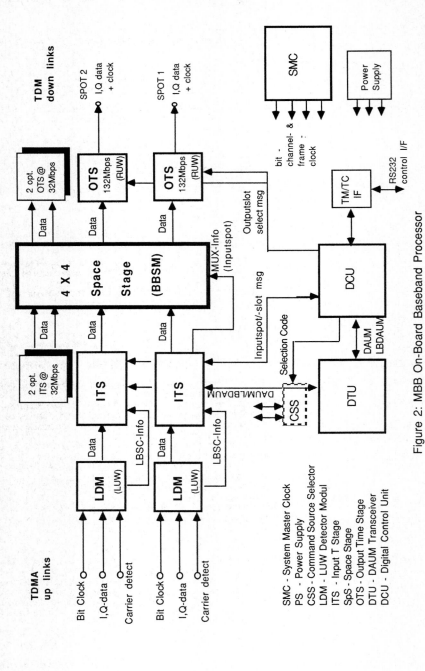

Figure 2: MBB On-Board Baseband Processor Breadboard Model

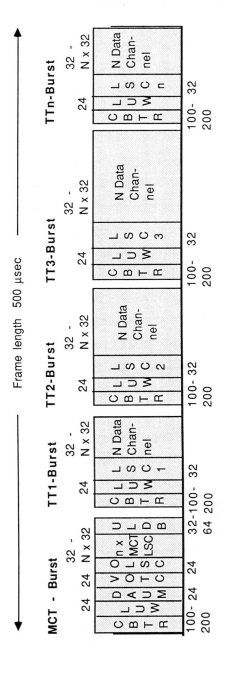

Figure 3: Example of TDMA Uplink Frame Configuration

testing under quasi operational traffic conditions via an experimental satellite link.(ESA/EUTELSAT Double-Hop experiment).

The communication between the TTs and the MCT via the BBP is controlled by a special protocol (ESA). This protocol serves as a common carrier for many independent communication channel covering user, administrative and operational information. Beside carrying the user information it delivers the demand assignment updates between MCT and payload. It also provides the feature of permanent ranging information mandatory for each TT to remain full synchronized with the system also in case of unavoidable attitude changes of the host satellite. In the bread board model up to now we have introduced only those elements which allow to establish the set-up and clear of call request.

CONCLUSIONS

This paper reported briefly on the design and development of a base band processor at MBB. The design of the available hardware and software has been performed in line with the ESA system concept to achieve a high degree of compatibility. This equipment provides a testbed for further development and investigations. It is scheduled that this breadboard model will also be used in the double-hop experiment of ESA/EUTELSAT where the operation of a TST/SS-TDMA system via real satellite channel will be tested.

REFERENCES

[1] G.Pennoni
"A TST/SS-TDMA Telecommunication System: from cable to switchboard-in-the-sky " ESA Journal 1984, Vol.8

[2] G.B.Alaria, et.al.
" On-board processor for a TST/SS-TDMA Telecommunications Sytem " ESA Journal 1985, Vol. 9

[3] G.Pennoni, G.B. Alaria
" An SS-TDMA satellite system incorporating an on-board switching facility: overall system characteristics and equipment description "
ISS 84, Firenze

[4] S.E.Dinwiddy
" Advanced on-board processing satellite system concepts " Graz 1985

COMPARISON OF DIGITAL TRANSMULTIPLEXER ARCHITECTURES FOR USE IN ON-BOARD PROCESSING SATELLITES

W. H. Yim, C. C. D. Kwan, F. P. Coakley, B. G. Evans.
University of Surrey, Guildford, Surrey, GU2 5XH, U.K.

A study of transmultiplexers for on-board processing satellites is presented. Different architectures are compared in terms of their architectural constraints and computational requirement. Results are given to indicate an approach to the design of a transmultiplexer.

1. Introduction

Future traffic requirements in mobile (land, maritime and aeronautical) and small fixed station business systems will demand new types of communication satellites.[1] For networks involving large numbers of small capacity, multi-service users, the conventional transparent-transponder satellite using FDMA or fixed TDMA access schemes is no longer efficient. Satellites employing multi-beam antennas, regenerative transponders and on-board processing (including switching) are needed to serve such networks economically. This means high satellite resource utilisation, hence low space segment tariffs, plus simple and cheap earth-terminals. In addition, studies of efficient access-schemes for such networks have demonstrated that low-cost earth-terminals result from the use of SCPC/TDMA on the uplink and TDM on the downlink[2]. This places an additional requirement on the satellite of transforming between frequency and time domains-hence the inclusion of transmultiplexers (TMUX). The proposed general architecture is shown in Fig.1, and has been referred to as the 'intelligent satellite'[3].

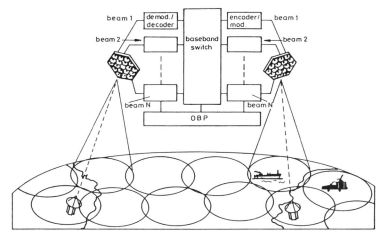

Fig.1 On-board processing satellite general architecture

In a previous paper[4] the on-board processing payload architecture and the overall design considerations for the baseband switch and transmultiplexer were examined. In this paper the design of a digital implementation of the transmultiplexer is considered in detail. A description is given of the simulation software which has been developed to facilitate investigation into the performance of various transmultiplexer architectures.

Comparisons can then be drawn between the various architectures with regard to the application areas already mentioned. The results of this study is presented, and should enable the appropriate selection and design of transmultiplexers for specific applications.

In this paper the term 'transmultiplexer' is used to apply to the demultiplexing function of the more general 'multi-carrier demodulator'(MCD). The latter adds the demodulation to the transmultiplexing function to produce the overall device (MCD) that would be used on board the satellite. For purposes of design, the functions (demultiplexing and demodulation) can be separately optimised, although in implementation they would be closely related, e.g. being in the same DSP chip.

2. Transmultiplexer Architecture

TMUX architecture/algorithms are largely based on the theory developed around terrestrial TMUX applications[5]. They can be widely classified into the two categories of transform and non-transform types[6,7]. However, the additional requirement of demodulation in the MCD also implies some important differences in the requirement for the TMUX in the MCD application[8]. This section outlines the main choices of architecture of TMUX and where possible indicates how the application affects the choice of such architecture.

2.1 Direct Method

This method performs filtering of the frequency multiplex by individual bandpass filters each assigned to a SCPC channel, as shown in Fig.2, and is similar to filtering by analog means.

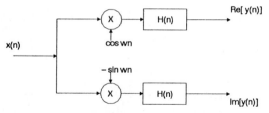

Fig.2 Direct method

It may shift the channels in frequency and perform lowpass filtering, as expressed in Eqn.2.1, or equivalently use bandpass filters that are frequency-shifted versions of the lowpass prototype[9], as in Eqn.2.2.

$$y(n) = \sum_{i=0}^{N-1} h(i).x(nM-i).e^{-j\omega_k(nM-i)} \quad 2.1$$

$$y(n) = e^{-j\omega_k nM} \sum_{i=0}^{N-1} h(i) e^{j\omega_k i} x(n-i) \quad 2.2$$

where $x(n)$=input sample, $y(n)$=output sample, $h(n)$=filter impulse response, $0 \leq n \leq N-1$. ω_k = nominal channel radian frequency, and M = decimation rate.

As will be seen later, this method is of little practical interest since its computational requirement is high, although it may be reduced by multistage decimation[10]. Symmetry of the FIR filter coefficients allows further reduction, although for complex filters the existence of this symmetry depends on its frequency shift from the lowpass prototype.

2.2 Tree Filter Bank

This method allows sharing of computation between channels by exploiting the symmetrical nature of the channel stacking arrangement. Each stage of a binary tree structure splits the input into a highpass and a lowpass branch, with decimation by 2. This process is repeated until channels are totally separated, as shown in Fig.3 for an eight channel case.

The frequency responses of the two branches at each stage may be produced from the same lowpass prototype as in Eqn.2.2, and their impulse responses are related by

$$h_{highpass}(n) = \phi(n).h_{lowpass}(n), \quad 0 \leq n \leq N-1. \qquad 2.3$$

$\phi(n) = (-1)^n$ if real filters are used as shown in Fig.4a, and $\phi(n) = j^n$ if complex responses as shown in Fig.4b are used.

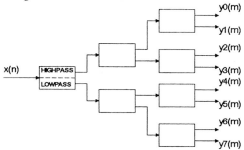

Fig.3 Tree filter bank

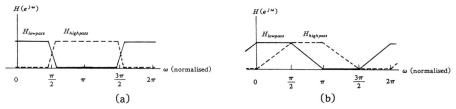

Fig.4 (a) real filters, (b) complex filters

Saving in multiplication may be made by using the same product terms $h(n).x(n)$ with sign-inversion and exchange of real and imaginary parts as necessary. Complex responses allow greater transition widths hence shorter filters, but the combination of the product terms is less straight-forward. It also requires post-filtering[11], although in a MCD this can be part of the demodulator and the additional computational effort is not significant. This was the approach taken in the simulation study.

2.3 FFT Filter Bank

The main principle of the FFT filter bank is to share the same lowpass filter amongst all of the channels. There are several variations of the FFT filter bank. The most straight forward is the polyphase-FFT method and is shown diagrammatically in Fig.5.

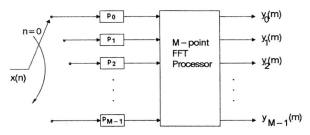

Fig.5 Polyphase-FFT filter bank

The structure is arrived at by polyphase decomposition of the lowpass prototype followed by algebraic manipulation of the variables[12], giving Eqn.2.4.

$$y(m) = \sum_{n=0}^{M-1} \sum_{i=-\infty}^{\infty} p_n(i) . e^{-j2\pi k \frac{n}{M}} x_n(m-i) \qquad 2.4$$

where $p_i(n) = h(nM-i)$, i.e. the $i\,th$ polyphase filter branch of the prototype h(n).

The constraint in this structure is that the outputs are critically sampled at the Nyquist rate of each channel, and that the decimation rate(M) equals twice the number of channels(K). This may not be desirable in a full MCD as the output sampling rate should be related to the symbol rate for convenience of demodulation. This implies arbitrary choices of decimation rate with respect to the number of channels, and for this purpose the weighted overlap-add structure is proposed. The generalised DFT[12] is applied if the frequency origin of the baseband multiplex is to be arbitrary, and in this study the odd-FFT is chosen for the required odd-channel stacking arrangement.

A brief comparison between the different methods in terms of their architectural constraints is given in Table 1, with the general assumption that each architecture is applied in its most efficient form. It is not, however, a comparison of their relative efficiencies.

Table 1.

Method	Uniform-bandwidth channels		
	Stacking arrangement	Number of channels	Decimation rate
Direct	no restriction	no restriction	no restriction
Binary tree	no restriction	power of 2	power of 2
Polyphase-FFT	uniform channel spacing	power of 2	2K
Weighted-overlap-add	uniform channel spacing	power of 2	no restriction

For non-uniform-bandwidth channels, the FFT-type methods are not efficiently applicable. A general comparison is not possible as it depends entirely on the specific channel arrangement in the frequency multiplex.

3. Comparison Of Architectures

3.1 Computer Simulation

In order to study the effect of TMUX distortion on the bit error rate (BER), a simulation procedure was developed which involves the generation of frequency multiplexed test signals, demultiplexing, demodulation and BER evaluation.

The input FDM consisted of independent 4-phase PSK channels generated from pseudo-random bit streams, at a rate of 64 kbit/s. Nyquist shaping filtering characteristic was equally shared between transmit and receive filters, with a rolloff factor of 40%. The number of PSK channels and the carrier spacing was varied to compare the computation complexity, storage and word length requirements of various TMUX architectures.

The BER at the demodulator output was estimated by the semi-analytic method, where the following assumptions were made:

(i) the signal and quantisation noise are uncorrelated,
(ii) the demodulator is ideal.

Hence the Gaussian noise power can be estimated from the equivalent noise bandwidth of the receive filter at the demodulator. The probability of individual symbol error was determined analytically from the distorted signal at the optimum sampling instant, and averaged over 100 symbols to obtain the BER. The TMUX's were compared with a target performance of 0.04 dB degradation at E_b/N_0 of 8.4 dB. In order to study the undesired signal distortion caused by demultiplexing, the test signal generation and demodulation algorithms were chosen such that the degradation due to interchannel interference, carrier and clock recovery was negligible. Fig.6 illustrates graphically the procedure of determining the word length requirements, in this case the analog to digital conversion (ADC) word length for the FFT TMUX.

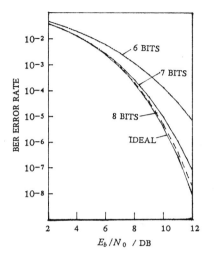 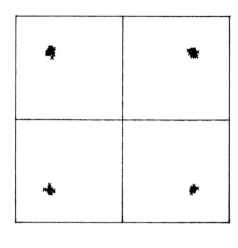

Fig.6 ADC word length effect on BER   Fig.7 Scatter diagram

The BER curve corresponding to the selected word length is not shown since the target degradation is close to ideal. The scatter diagram corresponding to the chosen design which included various finite word length effects is shown on Fig.7.

In the design of TMUX filtering bandwidth, a carrier uncertainty of ± 600 Hz was taken into account. Consequently, the simulation results cannot be directly applied to other bit rate systems, since the filtering bandwidth is not directly proportional to the bit rate.

3.2 Software

A process oriented package has been developed for the simulation of communication systems, specially tailored for multi-rate digital systems. The package is implemented in C language running on the Unix® operating system. All simulated sub-systems are represented by process-like functions and communicate through channels. The activation of sub-systems and synchronisation of data transfer are performed by the simulation kernel, which simplifies the programming task as new functions are required. Processes can be nested for building complex systems by using existing ones. The process features also enable the simulated system to be described in a data flow style resembling a block diagram description. The use of a conventional language retains the convenience of iterative simulation, required for optimisation etc.

In order to reduce the coupling between signal processing algorithms and hardware arithmetic units, basic operations at the heart of a sub-system are serviced through function calls. These functions are dynamically bound at different instants of the simulated system to different definitions, which represent particular hardware implementations.

3.3 Comparison

Fig.8 shows a comparison of the computational complexity as the number of PSK channels is increased, for a channel spacing of 64 kHz. The Direct method is unsuitable for

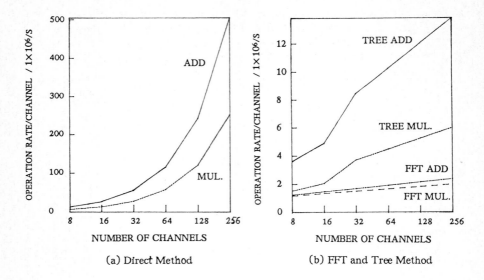

Fig.8 Computation rate comparison

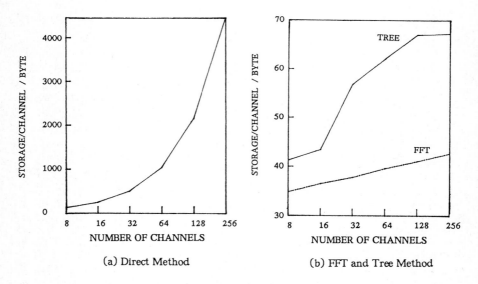

Fig.9 Storage comparison

implementation if the computation rate is the prime concern. The FFT method is the most efficient in all cases, whilst the Tree method remains competitive especially for small number of channels. It should be noted that the distortion due to the Tree method is significantly smaller than the other methods in many cases. This is due to the low order of the required half-band filters. The available choices of design often result in performance that are either unsatisfactory or much better than the design goal.

The corresponding storage requirement follows similar trend as the computation rate, which is shown on Fig.9. Table 2 summarises the various word length requirements.

Table 2. Word length requirements (including sign bit)

	Direct method						Tree method					
No. of channels	8	16	32	64	128	256	8	16	32	64	128	256
ADC	9	9	9	9	9	9	9	9	9	9	9	9
Filter Coef.	8	8	8	8	8	8	7	7	7	7	7	7
Filter Arith.	15	16	17	18	19	20	13	13	13	14	15	15

	FFT method					
No. of channels	8	16	32	64	128	256
ADC	9	9	9	9	9	9
Filter Coef.	8	8	8	8	8	8
Filter Arith.	11	11	11	11	11	11
FFT Coef.	6	6	6	7	7	7
FFT Arith.	11	12	12	13	13	14

The effect of channel spacing on the computation rate is shown in Fig.10, where the number of channels was 16.

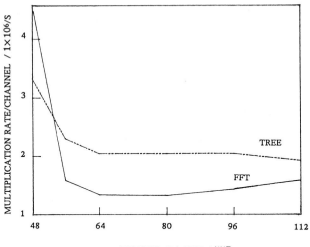

Fig.10 Effect of channel spacing on multiplication rate

The Tree method appears to be more efficient for narrow channel spacing. For low bit rate systems, the frequency deviation, due to Doppler effect etc, increases relative to the PSK signal bandwidth. In order to avoid degradation due to band limitation in the TMUX, a filter bank with narrower transition width is required to account for an effective increase in signal bandwidth. The result is similar to a decrease in channel spacing. Hence the Tree method is more suitable where a small number of low bit rate channels require high bandwidth utilisation.

In general, the sampling rate at the output of the TMUX is not suitable for the following demodulation process, where an integer number of samples per data symbol is required. A sampling rate conversion filter is necessary for interfacing between the demultiplexed channels and the demodulator. This can be integrated with the post-filtering stage of the Tree method with reduced computations. This comparative advantage over the FFT method is also highly dependent on the channel spacing etc, as mentioned above. Alternatively, all the per channel filtering operations can be integrated into the pulse shaping filter of the demodulator. The overall computational complexity involves demodulator design and optimisation of the whole MCD. In clock recovery schemes where the pulse shaping filter is implemented as an adaptive polyphase network[9], the additional integration does not introduce computational overhead, however, the control complexity is necessarily increased.

4. Conclusion

This paper has described the computational requirements for the FFT and Tree design for TMUX's. The 'best' technique depends on the number of channels and channel spacing with the Tree method probably better for small number of low bit rate densely packed channels. Other considerations, for example, ease of integration into silicon have not been considered — these may favour the Tree approach. Similarly the sharing of computation between the TMUX and a group demodulator may also influence the choice, but our flexible simulation approach allows a good compromise for a specific requirement.

REFERENCES

[1] 'Study of systems and repeaters for future narrowband communications satellites', ESA, ESTEC contract no. 5484/83/NL/GM(SC), Phase II final report, 1985.

[2] El-Amin, M. H., Evans, B. G., Chung, L. N., 'An access protocol for onboard processing business satellite system', Proc. 7th international conference on digital satellite communications, Munich, May 1986.

[3] Evans, B. G., 'Towards the intelligent bird', Int. J. Satell. Commun., Jul. 1985, No.3, pp.203-215.

[4] Evans, B. G., Coakley, F. P., El-Amin, M. H., Lu, S. C., Wong, C. W., 'Baseband switches and transmultiplexers for use in an on-board processing mobile/business satellite', IEE Proc., Vol.133, Pt.F, No.4, Jul. 1986, pp.356-363.

[5] Scheuermann, H., Gockler, H., 'A comprehensive survey of digital transmultiplexing methods', IEEE Proc., Vol.69, No.11, Nov. 1981, pp.1419-1450.

[6] Bellanger, M. G., Daguet, J. L., 'TDM-FDM transmultiplexers: digital polyphase and FFT', IEEE Trans. Commun., COM-22, No.9, Sept. 1974, pp.1199-1205.

[7] Tsuda, T., Morita, S., Fujii, Y., 'Digital TDM-FDM translator with multistage structure', IEEE Trans. Commun., COM-26, No.5, May 1978, pp.734-741.

[8] Kato, S., Arita, T., Morita, K., 'Onboard digital signal processing technologies for present and future TDMA and SCPC systems', IEEE J. Selected Areas Commun., SAC-5, No.4, May 1987, pp.685-700.

[9] Gardner, F. M., 'Onboard processing for mobile-satellite communications', ESA, ESTEC contract no. 5889/84/NL/GM, May 1985.

[10] Crochiere, R. E., Rabiner, L. R., 'Optimum FIR digital filter implementation s for decimation, interpolation and narrow-band filtering', IEEE Trans. Acoust. Speech Signal Process., ASSP-23, No.5, Oct. 1975, pp.444-456.

[11] Constantinides, A. G., Valenzuela, R. A., 'An efficient and modular transmultiplexer design', IEEE Trans. Commun., COM-30, No.7, Jul. 1982, pp. 1629-1641.

[12] Crochiere, R. E., Rabiner, L. R., Multirate digital signal processing, (Prentice Hall, 1983).

SESSION 5

USER-ORIENTED AND SPECIALIZED SATELLITE NETWORKS

Chairman: S. Tirrò *(TELESPAZIO, Italy)*

Present Status and Future Developments of Satellite
Business Services Networks in Japan

Yoshiteru Morihiro and Shuzo Kato

Nippon Telephone and Telegraph Corporation

NTT has been offering satellite business services all over Japan by using Japanese Communications Satellite, CS, and developing key technologies to realize these services cost effectively. NTT's new satellite business communication system development and planned launches of domestic satellites will provide more possibilities to enhance satellite business services in Japan.

1 INTRODUCTION

Japanese first domestic satellite communication systems were put into commercial service in June, 1983 by the successful launches of Communication Satellite-2s (CS-2a and CS-2b), in February and August, 1983. They are the world's first commercial satellite communication systems using the highest frequency (30/20 GHz) bands ever utilized.

In the beginning of services, the major roles of satellite communication systems were to establish trunk transmission lines between Regional Centers (RC) and relocatable transmission links of voice or video signals by transportable earth stations, in addition to providing transmission lines between the main island of Japan and remote islands.

On the other hand, The explosive growth of computer terminals and other non-telephone terminals has created a huge demand for data communication links. To meet these demands, NTT has been constructing the Information Network System (INS) which is composed of digital transmission lines and digital exchanges. Along with this target, NTT started a new broad band digital satellite service, which is called Satellite Digital Communication Service (SDCS) in 1984 and a new video satellite service, which is called Satellite Video Communication Service (SVCS) in 1986. These two services offer dedicated networks to customers by small earth stations.

In addition to these dedicated network services, satellite INS subscriber networks has been under research and development in NTT, which are a part of public switched networks (INS) and will offer economical broad band digital transmission links between subscribers and the nearest telephone offices to which the called party is connected.

Furthermore, with the planned launches of private company-owned domestic satellites late 1988 and early 1989, various types of satellite business services are under investigation.

2 SATELLITE DIGITAL COMMUNICATIONS SERVICES (SDCS)[1],[2]

The SDCS is a Multi Access Closed Network (MAC-Net), very close to a Local Area Network (LAN), making use of full advantages of satellite communication networks such as flexibility in setting up circuits, broadcast capability and multi-access capability.

2.1 MACNET

In the MAC-net, a satellite channel is allocated to each user group based on a pre-assignment mode. However, differing from conventional TDMA channel assignment, in the MAC-Net, a user group controls on/off of pre-assigned burst(s) by using signaling (S) bits. Therefore, there will be no bursts transmission to a satellite as far as "S" bit is "0". On the other hand, all users who belong to the same user group can always receive the transmitted signals to the dedicated channels. The logical channel configuration of the MAC-Net is shown in Fig. 1. As seen from this figure, by using the MAC-Net, it is easily possible to communicate in modes of (a) point to multi-point, (b) Multi-point to point and (c) Half duplex (changing transmitting points alternately) and so on. Since the MAC-Net has these characteristics, a terminal adapter is required for conventional both-way terminals to interface this network. A terminal adapter for this purpose is composed of an S-bit inserter, S-bit detector, controller and interface circuits(Fig. 2). By this terminal adapter, the MAC-Net satellite channels are used very efficiently.

2.2 CIRCUIT SWITCHED SYSTEMS

In addition to the above mentioned pre-assignment services, the MAC-Net offers circuit switched services for higher speed than 192 kb/s. This system requires as many demand assignment equipment(DAS) as the number of user stations and one demand assignment controller (DAC). A configuration for the SDCS with circuit switching capability is shown in Fig. 3.

2.3 SYSTEM CONFIGURATION

A configuration of the SDCS is shown in Fig. 4. The SDCS is composed of TDMA (Time Division Multiple Access) earth stations located in telephone offices with user interface bit rates varying from 64 kbit/s to 6144 kbit/s. This system employs 30/20 GHz frequency bands to avoid interference from/to terrestrial radio communication systems and QPSK (Quadrature phase shift keying) modulation and coherent detection. Moreover, convolutional encoding - Viterbi decoding as an FEC scheme is employed. This system has a transmission capacity of 160 channels (in 64-kb/s both way). Subscriber data ranging from 64 to 6144 kb/s are conveyed to an SLT (Subscriber line terminal) by radio- subscriber lines, optical fibers or metallic cables and are then multiplexed into 2.048 or 8.192 Mb/s signals by the SLT and interface to TDMA equipment. The major parameters of the SDCS are shown in Table 1.

2.4 EARTH STATION

In order to offer the SDCS cost effectively, a compact and high reliable 30/20 GHz earth station has been developed. The developed earth station and major parameters are shown in Fig. 5 and Table 2.

For TDMA equipment, to increase reliability and reduce hardware size drastically, LSI-implementation have been carried out for a synchronization unit (baseband signal processing part)[3]. By extraction of basic functions and assigning these functions to each LSI optimally, general-purpose six kind TDMA LSIs have been successfully developed(Table 3). By the advent of these LSIs, hardware size of TDMA equipment has been reduced to one - fourth of conventional TDMA equipment and the cost has been reduced drastically in proportion to hardware reduction(Fig. 6)[4]. It is noteworthy the developed six kinds of general-purpose LSIs can be applied various types of TDMA equipment with different frame formats, bit rates and so on. These are made possible by software parameter setting for each LSI and in-planted parallel processing capability of LSIs for higher clock rates than 25 MHz. For example, by employing QPSK modulation scheme, 200-Mb/s TDMA equipment can be very easily developed by four-parallel processing of the developed LSIs with drastically reduced hardware complexity and development time.

3 SATELLITE VIDEO COMMUNICATIONS SERVICES (SVCS)[5]

The SVCS is another version of the MAC-Net which provides NTSC video signals (4.2 MHz bandwidth) and voice signals between dedicated users. The SVCS has similar properties with the SDCS and satellite channels are available according to the on/off of "s" bit. In this SVCS, a relocatable transmission link is easily established among small earth stations. In the commercial systems, transportable earth stations have been used for transmission to provide flexibility and mobility to the systems since relocatable video transmission demands are very high.

There are increasing demands for video transmission such as intra-company video conferences with higher encryption in addition to satellite news gathering. To meet these demands digitalization of transmission links are under development, focused on high performance demodulators and forward error correction. For the demodulator, three monolithic ICs employing high speed silicon bipolar devices, two LSIs for baseband signal processing and automatic frequency control(AFC) automatic phase control (APC) have been developed in addition to three thick film ICs for automatic gain control, alarm detection and so on. Moreover, to achieve high quality signal transmission and to realize compact earth stations, a universal code rate Viterbi decoder has been developed with reduced hardware and general usability. This Viterbi decoder has been implemented in two full custom CMOS LSIs[6]. These LSIs are called NUFEC (NTT Universal FEC) and can operate up to a clock frequency of 25 MHz and various coding rate ((N-1)/N) FEC are easily realized by combining two types of LSIs. Moreover, two master slice LSIs have been developed for 32 Mb/s DPCM video decoders. A configuration of a digital video receiver and NUFECs are shown in Fig. 7 and 8.

EXamples of SDCS and SVCS users are shown in Table 4 and 5.

4 INS SUBSCRIBER NETWORKS

For realization of digitally integrated networks, INS, the most costly part is rural area which has sparse traffic. Therefore, to construct rapidly cost effective INS in nation wide, the networks must be supplemented with satellite networks. The satellite INS subscriber networks will offer transmission lines with (2B + D) interface between subscribers and the nearest telephone offices to which the called party is connected. From the view point of calling, users will not perceive the existence of satellite links: fully a part of digital public switching networks, INS.

5 OTHER PLANNED SERVICES

In addition to the government owned communication satellites, private company-owned communication satellites will be launched late 1988 and early 1989. By using these satellites, various types of direct user services such as CATV signals transmission, intra-company video transmission and so on are being planned.

6 CONCLUSION

This paper has summarized present satellite business services networks offered by NTT and future networks under development or planned in Japan. By using flexible and rapid network construction capability and other superior performances of satellite communications, more and more services will be offered by satellite communications.

ACKNOWLEDGEMENT

The authors would like to thank Dr. H. Fuketa, Executive Manager of Radio Communications Laboratories Dr. H. Yamamoto, Executive Manager of Communications Satellite Department of NTT and Dr. N. Ishida of NTT Network Systems Development for their guidance and useful discussions.

REFERENCES

(1) Y. Morihiro: "Satellite Digital Communication System for New Business Use", JTR, 26, 4, p.270, 1984.
(2) Y.Morihiro, H.Nakashima and S.Kato:"Satellite Digital Communication Services (SDCS), Review of the ECL Vol. 35 No.2 1987
(3) S. Kato, M. Morikura and M. Umehira:"TDMA System for Satllite Digital Communication Service", Review of the ECL, Vol. 35 No. 2, 1987
(4) S. Kato, M. Morikura, M. Umehira, K. Enomoto and S. Kubota :"General-purpose TDMA LSI Development for Low Cost Earth Station", ICC'86, 1987
(5) S. Yamakawa and K. Hidaka:"Video Distribution System by Using CS-2", JIECE Tech. Group SAt87-6, 1987 (in Japanese)
(6) S. Kubota and S. Kato:" A Proposal of Universal Coding Rate Viterbi Decoder - A Novel Encoding and Decoding Method for High Coding Rate FEC", ICC'87, 1987

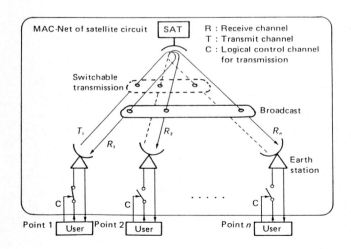

Fig. 1   MAC-Net configuration

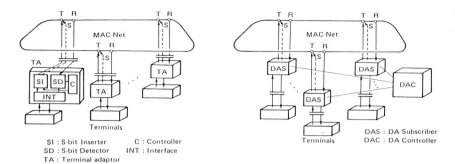

Fig. 2 MAC-Net PA service

Fig. 3 MAC-Net cericuit switched service

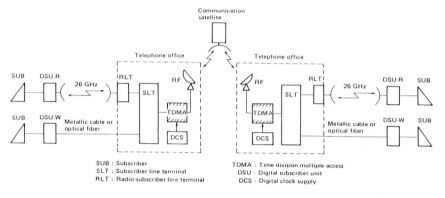

Fig. 4 SDCS system configuration

Table 1 Major system parameter

Freqeuncy	30/20GHz
Multiple access	TDMA
Earth stations	Reference stations: 2 Traffic terminals :50
Burst synchronization	Reference stations:closed loop Traffic terminals :feed-back loop
Modem	QPSK-coherent demodulation
FEC	Convolutional encoding (rate:½ constraint length:4)/Viterbi decoding
Clock rate	24.528MHz
Transmission capacity	160ch/transponder
Service bit rate	64, 192, 384, 768, 1536, 6144 kb/s
TDMA-SLT interface	2.048, 8.192Mb/s

Table 2 Major earth station parameter

Antenna (Offset Cassegrain)	Diameter		4.2 m φ
	Gain	TX	59.3 dB
		RX	55.4 dB
HPA (Klystron)	Output power		200 W
LNA (GaAs FET)	Noise temperature		330 K

Fig. 5 SDCS earth station

Table 3 General purpose TDMA LSI

CMOS master slice LSI(2μm)
Speed : 25MHz(max)

Function	Main Functions
1 UWD LSI	Unique word detection
2 ELB LSI	Elastic buffering
3 VTB LSI	Viterbi decoding
4 BCD LSI	Burst combining & dividing
5 TMC LSI	TX & RX timing control
6 CEB LSI	Compression and Expansion buffer control

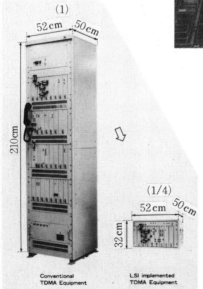

Fig. 6 LSI-implemented TDMA equipment

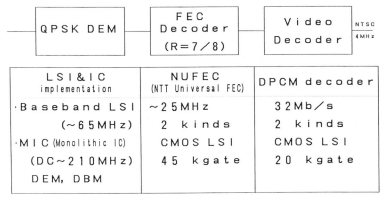

Fig. 7 Digital video receiver

Table 4 Example of SDCS user

User	Service bit rate	Access point	Mode/Use
A company	192Kb/s×2	Tokyo Osaka	
	64Kb/s×1	Tokyo Fukuoka Sendai Sapporo	Random access /computer net
B company	192Kb/s×1	Tokyo Nagoya	One way/ News delivery
C company	1.5Mb/s×2	Tokyo Fukuoka	Both way/ Video conference

Table 5 Example of SVCS user

User	Service class	Access point Transmit	Access point Receive	Mode
A company	NTSC (4MHz)	Heiwajima	4 cities (Kiryu etc)	Point to multi -Point
			4 cities (Hamanako etc)	
B company		Taira	3 cities (Fukui etc)	
		Kokura	6 cities (Matsuyama etc)	
C company		Tokyo	Fukuoka	Point to Point

NUFEC TYPE1 LSI
(R=1/2·K=7 Viterbi decoder)

NUFEC TYPE 2 LSI
(Path memory circuit)

R=3/4 : TYPE1×1 TYPE2×1
R=7/8 : TYPE1×1 TYPE2×2
R=15/16 : TYPE1×1 TYPE2×3

Fig. 8 NUFEC LSI for FEC

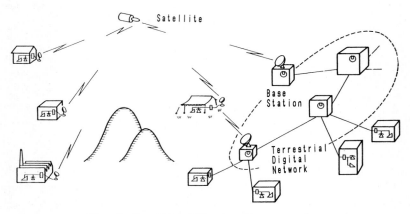

Fig. 9 Satellite INS subscriber network

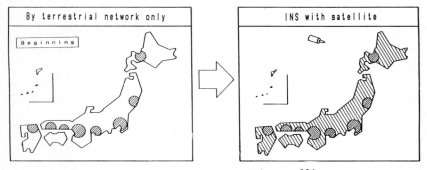

Fig. 10 Expansion of INS by satellite

ERCOFTACS PROJECT: A SATELLITE NETWORK FOR INTERACTIVE
2 MBPS COMMUNICATIONS

Sergio Benedetto* and Sebastiano Tirro'**

* Dipartimento di Elettronica, Politecnico di Torino,
Corso Duca degli Abruzzi, 10129 TORINO, Italy
** Telespazio, Viale Bergamini 50, 00136 ROMA, Italy

Summary ERCOFTACS is the acronym for a European project aiming at establishing an excellence Center for research and development of computer simulation of flow, turbulence and combustion research. The very large computing power available in the center (10 GFlops) will be made available to several subcenters located in various European countries through a high speed satellite telecommunication network. In this paper, preliminary hypotheses about the structure of the network are presented.

1. INTRODUCTION

Satellite communications are the only possibility for the implementation of high speed networks in the today's European reality. Experiments of data transmission at 2 Mbps have been succesfully performed as early as 1980 using the European experimental satellite OTS [1]. This paper discusses the possible implementation of an operational 2 Mbps network serving a European reasearch community.

2. ERCOFTACS: A BIT OF HISTORY

ERCOFTACS (European Research Center On Flow Turbulence And Combustion Simulation) is a project which started in early 1986 as the result of a common concern of a group of scientists and engineers from Universities, Research Centers and Industries with regard to the future competitiviteness of the European high technology industry involved in large scale, computer-based design with fluid flow turbulence, combustion and associated phenomena.
Aerospace, automobile, nuclear, chemical and metallurgical industries are related to the project.

The project proposes that a European Center of Excellence for Research and Development of Computer Simulation of Flow, Turbulence and Combustion be established and utilized by the participating countries.
The Center should be composed of
- a very large computer center containing a number of supercomputers to provide sufficient computational power to generate valid and advanced direct fluid-flow, turbulence and combustion simulations;

- a research institute of software specialists and experts in the field of turbulence and combustion sciences, applied mathematics and numerical analysis;
- a network of regional/national "satellite" sub-centers in the participating countries, equipped with computer hardware enabling them to be linked to the main computer center through high speed data flow connections.

It is envisaged that the funding of the Center would be largely by Governments, eventually through the European Community. Partial support by the industry can also be envisaged. Commercial use of the supercomputer facilities at the Center by industries and universities would substantially help the running costs of the Center.

The project is presently at the stage of preparing the final document containing detailed research programs for the first five years, the precise structure of the Center in terms of computer power, personnel and related costs, the costs-benefits analysis. This document will be presented to the Governments, National Research Councils and European Community to obtain the start-up funding. Encouraging preliminary colloquia have already taken place. In any case, being the project still in progress, the information contained in this paper should be considered as preliminary and subject to variations in the future developments.

A rough estimate of the annual costs of the entire project, with the exclusion of the sub-centers, is in the order of 25-30 MECU. It is foreseen to spend 1.5-2. MECU for the telecommunication network.

3. TELECOMMUNICATION USER REQUIREMENTS

The user requirements for the telecommunication network stem from the particular applications of the computing power located in the main Center on behalf of the regional sub-centers. It is initially foreseen to have a file transfer traffic originated from batch simulation programs sent to the main Center and big bulks of data obtained as results of simulations to be received and processed by the sub-centers.

To give an idea of the required data transmission rates, it must be noticed that a typical 3-dimensional fluid dynamics or combustion simulation problem can be characterized by the following parameters:

numbers of time steps	1000
spatial resolution	100x100x100
number of unknowns/node	6
number of matrix elements/node	162.

In order to avoid delay due to data latency, all operands need to be brought into first level memory, This makes the required size of the memory shared by the processors to be of the order of 8 Gbytes. Further considerations about the number of operations per time step lead to the conclusion that, for such a program to be executed in one hour of CPU the required computing power is of the order of 10 Gflops. Though these requirements cannot be met by any existing mono or multiprocessor system, they have inspired the design of the next generation of supercomputers, like CRAY 3, ETA 10 and the system which will result from the Japanese supercomputer project.

As concerns the size of data files to be transferred from the main Center to the subcenters, it can vary from a few Gbits to a hundred of Mbits depending on wether the entire raw bulk of the simulation results have to be transferred or a preelaborated version of them, in which only the most relevant informations are kept. At a transmission speed of about 2 Mbps the above figures lead to telecommunications session durations ranging from a few minutes up to hours. All in all, the choice of 2 Mbps seems a reasonable trade-off between users needs and costs, at least in a first experimental phase of the system operations.

As the communications are required essentially between the main Center and the sub-centers, a star topology of the network is envisaged. If available at very low extra-costs, a fully meshed network capability could bring some marginal benefits, like direct communication between two sub-centers needing to exchange data without passing through the main Center. Bidirectional links are required, with full-duplex capability. The largest transmission speed is required only in one direction, whereas the other connection could allow a speed of 64 Kbps as a return channel for protocol needs and/or for interactive applications that are foreseen in a second phase of the project.

4. THE TELECOMMUNICATION NETWORK

Around each regional or national subcenter, local area networks will allow communication exchanges between the center and users located in the neighborhoods of the subcenter. The technology will be ISO 8802.3 compatible (Ethernet type). Full connectivity between buildings and between remote subcenters is offered by a common backbone infrastrucuture. The backbone network acts as a bridge beteweeen the attached LANs and makes them to form a single logical LAN at the ISO-OSI level 2. The backbone network utilises fiber optic links for intrabuilding communication up to 140 Mbps capacity. As far as the 2 Mbps backbone network is concerned, the choice is restricted to a satellite-based network. In fact, neither terrestrial telecommunication facilities at the speed and with the required area coverage exist at this moment, nor are they foreseable in the near future. Only part of the network, at a regional or national level when available, could be supported by terrestrial links. The number of earth stations that will be required is of course dependent upon the availability of terrestrial links in some areas. So far, it can be foreseen that the number of earth stations be in the order of 10-15. A preliminary proposal of ubication is shown in Fig.1.

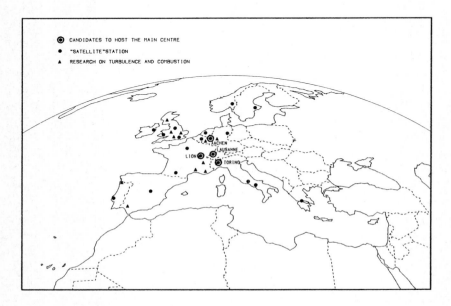

Fig.1 A preliminary proposal for ubication of the main center and subcenters

5. A POSSIBLE SPACE SEGMENT CONFIGURATION

5.1 Choice of the satellite

The required area coverage emerging from Fig.1 is well matched by the EUTELSAT I SMS beam. The receiving (RX) contours are shown in Fig.2, each successive concentric area having a 1 dB loss with respect to the adjacent inner one. The transmitting contours are only slightly different and are not reported here. Comparing Figs.1 and 2 it can be seen that most of the earth stations lie inside the -1 and -2 dB areas, and that no one is outside the -4 dB area. The situation would be even better with the foreseen EUTELSAT II SMS contoured beam coverage, shown in Fig.3.

An alternative to European satellites is represented by the INTELSAT V A IBS Atlantic 307° east satellite, which provides the spot beam coverage

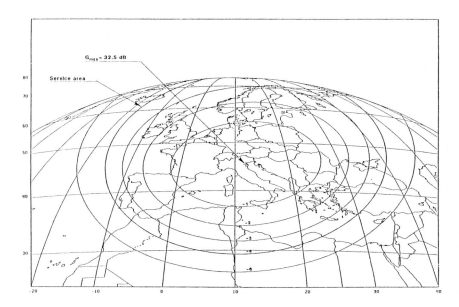

Fig.2 EUTELSAT I SMS beam - Rx contours

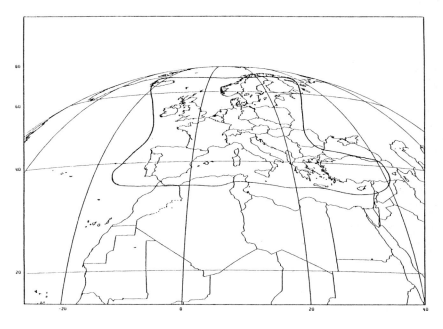

Fig.3 EUTELSAT II SMS SCPC coverage area

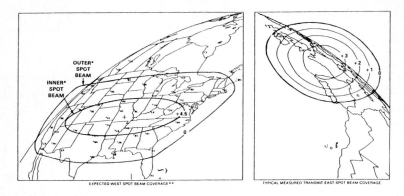

Fig.4 INTELSAT VA IBS spot beam coverage

shown in Fig.4. Although allowing a direct connection between ERCOFTACS centers and similar supercomputer centers located in the United States, its European coverage seems to require, in some regions, too expensive earth stations.

5.2 The access scheme

Considerations about the foreseen traffic made in Section 3 lead to a high channel occupation combined with low volume of traffic. This requires the availability of a full-time leased 2 Mbps channel. As to the access scheme, use of a random access is prevented by the aforementioned traffic considerations. Moreover, the average length of a session, of the order of 10-15 minutes, seems to discourage the use of an SCPC-based reservation scheme, due to the exceedingly long procedures to set up each connection.
The best solution to the problem at hand seems to be a TDMA channel access scheme with demand assignment based on a reservation procedure. Besides the previous considerations, a TDMA scheme would guarantee a high flexibility to the system. This aspect is very important for our application, which will require an initial test period to precisely assess the users telecommunication needs, with adaptive system readjustments.

The master station with central monitoring and a specialized network management will be located near the main center.

5.3 Earth terminal standard

Use of EUTELSAT SMS standard 1 and/or 2 are foreseable for both the master and peripheral stations. Their characteristics are as follows:

Standard	Antenna [m]	HPA Power [w]
1	5.7-6.5	100
2	3.7-4.0	300

Both standards can use GaAs FET low noise amplifiers. The achievable goal is the standard grade, i.e a bit error rate of 10^{-6} for 99% of the year among all stations of the network. The uniformity of earth stations standard could allow implementation of a fully-meshed network option at a marginal cost.

5.4 TDMA System structure

TDMA terminals with a flexible coverage of the 2-20 Mbps range are available on the market. For the application at hand, suitable characteristics would be:

  20 ms typical frame duraton

  4 μs guard time between adjacent bursts

  2.4 Kbps service channel from idle peripheral stations to master station allowing:
  -station synchronization
  -station and transmission monitoring
  -service communications

  64 Kbps channel between active peripheral station and master station to implement a return communication channel

  1.5 Mbps channel between active peripheral station and master station to implement the direct communication channel.

5.5 Network management

It is expected that the network management be part of ERCOFTACS organization located in the main center. It will interface both with the network and the users of ERCOFTACS supercomputers, agreeing with them a reservation scheme and, consequently, a time plan production and a cost sharing for the space segment. Initially, weekly time plans can be foreseen.
The functions of the network management with respect to the network will consist of time plan implementation, earth terminals control and monitoring, and transmission quality monitoring.

6. INSTITUTIONAL ASPECTS

The satellite network requires the leasing of the space segment and the earth terminals. As is known, direct negotiations between EUTELSAT organization and users are not possible. They must be mediated by the PT administrations of the participating countries. It is expected that the leasing of the satellite channel be agreed between ERCOFTACS and the PT administration of the country hosting the main center, leaving to the network management the issue of the cost sharing among users. As far as the earth stations are concerned, individual agreement between subcenters and national PT administrations will be probably required, although a single point of contact between ERCOFTACS and a designated committee representing all the PT administrations would be highly desirable. As a matter of fact, this would simplify and shorten technical discussions, contract negotiations and would allow uniformity of the tariffs for the earth stations leasing.

6. COST ESTIMATES

Being the project still under its way, only a rough estimate of the costs is possible at this stage. Based upon the use of EUTELSAT I SMS, the cost of the satellite channel is in the order of 450 KAU/year (Kilo Accounting Unit per year), whereas the yearly service charge for each earth terminal, inclusive of installation, maintenance and operations, is estimated in the order of 150 KAU.

REFERENCES

[1] A. Marzoli, "Telespazio's contribution to SMS-type experiments via OTS," EUTELSAT SMS Symposium, Rotterdam, June 1985

[2] European Research Centre on Flow, Turbulence and Combustion Simulation, proposal for ERCOFTACS, London 1986

[3] A. Endrizzi and R. Gruber, Technical Report ERCOFTACS, Lausanne 1987

THE EUROPEAN SATELLITE BUSINESS SERVICE: EUTELSAT SMS

Ms. Martine Papo, Associate Strategic Planner

EUTELSAT, Tour Maine-Montparnasse, 33 avenue du Maine
75755 Paris Cedex 15

Ariane's launch was a full success this morning. This is great news for the entire satellite telecommunications industry which will resume its high level of activity. For EUTELSAT, it means that within a few weeks, the long expected three operational satellite configuration will be in place. The EUTELSAT space segment presently consists of the EUTELSAT I F1, F2 and soon, F4 satellites as well as capacity leased on the French TELECOM 1 satellite for certain international applications.

The EUTELSAT I F1, F2 and F4 satellites were launched on 16 June 1983, 4 August 1984 and today, 15 September 1987. F1 and F4 are to be used exclusively for distribution of TV programmes within Europe on an international or domestic basis. F2 is being used for the distribution of the European Broadcasting Union Eurovision TV programmes, for domestic TV leased services and transmissions on an occasional basis, for satellite multiservices system (SMS) applications and for telephony services.

EUTELSAT I F2 Satellite - allocation of capacity per service

EBU	2 transponders
TV lease	1 transponder
Occasional use	2 transponders
Telephony	3 transponders
SMS	1 transponder

Current operational plans call for EUTELSAT I F1 to be replaced by F5 when the former reaches its end of life. The F5 satellite will be ready in September 1987 and will be kept in storage until its launch in the middle of 1988.

EUTELSAT Satellite Type	Location (°E longitude)
I	7,10,13,16
II	7,10,13,36

Launch Schedule

EUTELSAT I F4	15 September
EUTELSAT II F5	June 1988
EUTELSAT II F1	January 1990
EUTELSAT II F2	March 1990
EUTELSAT II F3	September 1990
EUTELSAT II F4	1991

EUTELSAT's endeavour is to react quickly and efficiently to new customer demands, while at the same time provide the basic telecommunications services to all its users on a non-discriminatory basis. With this in mind, EUTELSAT has taken a number of actions to ensure that its facilities are efficient, economic and well-adapted to the business customers' requirements.

BUSINESS SERVICES

EUTELSAT has established its Satellite Multiservice System (SMS) for business service applications in Europe. It is designed to operate with small-dish earth stations that can be installed close to, or on the users' premises, thus avoiding the need for long terrestrial extensions. Use of the 12 GHz downlink frequency band greatly facilitates the location of earth stations everywhere within the system coverage area.

The system offers unidirectional or bidirectional, point-to-point or point-to-multipoint digital circuits of various bit rates, alloted on a full-time or part-time basis, in a fully digital integrated manner, of speech, data, text and compressed video images.

The EUTELSAT SMS comprises two distinct networks with different transmission methods, which at present use different space segments facilities. One network uses the 14/12 GHz SMS package on-board the EUTELSAT I satellites, and the other uses part of the 14/12 GHz capacity of the French TELECOM 1 system, made available to EUTELSAT under an arrangement with the French Administration.

The service areas of the present EUTELSAT I and TELECOM I networks are shown in Figure 1.

SMS capacity became available in 1984 with the successful launches of EUTELSAT I F2 and TELECOM 1A. The early use of this capacity, however, was lower than anticipated and therefore a number of actions were rapidly taken in order to stimulate the demand for this capacity.

1. The SMS tariff structure and charge levels were completely revised with an overall *30% reduction of the tariffs*.

2. Efforts were made to facilitate the utilization of *closed network* business applications, and the appropriate booking and charging rules were approved for these applications.

3. Finally, approval was given for the introduction of smaller, simpler and cheaper earth stations with the adoption of *second and third standards* for SMS earth stations.

SMS Standard Earth Stations

	antenna size	G/T
Standard 1	5 meters	29.9 dB/K
Standard 2	3.5 meters	26.9 dB/K
Standard 3	2.4 meters	23.9 dB/K

The largest utilization of *very small aperture antennas (VSATs), or microterminals* in the EUTELSAT system is also under study. It is expected that the use of these antennas will increase considerably in Europe in the next few years and the Organization is preparing to facilitate their implementation on a large scale.

The advent of VSATs on the market enlarges the potential customer base of the satellites. It is estimated that 1000 companies could benefit from space communications in Europe today.

It is, therefore, expected that the one transponder available on EUTELSAT I F2 for SMS use will be saturated by 1990, and that a second one will need to be introduced at that time.

It is envisaged that by 1991, the EUTELSAT II satellites take over the SMS services from the TELECOM 1 and EUTELSAT I satellites; the two networks would then co-exist on the same satellite. The service areas of the

EUTELSAT II satellites for the SCPC and TDMA open networks are illustrated in figure 2. At the end of 1994, the number of transponders allocated to SMS is expected to be five.

EUTELSAT SMS offers a basic open system network but closed user networks may also be developed according to user needs. The FDMA utilisation of the transponder permits an easy allocation of band and power to create these closed networks according to requirements.

THE OPEN SYSTEM

In the open system, all transmissions fully comply with specified character- istics concerning modulation, coding, framing, earth station performance, etc. This permits equipment standardization and facilitates the inter-linking of system users and the integration of satellite links in the public terrestrial digital network.

In the EUTELSAT I SMS network, the earth stations access the satellite in FDMA/SCPC mode, while the TELECOM 1 part uses a 24.5 Mbit/s TDMA method of access.

Although of different technical design, the two networks offer essentially identical service capabilities:

Service Characteristics:

a) customer bit rate

2.4, 4.8, 9.6, 48 kbit/s
64 and n x 64 kbit/s, where n = 1, 2, 3 ... up to 31

b) earth station standard

Three different standards of earth station can be used in the EUTELSAT I-SCPC network:

- the standard 1 earth station which requires a 5.5 m dish,
- the Standard 2 earth station requiring a 3.5 m antenna,
- and the standard 3 requiring a 2.4 m antenna without tracking, which is of particular interest for applications on the customers' premises.

c) allotment modes

Full-time, subscription, occasional

d) circuit configuration

point-to-point, multi-point

CLOSED USER NETWORK SERVICES

Special customer requirements led to the consideration of "customer-tailored" closed networks, for which transmission parameters and earth station characteristics would depart from the specified SMS system and provide a better response to the needs of individual users or closed user groups. Virtually any modulation or accessing technique can be envisaged, as well as a choice of user bit rates and link quality.

Applications using microterminals are a typical example of closed networks; in Europe today, such a network would in most cases have the following characteristics:

- a star network configuration
- hub station owned by PTT
- remote microterminal owned by user if receive-only
- remote microterminal owned by PTT if receive/transmit

The first operational network using microterminals in Europe uses the EUTELSAT SMS space segment. Its owner is POLYCOM, a French company partly owned by the French press agency AFP, which distributes news in text and images on a permanent basis to nearly one hundred terminals.

EUTELSAT II PROGRAMME

The procurement contract of a second generation of EUTELSAT satellites was awarded in May 1986 to the European consortium led by Aérospatiale. Three satellites were ordered immediately, for delivery starting in the autumn of 1989, with an option of up to five additional satellites. The purchase of a fourth flight model was confirmed in June 1987.

A four operational EUTELSAT II satellite in-orbit configuration is planned. F1 and F2 providing mainly non-preemptible high e.i.r.p. leases; F3 and F4 operating SMS, telephony, service to EBU and preemptible leases.

The EUTELSAT II satellites will provide substantially improved communication performance and capacity compared to the first generation.

Each satellite will have 16 transponders, each one radiating 50 W RF permanently, including during eclipse. A high reliability is achieved through the use of a ring redundancy scheme whereby RF channels are arranged in two groups of eight active channels out of 12. With this scheme, up to four failures in each group can be tolerated without loss of capacity.

A contract was placed in June 1986 with Arianespace for three satellite launches, plus two option flights. EUTELSAT has felt it to be more prudent to adopt a policy of also having a back-up launcher for the EUTELSAT II satellites to ensure the timely availability of satellites and continuity of services. The Atlas Centaur launcher of General Dynamics was selected. A contract was signed for the launch of two EUTELSAT II satellites and an option for the launch of a third satellite, in the event of the unavailability of the Ariane launcher.

CONCLUSIONS

For business communications, satellites at 14/12 GHz offer ideal opportunities:

- wideband channels, suitable for digital as well as analogue transmissions
- immediate availability within coverage
- access to/from small dishes
- flexibility to reassign capacity between locations or services
- broadcasting capability.

These unique features, however, can only be taken advantage of, if the proposed satellite network offers:

- satifactory in-orbit protection
- sound system control and management
- availability at European level
- continuity in the long term.

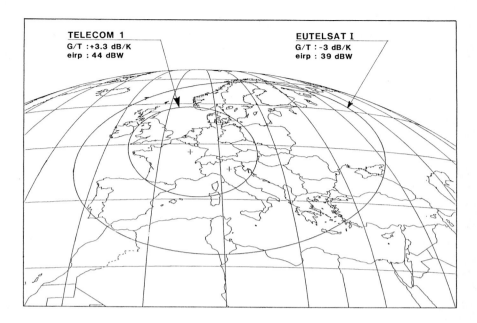

EUTELSAT I / TELECOM 1 SMS SERVICES AREAS

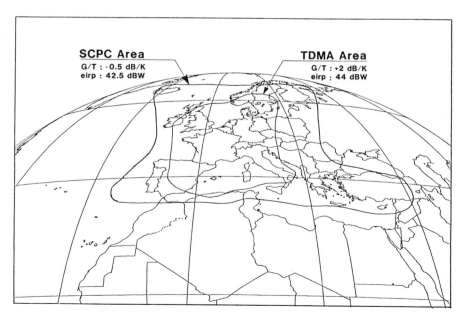

MINIMUM SMS TDMA/SCPC COVERAGE AREAS ON EUTELSAT II

BUSINESS SERVICES VIA SATELLITE: ITALIAN PROGRAMS

S. DE PADOVA*, A. PUCCIO°

*SIP - Rome (Italy)
° TELESPAZIO - Rome (Italy)

ABSTRACT

In countries with a developed network the satellite medium is very useful mainly for two reasons:
- Quick implementation, at relatively low cost and financial risk, of a business network with the desired connectivity.
- By means of a high degree of penetration of the satellite system, the whole network is given the flexibility required by an advanced network management

This paper deals with the technical programs leading to the implementation of a domestic satellite network and with the implementation plans, mainly in the short terms, based on the use of existing business satellite systems IBS and SMS.

1. INTRODUCTION

In the today telecommunications world the emphasis has been shifted from the system, technical and technological aspects of the networks towards the services to be supplied by the networks to the users.

The use of informatic and videomatic tools is becoming more and more popular both at business and residential subscriber level. This will dramatically impact on the communication needs of the people and of the organizations.

Therefore the present problem is not meeting the subscriber needs in the near future but defining these needs well in advance to the developments of the networks. In practice this is a huge marketing problem.

In the light of the foregoing the satellite systems are very attractive also in developed countries to implement quickly, at relatively low cost and financial risk, a domestic network for business services providing the desired connectivity. This network can be used for an early provision of business services and as a test market tool. At a later stage satellite systems can be fully integrated in the coming ISDN (Integrated Services Digital Network) to exploit their well-known flexibility and to provide wideband digital links either point to point or point to multipoint that can be assigned on demand, either on reservation or on call-by-call basis.

Short terms implementation plans of a domestic business network in Italy are also based on the available capacity in the INTELSAT and EUTELSAT systems.

The Intelsat Business Services (IBS) system is operational via Intelsat V since October 1983 to satisfy a wide variety of the business users emerging requirements. The IBS allows the operation both of open and closed networks: in the latter case only RF specifications need to be met. A wide range of transmission rates is possible (from 64 Kbps up to 8.448 Mbps); "basic" (BER better than 10^{-6} for more than 99% of the year) and "super" (i.e. ISDN standard) quality is offered.

The Eutelsat SMS (Satellite Multi Services) system uses the same standard as the IBS system; a further system is offered via the Telecom 1 satellite over a limited coverage of Europe.

Medium terms implementation plans of the domestic business satellite network in Italy will be also based on the exploitation of the capacity offered by the Italsat system. A significant technical and technological development effort has been devoted to the Italsat programme in order to implement a system fullfilling the requirements of the Italian network.

2. THE ITALSAT PROGRAMME

Italsat is a domestic experimental - preoperational satellite to be launched in the year 1990; the related system shows the main features* listed in the following /1/ /2/ /3/:

- Use of the 20/30 GHz bands where the large bandwidth (2.5 GHz) available to the fixed satellite services can be exploited to implement high capacity systems and therefore to recover the high cost of the common parts in the earth and space segment;
- The capacity is offered by means of a multibeam coverage system with six circular spots 0.5° wide covering all switching nodes at district level of the future italian digital network and a global beam system with an elliptical beam 1.5° x 1.1° wide;
- The multibeam system provides a capacity of 6 regenerative (coherent QPSK modemodulation) 83 MHz transponders, interconnected by an onboard baseband switching matrix and acceded by an SS-TDMA (Satellite Switched TDMA) technique with a 147.5 Mbps bit rate;
- The global beam system provides a capacity of three 36 MHz transparent repeaters allowing a TH-TDMA (Transponder Hopping TDMA) access at 24.5 Mbps with coherent QPSK modulation; obviously this capacity is compatible also with more conventional access techniques and with the analog transmission;
- use of roof-top antennas (diameter = 3.5/5 mt according to the climatic zone) to be installed in general at the switching centres premises; a power of about 50 W and 100 W is in general requested to the HPA system in the multibeam and global beam respectively;

---

* The results given in this section have been obtained in the frame of a system study performed by Telespazio under contract to the Italian National Council for Research (CNR). The ideas reported here are not necessarily those of CNR.

- A modular frame with a 32 msec length is adopted in the global and multibeam system and only two modules (e.g. burst sizes) are possible, e.g. one or four 32 Kbps channels per burst;
- Use is made both of DSI and ADPCM techniques in the part of capacity handled by four channels per burst modules where it is implemented the demand assignment on a call by call basis at destination spot level (e.g. traffic originating in one station and related to all the stations operating in a spot is carried by means of a single trunk group);
- Use is made of DDI (Direct Digital Interface) to handle the part of the frame with one channel modules, this capacity being used to carry the traffic overflowing from the capacity handled with DSI and ADPCM; in this part of the frame it is implemented a full demand assignment technique;
- Dynamic traffic re-routing is feasible: the earth stations or the associated terrestrial exchanges transmit the activity status of their trunk groups to the control centre which in turn updates the assignment of the satellite system capacity and therefore updates the time plan of the stations and the onboard matrix.
Traffic re-routing can be implemented in order to cope also with periodic traffic matrix variations;
- Both global and multibeam systems have been dimensioned according to the quality criterion:

    BER better than 10^{-6} for more than 99% of the year

    and according to the availability criterion due to the propagation effects:

    BER better than 10^{-3} for more than 0.2% of the year.
- Both the multibeam and global system has been designed to offer telephony, business services and TV programs transmission.

2.1. Business services

Business users could be connected to the Italsat earth stations by means of 2 Mbps links with synchronism and signalling signals in the TS 0 and 16 respectively. These links will be provided on dedicated lines and therefore will not undergo to the switching function of the exchanges.

Data Services

Switched telephony and data (nx64 Kbps where $n = 1 \div 4$) services can be offered in an integrated manner; the system study has highlighted that, because of the low volume of the data traffic, it is convenient to bundle this traffic with the telephony one in order to increase the efficiency (erlang/circuit) of the data circuits. No overdimensioning results because it is possible to size the trunk groups on the basis of the telephony grade of service (if n is limited to a low value, the resulting blocking probability of the data calls is still acceptable).

These data services can be offered on a reservation or a switched basis. In the latter case, when an incoming data call is detected in the satellite system by means of the signalling analysis, the ADPCM function is obviously disabled in the seized circuit; DSI also must be disabled because the speech detector, designed for the detection of the voice, could misinterpret certain data sequences as an absence of speech activity and this fact can cause the release of the DSI channel during the connection.

Data streams are transmitted using the 32 Kbps channels available in the system. Multislot services, e.g. nx64 Kbps services, require that the TSSI (Time Slot Sequency Integrity) is maintained at the DTE (Data Terminal Equipment) level. This means that the order in the frame of the n Time Slots (T.S.) of the digital stream is not changed and there is no differential delay between these Time Slots.

At present no definitive recommendations have been issued by CCITT. Studies are in progress about the opportunity that the TSSI be guaranteed by the network or recovered by properly modified Data Communication Equipments (DCEs). In the former solution it is needed in the exchanges to implement a proper path finding algorithm, to double the T-stages memories in order to avoid the differential delay and to route the various TS's on the same transmission resource. In the latter solution, for instance, a test pattern can be transmitted during the link set-up phase in order to align the transmitting and receiving terminals (i.e. to recover the TS's order and the possible differential delay). Anyway Italsat will operate according to the specifications that will be recommended for the terrestrial network.

As to the 64 Kbps data, TSSI is maintained because the DSI/ADPCM equipment, the function of which is disabled in the data circuits, transmits in the proper order the incoming data stream by means of two consecutive TS's. However in the case of 64 Kbps data handled by a DDI, the T-stage memories in the DDI need to be doubled.

As to the link performance, the error protection and encryption functions, if required to meet specific customer requirements, could be carried out either in the interface between the earth station and the terrestrial network or could be allocated to modified DTE/DCE equipment.

Videoconferencing

A videoconference can be defined as a substitute of a meeting rather than an upgrading of the telephony; accordingly the videoconferencing service is offered on a reservation basis.

No forecast data are available on the percentage of multipoint videoconferences; however a reasonable estimation is that 75% are point-point, 20% are point-two points and 5% point-three points.

It is possible to implement a switching protocol in case of multipoint videoconferences in order to save capacity mainly in the multibeam system /4/.

The video signals can be controlled in order to allow only the transmission of the new speaker video and, if required, of a further studio, either the previous or the preferred one. In addition a permanent capacity must be assigned to each earth station in order to transmit audio, data and codec-to-codec signalling, if the continuity of these signals is requested.

The switching protocol is based on the possibility to freeze the image on the screen during the reallocation of the space resources. No synchronous procedure is needed. On the basis of the signalling sent by the chairman, the master station sends freeze and afterwards switching commands to the earth stations and to the satellite, enabling or disabling the transmission of the new and old video signal respectively. After the reception of all the acknowledgments, the master station sends a fast update request to the new speaker codec, allowing all the codecs to recover rapidly the new image.

The total switching time is about 1.5 seconds and therefore it can be considered acceptable to the customers also because audio continuity is guaranteed during the switching of the video.

3. IMPLEMENTATION PROGRAMS

The use of satellite systems in Italy and the relevant application areas are differentiated according to the opportunities offered by the present or shortly operating satellites and by the future italian satellite Italsat.

As to the operational satellites, we are referring to the use of Intelsat and Eutelsat systems to provide digital high bit rate (64 Kbps - 2.048 Mbps) links for business services, as anticipation and integration of terrestrial network facilities.

In order to cope with the potential customer needs, the Italian Telecommunication Operating Company (SIP) has planned to install community earth stations in the main metropolitan areas of the country, where there is a high concentration of business organizations.

This choice derives mainly from a trade off between two opposite requirements: on the one hand the need to install the earth stations near the user premises because of possible lacks of terrestrial digital links and on the other hand the necessity to share the relatively high costs of the earth stations among several users. That, at least as to the costs of the present standard earth stations specified by the space segments Operators, Intelsat and Eutelsat.

The community earth stations will be the backbone to provide european and/or intercontinental digital links via satellite; however for peculiar customer requirements, solutions based on single user earth stations could be carried out. That, obviously, according to the evolution of the terrestrial digital network and to the opportunities offered by the use of non standard low cost earth stations and by new generations high powered satellites.

According to the planning, two standard 1 ECS/SMS community earth stations will be operational respectively in Rome and Milan by 1988 when the National Control Centre in

Fucino will be operational too.

Recently new systems for data applications have begun to mature. They are based on the use of very small aperture terminals (VSAT). Typically VSAT systems are configured in a "star network" with one large central hub station and numerous remote VSAT stations, installed at geografically dispersed sites. The hub station may be dedicated to one VSAT network or shared among several VSAT subnetworks. Moreover a System Control Centre, associated to the hub station, is responsible for the control and monitoring of the entire network.

The VSAT market has gone off in USA, where different systems are currently in use by large companies for one and two-way data trasmission. At present there are already several companies manufacturing VSATs and related subsystems, which are characterized by different satellite link protocols, transponder access protocols, network bit-rate, etc. Therefore, in general, these systems cannot interwork.

Recently also European PTT's have shown interest in the use of VSAT's and field trials are in progress. The common approach of the European PTT's is to use a community hub, shared among a number of separate customer networks; thus the high costs of the hub can be shared among a large universe of peripheral earth stations. According to this approach the Italian Telecommunication Operating Company (SIP) is planning to integrate two hubs in the SMS earth stations in Rome and Milan.

Moreover in order to cope with specific customer requirements, several projects, based also on the use of narrow-band TDMA for meshed satellite networks, have been carried out with the cooperation of Telespazio. For some of these projects, experimental field trials are in progress with the customers.

4. CONCLUSIONS

For the early start-up of the business services in Italy also the available capacity in the IBS and SMS systems is exploited to satisfy customer needs.

At a later stage the Italsat system, a domestic satellite network for telephony and business traffic, will be fully integrated with the future Italian terrestrial network.

REFERENCES

/1/ S. Tirrò - "The Italsat preoperational programme" - 6th ICDSC, Phoenix, 1983

/2/ F. Marconicchio, F. Valdoni, S. Tirrò - "The Italsat preoperational communication satellite program" Acta Astronautica, Feb. 1983

/3/ G. B. Alaria, S. De Padova, M. Tommasi, A. Vernucci - "System architecture, services and performances of the Italian domestic satellite Italsat" - ISS '87 - Phoenix

/4/ A. Perrone, A. Puccio, S. Tirrò - "Optimization of connection techniques for multipoint satellite videoconference" - Space Communication and Broadcasting 3 (1985)

USERS AND ECONOMICAL VIABILITY FOR THE TELE-X BUSINESS SERVICES

Lars Backlund, Head, Business Development

Swedish Space Corporation
Box 4207, S-171 04 SOLNA, SWEDEN

The first Nordic communications satellite, Tele-X, is planned to be launched in 1989. Tele-X covers Denmark, Finland, Norway and Sweden and is designed to provide new data and video services for the business community using small rooftop antennas and two channels of direct broadcasting. During spring 1985 the Swedish Space Corporation concluded a market and applications study for the Nordic countries with very encouraging results. For the purpose of the study inception of service was assumed in 1986 and a forecast for the number of stations, traffic and revenues was made for a 10 year period to demonstrate viability for a commercially operated system based on a series of Tele-X satellites. It was concluded that the number of data/video stations would grow to around 2000 during the 10 year period and that the system would earn significant profit for its operator.

1. THE INFORMATION SOCIETY - A DEMANDING TECHNICAL ENVIRONMENT

The evolutionary growth of the information society puts ever increasing demands on efficient and flexible telecommunications services. The number of office computers is increasing extremely rapidly in all types of companies and computers are penetrating all aspects of modern society. This leads to an increased demand for exchange of large volumes of information between different users. Paper is replaced by digitally stored information, which is currently growing in volume by 40 % to 50 % annually.

Filing, retrieving and dissemination of information and automation of processing, leads to increased efficiency and more rational ways of working.

The conception of the electronic office comprises a broad menu of new services. In addition to various forms of data communications, new types of conferencing services based on graphics and image communication are expected to become an important growth area. The videoconference may become an attractive alternative to some business travel. High speed facsimile may radically change the way of distributing drawings, images, print originals, newspapers etc. Education can be offered in new ways when geographical distances and the traditional lecture are no longer limiting factors. New telecommunications services for data and video represents a vast area for innovation and growth. The availability or lack of these services will presumably affect a nation's ability in worldwide competition in the same way as the availability of roads, railroads and telephones do. The various services can be characterized by different levels of maturity as illustrated in figure 1.

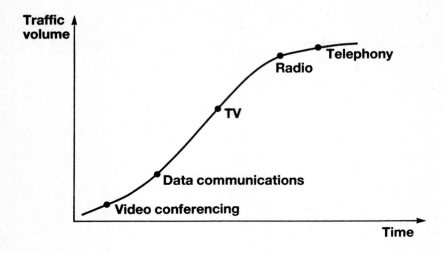

FIGURE 1
Maturity curve.

Telephony is a mature service which can not be expected to grow significantly with time. Radio is somewhat below on the curve and there is still room for more diversification of programs. Television has a substantial growth potential and the subscribers can be expected to have a significant interest if additional programming can be offered. Datacommunications is a relatively new service which in Sweden accounts for about 10% of the traffic in today's terrestrial network. The rate of growth dominates over other services and accounts for about 60% of the total traffic growth in the network.

Videoconferencing finally, is not commonly available today and finds itself in an early phase of growth and introduction. The new services will find their applications in the business community and their characteristics will require high data rates, flexible access and good geographical coverage. A telecommunications satellite system, featuring full interconnectivity between user-located earth stations, offers an efficient alternative to introduce the new services. From the moment of inception, service can be offered throughout the entire service area. A user can communicate immediately with everyone else having installed an earth station. Rural and remote areas are given the same priority as urban areas.

2. THE CONCEPT OF DATACOMMUNICATIONS VIA HIGH-POWER SATELLITES

Business communications via satellite was introduced in the USA in 1980 by the Satellite Business Systems (SBS) Company. Other systems have followed: the TELECOM-1 system in France, the IBS system offered by INTELSAT and the SMS system offered by EUTELSAT. These systems are all using satellite transponders with around 20 Watt saturated output power and are designed to be bandwidth-efficient. Hence relatively expensive and large Earth stations are needed. For the SMS and IBS systems this means 5-8 m diameter antennas in Europe with a total earth station cost exceeding 1 MUSD. Obviously that type of earth station is generally not well suited for installation at individual user premises. The high costs also leads to high fixed yearly charges giving high operating costs to the user even if the traffic volume in the beginning is limited. Thus, for this type of system, there is a high threshold in initial cost to be paid by the user before the possible benefits of high speed datatransmissions have been demonstrated to him. The limited success in finding customers for the systems mentioned above has been a strong indication that the threshold has been too high.

Obviously, another technical concept has to be implemented in order to attract customers. Key factors in this concept must be to decrease cost, facilitate equipment and adapt to traffic patterns required by the individual users. In a satellite system this can be done by increasing satellite power in order to decrease the size and cost of the earth stations, avoiding complex features like TDMA and dynamic assignment of satellite capacity.

One of the basic design goals when the Nordic Tele-X system was conceived in the beginning of 1980 was to introduce such a system operating with small and low-cost earth stations installed on user premises offering full interconnectivity between all users in the system. The system design was based on technology advances during the 70's which made it possible to implement these ideas.

3. NEW TECHNOLOGIES AVAILABLE DURING THE 70-IES

During the decade 1970-80 some important advancements were made in available technologies for spacecraft and earth station design and launching of large satellites. Large solar arrays generating electric power in the kW-range and high power spacecraft transmitting tubes with a saturated power exceeding 200 W were developed and qualified. Steerable narrow-beam spacecraft antennas capable of maintaining high-precision pointing were designed. Transponders operating in the 12.5-12.75 GHz band became available to take advantage of the exclusive allocation of that band to fixed satellite services. Using this band for business communications dramatically reduces the frequency coordination problems for the earth stations compared to the 4 and 11 GHz bands where the satellite service is shared with radio relay link services and consequently all Earth stations have to be coordinated with existing radio relay links to avoid mutual interference.

The development of Ariane and the Space Shuttle meant that physically big and heavy satellites could be launched into geostationary orbit with large solar arrays and large antenna dishes.

For the earth stations low noise receivers and solid state power amplifiers became available. Microcomputers made it possible to design advanced error correction, switching and access systems built into the individual stations.

When the Tele-X datatransmission system was conceived in the beginning of 1980, the new technology available was taken into account for the system design.

4. THE NORDIC TELE-X SYSTEM

4.1. The basis for the Tele-X system definition

The definition phase for Tele-X started out in 1979 with the conception of a scenario describing the information society and the associated telecommunications services that could be envisaged. A fairly long list of data and video services was compiled and an assessment of the profitability for the users to bring the new services into their organizations was made.

Even if a probable set of services was defined at the beginning of the Tele-X program the various services have been constantly analyzed and iterated from the start of the program. A lot of attention has been put on user aspects on the various services. As a result the interest has gradually focused on videoconferencing and data communications at 64 kbps and 2 Mbps.

A general goal formulated for the definition of the Tele-X system was that the system should be a complement, not a competitor, to other existing or planned datatransmissions systems. This was fulfilled by limiting the coverage to the Scandinavian area and by optimizing the transmission system for 64 kbps and higher.

Limiting the coverage to the Scandinavian area (figure 2) implies the use of a high-gain, narrow-beam satellite antenna. In combination with the 230 W tube, originally developed for TV-broadcasting, a very high flux density could be obtained on the earth, taking full advantage of the absence of a general flux density limitation for the frequency band used. In this way small 1.8 m antennas could be used for data rates up to 2 Mbps (figure 3).

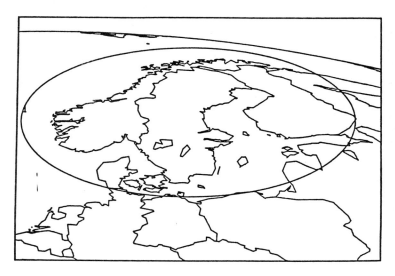

FIGURE 2
Tele-X coverage area. (BER <10E-6 for 99% of worst month)

FIGURE 3
1.8 m antenna

Another very important decision to be taken during the system definition phase was the choice between TDMA and SCPC for the transmission system. The inherently much lower price for SCPC stations as compared to TDMA was here a decisive factor. In addition to this an SCPC system can operate with more earth stations than a TDMA system. By the use of a demand assignment system for the SCPC-channels with an ALOHA-type of signalling, earth stations can access the system without the requirement for accurate, continuous synchronization that is necessary for a TDMA system and which limits the number of operating earth stations. Another advantage of an SCPC system is that the bandwidth for the Earth station receiver can be limited to what is necessary for a single 2 Mbps datachannel while the bandwidth in a TDMA system must correspond to the total capacity of the system. Again the use of smaller antennas for the SCPC system is facilitated.

4.2 System characteristics

The Tele-X communications system for data and video services consists of a satellite with linearized high power transponders, a traffic control station and a large number of small low cost earth stations installed at user premises (figure 4). The access system is able to handle up to 5000 earth stations. Up-link Power Control and Forward Error Correction is implemented in the system to maximize system efficiency and to decrease the vulnerability to intersystem interference and to facilitate satellite system coordination. The satellite will also provide two direct broadcast TV-channels.

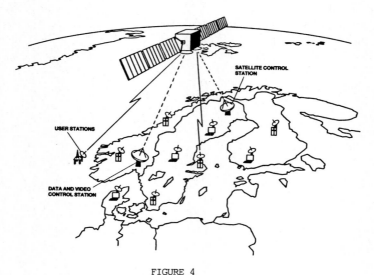

FIGURE 4
Tele-X communications system.

5. THE MARKET STUDY FOR TELE-X

5.1. Introduction

As one of the steps in investigating the viability of Tele-X and follow-on satellites in the Nordic countries a study of future needs for data communications and videoconferencing services within large companies was undertaken in the time period February to June 1985. The study dealt with data rates from 64 kbps to 2 Mbps in a 10 year forecast. For the purpose of the study inception of service was assumed in 1986.

The study comprised only large industrial companies in Norway, Finland and Sweden and their communications needs within the coverage area of Tele-X. The needs of governmental and other public sector organizations as well as banks, insurance companies, trading companies and investment companies were intentionally excluded to limit the scope of the overall study to a reasonable volume.

5.2. Interviews

In order to get accurate results from the interviews a complete "scenario" describing all relevant conditions was worked out in detail. It was considered of utmost importance to find and meet the right persons in the organization. First the full scenario of the satellite communications system was carefully presented. After a short discussion to check that all facts were correctly understood the interviewed persons were requested to quantify and describe what the new communications services could mean to their corporation. As this would require good understanding of the strategic plans for the coming 10 years it was most important to trace persons on corporate level having good knowledge of the future plans of the corporation and a good technical/economical understanding of communications services. The following companies were visited:

Finland:
- Kone OY
- Neste OY
- Nokia OY
- Valmet OY
- Wärtsilä OY

Norway:
- Det Norske Veritas A.S.
- Kvaerner A.S.
- SAGA Petroleum
- Statoil A.S.

Sweden:
- ASEA
- Ericsson
- Philips
- Saab-Scania AB
- Volvo AB
- Volvo Flygmotor AB

5.3. Forecast

Based on the results of the interviews a forecast for the use of Tele-X was made taking into account all potential industrial users in Norway, Finland and Sweden.

The forecast was done according to the following principles.

After a careful analysis of the interviews, a model assuming three basic utilization modes of the satellite services was developed. Each potential user was either identified with one of the three modes reflecting its typical use of the earth stations or completely omitted as potential customer for the satellite services.

The grouping of the companies was made with regard to
- type of activities
- geographical distribution of activities
- sales
- number of people employed.

It turned out, however, that the yearly sales figure was the most useable single factor available after detailed study of a great number of yearly statements.

The forecast finally included 77 companies in Finland and Norway, 79 in Sweden, in total 156. This includes the companies visited. No company with sales 1984 less than 1000 MSEK (approx 140 MUSD) was included in the forecast.

5.4. Results

The forecast for Finland, Norway and Sweden gave the results summarized in the tables below:

TABLE 1
Number of stations

Type of station	Number of earth stations		
	1988	1990	1995
64 kbps data	180	420	790
2 Mps data	180	530	810
Videoconf.	200	450	600
Total	560	1400	2200

As a result of the interviews a forecast of the typical use of a communication channel has been made. Regarding data communications, usage ranges from short bursts with limited amounts of data at 64 kbps up to long connection times at 2 Mbps for transfer of data bases.

Average values have been estimated as shown below and these have been used in the traffic model.

TABLE 2
Traffic characteristics

Type of service	Number of calls per 24 hour day			Average time per call minutes		
	1988	1990	1995	1988	1990	1995
64 kbps	3	3.5	4.5	7	10	11
2 Mbps	1	2	3	4	10	13
Videoconf	0.5	0.7	0.8	60	60	70

The table on average use indicates the importance of the 2 Mbps data traffic, where both the duration of the calls (from 10 to 13 minutes 1990 to 1995) and the number of calls per day (from 2 to 3) are increasing. The result is a doubling of traffic from 1990 to 1995.

Income from earth station lease charges and fees for traffic is given in the following table (values in MUSD):

TABLE 3

Income

Type of service	1988		1990		1995	
	Lease	Traffic	Lease	Traffic	Lease	Traffic
64 kbps	2.5	1.6	5.6	5.4	10.7	15.9
2 Mps data	3.6	1.8	11.0	27.3	16.7	77.3
Videoconf.	5.5	7.1	12.2	23.1	17.1	40.2
Subtotal	11.6	10.5	28.8	55.8	44.5	133.4
Total income MUSD	22.1		84.6		177.9	

From this summary of the forecast it can be seen that:
- although the number of 64 kbps data stations almost equals the number of 2 Mbps stations, the traffic revenue from 64 kbps is only about 17 % of the total data traffic revenues 1990 and 1995!

- video conferencing will give the main revenue in the beginning but later on 2 Mbps will take over this role.
- 2 Mbps data communications certainly has the biggest growth potential during the 1990's.

6. MAIN USER GROUPS AND SERVICES OFFERED

6.1. General

The need for communicating data is increasing at an ever increasing pace. The main reason is that the efficient utilization of information is becoming a very important competitive factor. There is a rapidly growing awareness of this factor in the industrial companies. That means that the possibilities created by new technologies will be more and more integrated in the strategic thinking of the management. As an example it is quite usual that the growth of the amount of stored data is in the order of 40-50% annually for industrial companies. The result is a growing need of data communications. It can be noted that a doubling in data traffic has occurred for the Swedish Telecommunications Administration between 1984 and 1985.

6.2 Data communications

Manufacturing industries will certainly be a major user group for data communications by satellite.

The process of design will need more and more computerized support. Computer Aided Design (CAD) is since years an established technology, but the use will increase rapidly. Computer Aided Manufacturing (CAM) is also well introduced in the manufacturing industry, but a significant integration between the two methods will take place in the coming years. A tool in this integration will be Computer Aided Engineering (CAE). All this methods will result in very large databases. In practice it will not be possible to work with interactive traffic between a large central database and non-intelligent distributed work stations. Instead large files have to be transferred from the central database to a local database, where the person at the work station can work without long response times. When the job is finished, the updated part of the data base is again transferred to the central location. In this way an updated database can always be maintained even in large systems.

In the production flow it is increasingly important to keep the quality high and the amount of stocks low. In both cases local databases best serve the purpose, but they have to be updated one or several times in a 24 hour period. This creates very high demands for fast and reliable communication.

The use of knowledge based systems (expert-systems) is not yet spread in the industry, but certainly they will come and demand fast communication between different data bases.

As a general trend, computer graphics will gain greater importance every year. The time when big heaps of printed lists were distributed will soon have passed. Instead graphics on desk screens will be the normal way of presentation. This means that the time when information flow was counted in number of bits to fill a certain amount of A4 pages with text and numbers is passed.

This is not only valid for CAD, CAM and CAE in the manufacturing industries. It will also be the case for administrative data handling, where business graphics is rapidly growing. Consequently also the trading companies will represent a fast growing market for high speed data communications.

The need to keep computer software well updated, so that various computer centers can work on the same program release version will require high speed data transmission. This is another reason why high speed communication from one computer to another will grow rapidly.

Many of these services will require 2 Mbps speed in order to transfer the information in an acceptable time. However, the lower speed, 64 kbps, will also have a market among the above mentioned user groups.

Relatively few telephone calls result in a contact with the person called. Letters take quite some time - time to write and time to be transferred. For these reasons, the use of electronic mail is growing very fast among the industries who have introduced such systems. For some applications 64 kbps will be sufficient but as the amount of data and messages grows it will be necessary to use 2 Mbps to handle this traffic.

Furthermore, interactive polling traffic with limited amount of data will remain a market in the future as well.

6.3. Video conferences

The interest in video conferences among the visited users was of a more complex nature than for data communications. One of the positive elements was the possiblity to get a video conference room installed on the premises of the plant. The general attitude was that if a trip (e g to a city center) had to be undertaken to reach a studio, much of the interest would be lost. Several companies with large premises foresaw the need of having more than one video conference room installed per plant.

One reason for hesitation was that audio conferencing has been tried out by some of the organizations with limited success. However, the possibility to show documents, pictures etc was considered as a major and very positive difference. Further, the improved personal contact that is created in a video conference made the potential users more positive.

A negative element in the introduction of video conferences is that it requires more of a change in culture in the organization than the introduction of fast data communications. As examples of change of culture it could be mentioned that people have been accustomed to a certain travelling pattern and a rapid change will probably be resisted. It is well known from USA that certain persons that are dominating a "live" conference have difficulties when it comes to video conferencing, where instead other persons more easily play an important role.

The responsibility for promotion of use of video conferencing in a large organization is not easily identified. People responsible for administration, organization, education, training and communications have to involved. Top management certainly has to play an important role, as user as well as promotor.

The uncertainty in the forecast for video conferences is bigger than for data communications and data from the interviews have been used very conservatively in the traffic forecast.

6.4. Other services

Other services possible via satellite communications as OB-transmission, data broadcast etc were shortly discussed during the interviews. Industry did not show significant interest in this form of services. In some cases, like

OB-transmission and event-TV with big needs for satellite capacity, it was feared by industry that these applications could seriously affect system capacity and cause temporary blockage.

The recommendation is to limit the use of satellite communications to switched high speed data communications and videoconferencing for the ten year period ahead.

6.5 General observations

It is very obvious that all industry and trade companies will require that all information sent by satellite must be protected against eavesdropping. This is valid for both video conferencing and data communications. All forecasts are based on the existence of efficient encryption and decryption methods. It is also very evident that the organizations want to handle this matter themselves. It was therefore fully accepted that such equipment is not included in the prices of the earth stations. However, the satellite organization must have a deep knowledge of these matters, in order to be able to give advice, especially to the smaller organizations.

Equipment for in-house switching of different user channels is neither included in the earth station. This was also well understood. The cost for such equipment was asked for. It should be in the interest of the operating company to give advice on this point as well.

Flexibility of a system with small aperture antennas located on the user premises was highly esteemed by industry. Communication can be established on short notice to any location within the service area. There is no dependence of the planned growth of the terrestrial network. High speed data communications can be established to subsidiaries, subsuppliers and other business partners independently of their localization. Engineering companies can easily establish temporary communications to various construction sites and establish their own computer systems on client premises.

The low bit error rate in the communications channel was regarded as a very important factor.

Satellite communications as outlined in the scenario, is regarded to be complementary to the services offered by the Telecommunications Administrations, and as such, of great importance to the further development of industry in the countries concerned.

7. FINANCIAL ASPECTS

In the original Nordic study a detailed financial analysis was included to show the viability and profitability of the Tele-X system with users in Finland, Norway and Sweden. To make such an analysis a large number of assumptions have to be made which are particular to the system being looked at. For the purpose of the Nordic study the satellite system was assumed to be operated and owned by a limited company called Nordcom. Revenues from operations covered insurance and replacement of satellites in order to support a two-satellite system in orbit. The analysis also included all other cost associated with the system like operations and dividends to the owners.

With all the assumptions taken into account complete balance wheets for a ten year period were computed. The analysis (illustrated in figure 5) showed that the company, measured over the whole period, would perform very satisfactory.

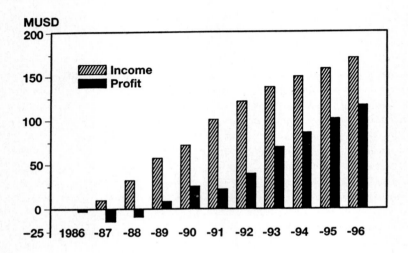

FIGURE 5
Income and Profit for Nordcom

8. CONCLUSIONS

8.1. System viability

From the market study in the Nordic countries it can be concluded that a system implemented along the ideas presented in this paper is technically and economically viable. Interviews with potential users have shown that one of the conditions to make a success is that the system is established with the relatively low yearly costs made possible with the system and that users are allowed to install stations and connect to the system without inhibiting

regulation and bureaucracy. Interconnection between the different areas must
also be provided. In a longer perspective the users must also be given the
possibility of gateway traffic and interconnection to other networks i.e
across the Atlantic.

8.2. Overall aspects

From the market study in the Nordic countries it can be concluded that the
interviewed corporations clearly identify a number of new services that
require a high speed digital communications network. The typical access
pattern of a 2 Mbit channel is approximately 10 minutes three times a day for
data traffic. If an attempt to squeeze the same service into a 64 kbps channel
should be made, the same amount of data requires approximately 16 hours of
transmission time. The corresponding user pattern for videoconferencing is one
hour once a day at 2 Mbps or 512 kbps.

The user pattern clearly indicates the necessity of a fully switched network where a great number of users can share the available channels. A system
using fixed channel allocation for point-to-point communication would exhibit
a much lower total capacity as most channels would be idle for long periods of
time for the type of traffic described in this study. The potential earning
capacity of a fixed allocation system would be much lower. Consequently the
tariffs in such a system would also have to be significantly higher or even
"prohibitive" to reach the necessary revenue. The conclusion is obvious:

If a high speed data network which is easily accessible, reasonably costed
and adapted to the traffic pattern that has been identified is not implemented, the new type of applications will and can simply not emerge.

The implications in a longer perspective can be serious. An important piece
of telecommunications infrastructure could be missing and thus depriving
European industry of a tool that could provide an edge in worldwide competition. Development in commmunications technology and industrial efficiency
could be hampered. Nations outside Europe could be given the opportunity to
develop and capture a leading position in areas such as Data Base Management
and applications of Computer Aided Design, Engineering and Manufacturing.

In several European countries the Telecommunications Authorities are
replacing the old analog networks with switched digital networks which
eventually will develop into nationwide full ISDNs (Integrated Switched
Digital Networks) providing voice and data communications services based on 64
kbps digital transmission channels. These projects run over long periods of
time and it will take decades to establish networks in and between all European countries. The plans to extend into higher data rates are very diffuse
and the time to establish switched services at 512 kbps and 2 Mbps will reach
well into the next century.

The technology for videoconferencing and high speed and high volume data communications is already existing and available on relatively short notice as far as user equipment and user installations are concerned. The limiting factor today is the lack and unavailability of transmission capacity at resonable tariffs. Satellite technology offers a possibility to establish the necessary transmission network and offer attractive tariffs in a near future and it would be unfortunate if the applications of new services were unnecessarily delayed.

Some reflexions can also be made on the terrestrial implementation of a communications network offering the services given in this study. In the Nordic area the forecast indicates approximately 2000 user stations 10 years after inception of service. Each station will access the system for short bursts of high speed data and for relatively short periods for videoconferencing. The expected traffic can be supported by a fully switched demand assignment communications network having a capacity in the order of 100 simplex 2 Mbps channels. To build a nationwide terrestrial network or to expand the ISDN network into higher data rates only to provide a few hundred simultaneous simplex channels to several thousand stations becomes very expensive. The technical solution leads to a small number of exchanges with very long subscriber connection lines.

In conclusion, a satellite system provides a very well balanced solution in terms of viability, cost, capacity and technical complexity for the particular applications dealt with in this study.

REFERENCES

1. - NORDCOM - A communications system for data, video and TV in the Nordic countries.
   - Forecast for new sevices in the Tele-X system
   - Prospectus for investors in Nordcom

   Swedish Space Corporation, September 1985
   (These three documents are written only in Swedish).

2. Tele-X, The First Step In A Satellite Communications System For The Nordic Countries. AIAA-84-0713.
   L Backlund et al, March 1984.

3. Pan-European Business Services By Satellite - A Study Of The Potential Future Demand. ESTEC contract report 6704/NL/DG.
   SSC, L Backlund et al, September 1986.

SESSION 6

TRANSMISSION TECHNIQUES AND INTERFACES

Chairman: G. Tartara *(Politecnico di Milano, Italy)*

Satellite Integrated Communications Networks
E. Del Re, P. Barthelomé and P.P. Nuspl (eds.)
© Elsevier Science Publishers B.V., 1988

AN ACQUISITION AND SYNCHRONISATION UNIT FOR A SS-TDMA NETWORK

KRG FOWLER and MRW MANNING

*British Telecom Research Laboratories, Martlesham Heath, Ipswich
IP5 7RE, England

ABSTRACT

This paper describes an acquisition and synchronisation unit (ASU) which is being designed and constructed at British Telecom Research Laboratories (BTRL) and is intended to interface with the reference terminal of a commercial TDMA system to thus permit SSTDMA operation via the specialised services payload of the ESA OLYMPUS-1 satellite.

1. INTRODUCTION

It is intended to take the opportunity afforded by OLYMPUS-1 to examine the operational and technical features of SS-TDMA, practically and at modest cost. In particular, the problems of acquiring and synchronising with the on-board SS-TDMA switch are to be studied. It is believed that the best approach to an SS-TDMA implementation is to provide appropriate interfacing to facilitate the use of standard proprietary TDMA equipment. It is also desirable that the interfacing equipment (ASU) is capable of adaptation to a variety of proprietary TDMA systems. These objectives are reflected in the modular design of the ASU, which allows maximum flexibility in the changing of parameters.

Apart from British Telecom, two other main parties have an interest in this experimental programme: DBP/FTZ (West Germany) and Telecom Denmark.

2. OLYMPUS-1 SATELLITE

The ESA OLYMPUS-1 satellite is due for launch in 1989. The 12-14GHz specialised services payload, which includes a steerable multiple-beam transmit-receive antenna and an IF switch matrix, is intended for communications experiments between small earth terminals [1].

Figure 1 shows the coverage zones of the five beams at nominal pointing over Europe; the footprints include 0.2 degree beam pointing error, and are for the uplink at 14GHz. The numbers on the contours are dB related to the beam peak. The whole five beam cluster can be moved to any point over the visible earth; not all coverage zones can be served simultaneously.

The satellite contains a 4x4 IF switch. The switch matrix operates a 256 sequence of switch connections, repeated every 20ms. It is apparent that the

---

*BTRL are sponsored and supported in this work by British Telecom International.

timing of all bursts from TDMA earth stations must be synchronised with the on-board switch (otherwise bursts will be misdirected to the wrong beam, or truncated by the switch). The ASU is a mechanism which synchronises the TDMA frame period to that of the on-board switch.

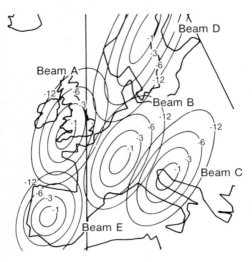

FIGURE 1
OLYMPUS-1 Specialised Services Coverage

3. ASU DESIGN

Figure 2 shows the interfacing between the ASU, the commercial TDMA equipment and the earth station equipment. The ASU is inserted between the TDMA equipment and the earth station, and comprises an acquisition and synchronisation module (ASM) and an ASM to TDMA interface (ATI). The function of the ASM is to establish knowledge of the on-board switch timing, while the function of the ATI is to interface signals between the ASM, reference terminal equipment (RTE) and the earth station equipment; the ATI also synchronises transmissions from the reference station to the on-board switch.

Adaptation of the ASU to a variety of TDMA systems is made easier by this modular design approach. The ASM is largely independent of the commercial TDMA equipment, although reconfiguration of the ATI will be required when alternative TDMA equipment is used.

For the TDMA systems under consideration, all remote sites synchronise their transmissions to the burst transmitted from the reference station. Therefore, synchronisation of the reference station transmissions to the on-board switch (by the ASU) will synchronise all stations to the switch.

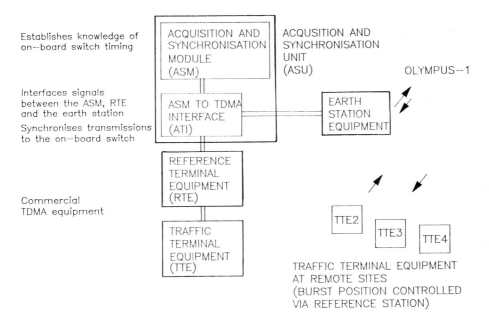

FIGURE 2
ASU Interfacing

The ASM synchronises to the on-board switch using a window technique [2] as shown in figure 3. A special word is configured which comprises a carrier and bit timing recovery sequence (CBTRS), a unique word and a metric. The pattern and length of the special word is programmable, to some extent, to facilitate use with alternative TDMA equipment; Telecom Denmark have commissioned software simulation studies to determine optimum sequences for the unique word and metric [3].

The optimum CBTRS pattern and length is dependent on the modem. The modem interface is an ATI function; initially the modem associated with the TDMA equipment will be used to transmit and receive the special word (in addition to all other reference station bursts), but later development may provide a dedicated modem for special word transmission-reception, which would make the ASM completely independent of the TDMA equipment.

There must be some interval during the 20ms OLYMPUS-1 frame period when transmissions from the beam in which the reference station is located are directed back to the reference station beam by the on-board switch. This interval is here termed the reference station beam loop-back window (RSBLW). When correctly positioned, the special word is truncated such that exactly half of the metric is received at the reference station. The transmission instant of the special word is adjusted such that synchronisation is maintained.

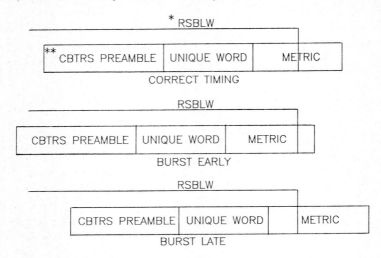

* RSBLW   REFERENCE STATION BEAM LOOP—BACK WINDOW
** CBTRS   CARRIER AND BIT TIMING RECOVERY SEQUENCE

FIGURE 3
Synchronisation Using the Window Technique

The ATI receives frame timing information from the on-board switch via the ASM, and frame timing information from the TDMA equipment via the RTE. The re-timing of the bursts from the reference station is arranged such that the on-board switch and TDMA frames coincide; the re-timed bursts from the RTE, together with the special burst from the ASM, are fed from the ATI to the earth station for transmission to the satellite.

Upon system initialisation, the ASM must first acquire synchronisation to the on-board switch; this is here termed the acquisition sequence. Signals from the reference station TDMA equipment are then accepted for transmission and synchronised to the switch; remote stations then access the satellite in accordance with instructions from the reference station in the normal manner.

4. ASU OPERATION

The Z8000 series microprocessor is employed in both the ASM and the ATI. The prime reason for this choice is that the ASU design uses burst chips which are intended as peripherals for the Z8000. The burst chip is a semi-custom IC, previously developed for a multi-point radio system [4].

Figure 4 is the ASU block diagram in which a specific design of ATI is shown. In this case it is assumed that the first reference unique word (RUW), which defines the start of the TDMA frame, is transmitted by the TDMA equipment at an arbitrary point in time; it is also assumed that there is no easy method of controlling the position of the RUW by direct access to the TDMA equipment.

The (burst) data and burst enable signals, which are normally fed directly from the RTE to the modem, are interrupted by the ASU; the received data, normally fed from the modem to the RTE, is monitored (but not interrupted) by the ASU.

The ATI incorporates a FIFO buffer circuit to delay transmissions from the reference station such that synchronism at the satellite between the TDMA and on-board frames is achieved and maintained. It is apparent that this buffer will eventually overflow is the OLYMPUS-1 clock is not occasionally corrected. The clock timing information is available, for external use, at the ASM; Telecom Denmark are investigating methods of on-board clock control.

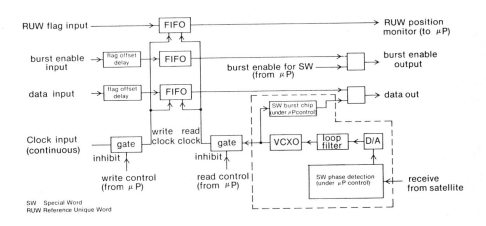

FIGURE 4
ASU Block Diagram

4.1. Steady-State Operation

The special word is assembled by the special word burst chip and output to the modem for transmission. The burst clock is derived from a voltage controlled crystal oscillator (VCXO), and is a multiple of the OLYMPUS-1 clock. The modem employed for special word transmission-reception for the initial experiments is that of the TDMA equipment, therefore the VCXO frequency is (approximately) that of the TDMA system.

The special word is interrogated within the special word phase detection circuit and the average of a number of metric correlations calculated. The output from the phase detector is fed via a digital-to-analogue converter and loop-filter to the VCXO. The action of the phase-locked-loop (PLL) is to modify the VCXO frequency (if necessary) such that exactly half of the metric is received; the VCXO is thus locked to the frequency of the on-board switch.

A continuous clock derived from the RTE is fed, via a gate, to the write inputs of a set of FIFO buffers; the gate is enabled during steady-state operation. The read inputs are fed with a clock derived from the VCXO, also via a gate which is enabled during steady-state operation.

Data bursts from the RTE are input to a FIFO buffer (via a delay circuit). The bursts are thus re-timed to a frequency which is synchronous with the on-board clock. The burst enable signal, normally fed directly to the modulator, is similarly re-timed.

4.2. Acquisition Sequence

The TDMA equipment (with the exception of the modem in this case) is not activated until the completion of the acquisition sequence; in fact, during the acquisition sequence, the ATI inhibits any transmissions from the reference station TDMA equipment.

There may be several RSBLWs per frame, depending upon the time-plan; the metric must be placed at the end of the correct RSBLW. The method chosen is to first transmit a number of words per frame, each with a unique identifier. The burst frequency is normally controlled by the PLL, but during this acquisition sequence the VCXO is free-running. The received words are then compared with a map based on the time-plan (this map can be readily amended for alternative time-plans). The correct RSBLW is then identified and a single unique word followed by a long metric transmitted, such that the metric is truncated by the correct RSBLW. Interrogation of the received metric enables the precise position of the end of the RSBLW to be determined, and the special word is then transmitted at the correct position. The PLL is then engaged, and the VCXO tracks the on-board clock as previously described.

The TDMA equipment is then activated. It is assumed that the reference station TDMA equipment transmits the first RUW, which defines the start of the TDMA frame, at an arbitrary instant such that it is necessary to delay this transmission within the ATI in order to obtain frame synchronisation with the on-board switch. It is assumed that a RUW send flag is available from the RTE (otherwise a suitable RUW detect circuit must form part of the ATI). When this flag is received at the ASU the write clocks on the FIFOs are enabled. It is necessary to insert a flag offset delay in the data and burst enable input lines because of the time delay between the start of the RUW burst and the RUW flag received from the RTE.

The reference burst and the associated burst enable signal are delayed in the FIFOs until the time interval between the RUW and special word transmitted from the ASU is correct (the correct delay value is a function of the time-plan and is programmed into the ASU; it can readily be changed). The read clocks on the FIFOs are enabled at the approriate moment. Both read and write clocks are enabled for the remainder of the system operation. A third FIFO is inserted such that the correct positioning of the transmitted RUW and special words can be monitored.

5. CONCLUSIONS

A design for an ASU has been presented which allows SS-TDMA operation to the OLYMPUS-1 satellite using standard proprietary TDMA equipment. A modular design is adopted; reconfiguration of the ATI enables operation to a variety of TDMA systems, but the ASM is largely independent of the TDMA equipment. The ASM also provides a terrestrial marker of the on-board frame period which can form part of an on-board clock control system.

ACKNOWLEDGEMENTS

Telecom Denmark are contributing to this work by providing theoretical assistance in the choice of ASU design parameters.

DBP/FTZ (W Germany), Telecom Denmark and the University of Surrey (UK) are collaborating with BT in various aspects of the experimental programme.

REFERENCES

[1] ESA OLYMPUS users guide, UG-6-1, part 2, specialised services payload.
[2] SS-TDMA frame synchronisation, S Joseph Campanella and Thomas Inukai, IEEE International Conference on Communications, June 1981.
[3] A Model for a Frame Synchronisation Unit, Karsten Olsen, Ibidem.
[4] Aspects of Design in a Semi-Custom IC Development of a Multi-point Radio System, Forty, P.J., Ballance, J.W., Edwards, S.J., Reger, J.D., 6th International Conference on Custom and Semi-Custom ICs, London, November 1986.

A MODEL FOR A FRAME SYNCHRONISATION UNIT IN SS-TDMA

Karsten Olsen

Aalborg University, Institute For Electronical Systems, Strandvejen 19, 9000 Aalborg, Denmark[*]

ABSTRACT

This paper describes a frame synchronisation unit for a reference station in a SS-TDMA system. It gives a mathematical model for the method proposed in [1]. The model allows the performance to be predicted in terms of acquisition time and phase jitter. Two methods to determine the location of the truncation are described. The first method gives a bias free estimate but requires knowledge of the bit error rate. The second is simple to implement but gives a small bias in the estimation of the location of the transition.

1. INTRODUCTION

SS-TDMA introduces a number of new problems. Most of them concern synchronization. An important problem is the earth stations synchronization to the on board switch matrix. Using a centralised system with a reference station, the earth stations synchronization to the switch matrix can be performed by locking only the reference station to the transition times of the switch matrix. The remote stations are then synchronized with the switch matrix when they are synchronized with the reference station, and can operate as ordinary TDMA remote stations.

This paper concerns the reference station's synchronization to the switch matrix. In [1] a metric truncation method is proposed. The reference station transmits a set of special bursts with a format as shown in figure 1.

carrier and clock recovery	unique word	metric

Figure 1. Format of metric burst.

The metric field consists of logic ones. When a metric burst is sent with the correct timing, half of the metric is received by the reference station, and the other half is truncated by the synchronization window. The received bits

---

[*]This work was carried out at British Telecom Research Labs. Ipswich in a cooperation project between British Telecom, Telecom Denmark and Aalborg University concerning SS-TDMA experiments with Olympus 1. Karsten Olsen is usually located at Telecom Denmark.

will eventually be corrupted with bit errors. It is assumed that the truncated bits occur as random bits in the receiver, with equal probability of one and zero.

The estimation of the location of the truncation is based on the average of several metric burst samples, in order to get a estimate with a small error. The number of samples must be sufficiently small to allow tracking of the frequency offset between between the local oscillator in the receiver and the the on-board clock, and the Doppler shift due to the distance change between the satellite and the reference station.

The block diagram for the synchronization is shown in figure 2. The phase error (measured in units of bits) is determined in the phase error detector. The phase error is used to adjust the transmitting times for the next set of metric bursts.

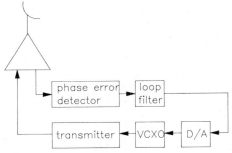

Figure 2. Block diagram of the synchronization unit.

2. MATHEMATICAL MODEL

The structure of the synchronisation unit shown in figure 2 is similar to a phase lock loop and it can also be described with the digital phase lock loop as shown in figure 3.

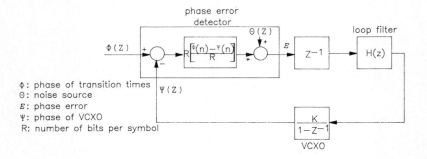

Figure 3. Mathematical model of frame synchronization unit.

All frequency dependent terms are neglected.
Due to the random nature of the phase error estimate a certain amount of phase jitter is introduced, here represented with a noise source θ.
Due to the round trip delay it is not possible to update the VCXO immediately after the transmission of a metric burst set. By assuming that the duration of one set of metric bursts is longer than one round trip delay, the influence of the round trip delay can be represented as a single sample delay z^{-1}.
The sample time is NT_F where N is the number of burst in a set and T_F is the frame length in seconds.
The non-linearity in the phase detector is ignored, with the exception of points discussed in section 6 "Simulation with non-linarities".
Let K_V (Hz/v) be the gain of the VCXO so that the phase of the VCXO is given by

$$\psi(nNT) = \psi((n-1)NT_F) + \omega_0 NT_F + K_v \int_{(n-1)T_F}^{nT_F} u(nNT_F)dt$$

$$= \psi((n-1)NT_F) + \omega_0 NT_F + K_v NT_F u(nNT_F) \qquad (1)$$

The relation between K and K_V is given by:
$$K = K_v NT_F \qquad (2)$$

3. PHASE DETECTOR

The following notation is used:

- Pe: bit error rate
- L: numbers of bits in the metric
- N: number of transmitted metrics
- m_{ij}: equal to 1 if the jth bit in the ith metric is logic one otherwise equal to 0
- M: the N by L matrix with elements m_{ij}
- δ: $L/2+\delta$ is the position of the transition, so the bits $1..L/2+\delta$ are received as ones and the rest are random bits.

The estimation of the truncation times must be based on M. Ideally estimation must take into account the possibilities of bit errors in the returned part of the metric. Following estimate can be used :

$$ES(M) = \frac{1}{N(1/2-Pe)}\left(\sum_{i=1}^{N}\sum_{j=1}^{L} m_{ij} - NL(3/4-Pe)\right) \qquad (3)$$

This is a bias free estimate because if it is assumed that the transition occurs after $L/2 + \delta$ bits we have:
$$E\{ES(M)\} = \delta \qquad (4)$$

On average we can expect a correct estimate of the location of the truncation. The variance of θ when $\delta=0$ is:

$$E\{ES^2(M)\} = \frac{L(1+4Pe-4Pe^2)}{2N(1-4Pe+4Pe^2)} \qquad (5)$$

This shows that the variance increases with the length of the metric and with increase in the bit error rate, and decreases with the number of transmitted metrics.

The noise is assumed to be white with the variance given in (5) and uncorreleted with ϕ. It is a reasonable representation in stationary mode but represents a small approximation during acquisition because the variance depends on δ.

The estimate in (3) requires a knowledge of the bit error rate. It is possible to get an estimate of the bit error rate by counting the number of errors in the unique word. A simpler estimation of the transition times is given in (6) which does not require knowledge of the bit error rate. This estimation is obtained from (3) by setting P_e to 0.

$$ES'(M) = \frac{2}{N}\left(\sum_{i=1}^{N}\sum_{j=1}^{L} m_{ij} - 3\frac{NL}{4}\right) \qquad (6)$$

With this estimate bias is introduced:

$$E\{ES'(M)\} - \delta = -Pe(L+2\delta) \qquad (7)$$

Thus with a small bit error rate and a small length of the metric the bias is small.

The length of the metric must be long enough to allow for drift between the VCXO and the oscillator on-board the satellite, and the Doppler drift due to distance changes between the reference station and the satellite. The Doppler drift is in the order of 10^{-8} for a geostationary satellite so it is of minor importance. Unfortunately there exists no accurate model for the short term drift between oscillators, so it cannot be included here.

4. LOOP FILTER

The transfer function for the loop filter is chosen so that it is possible to track a frequency offset with a stationary phase error. When a frequency offset occurs, the input phase is a ramp function:

$$\phi(n) = nu(n) \qquad (8)$$

or in the z-domain:

$$\Phi = -\frac{1}{(1-z^{-1})^2 z} \qquad (9)$$

The close loop transfer function is given by:

$$F(z) = \frac{E(z)}{\Phi(z)} = \frac{1}{1 + z^{-1} H(z) \frac{K}{1-z^{-1}}} \qquad (10)$$

The infinite value theorem is:

$$\lim_{n \to \infty} \epsilon(n) = \lim_{z \to 1} \left(1 - z^{-1}\right) E(z) \qquad (11)$$

by using the infinite value theorem on (9) and (10) we get:

$$\lim_{n \to \infty} \epsilon(n) = \lim_{z \to 1} F(z) \Phi(z) \left(1 - z^{-1}\right)$$

$$= \lim_{z \to 1} -\frac{1}{\left(1 - z^{-1}\right)^2 z + H(z) K}$$

$$= \lim_{z \to 1} -\frac{1}{H(z) K} \qquad (12)$$

And from this we get:

$$\lim_{n \to \infty} \epsilon(n) = 0$$

↓

$$\lim_{z \to 1} H(z) = \infty \qquad (13)$$

This means that the loop filter must have a pole at z=1, for tracking a frequency offset with a zero phase error.
The loop filter could be chosen as :

$$H(z) = \frac{K_1}{1 - z^{-1}} \qquad (14)$$

The root curves for this loop filter are shown in figure 4.

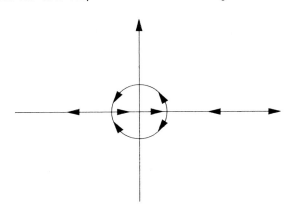

Figure 4. Root curves for the loop filter.

When KK_1 is ∞ the loop has a pole at $z=\infty$ and at $z=0$. When KK_1 approaches 0 both poles approach 1. The two poles continue on the unit circle, following each half circle until they meet at $z=-1$. One of the poles then continues to $z=-\infty$ and the other contines to $z=0$. At no time do both poles lie inside the unit circle so the loop is inherently unstable. By including a zero in the loop filter :

$$H(z) = K_1 \frac{K_2 - z^{-1}}{1 - z^{-1}} \tag{15}$$

The transfer function becomes:

$$F(z) = \frac{(z-1)^2}{z^2 + (KK_1K_2 - 2)z + 1 - KK_1} \tag{16}$$

and for a desired pole location corresponding to:

$$F(z) = \frac{(z-1)^2}{z^2 + \alpha z + \beta} \tag{17}$$

we shall select:

$$K_1 = \frac{1-\beta}{K} \quad \wedge \quad K_2 = \frac{\alpha + 2}{1 - \beta} \tag{18}$$

with (15) it is possible to select the poles freely inside the unit circle. (15) can also be written as:

$$H(z) = K_1 \left(1 + \frac{K_2 - 1}{1 - z^{-1}}\right) \tag{19}$$

The complete model in the time domain is shown in figure 5.

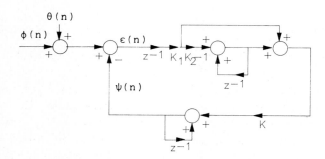

Figure 5. Time domain representation of the synchronization unit.

5. PHASE JITTER

The phase jitter can be expressed as

$$\sigma_\epsilon^2 = \frac{1}{2\pi} \int_{-\pi}^{\pi} |F(e^{j\omega})|^2 P(e^{j\omega}) d\omega \qquad (20)$$

where $P(e^{j\omega})$ is the power spectrum of θ. Under the assumption that θ is white (20) can be expressed as:

$$\sigma_\epsilon^2 = \frac{\sigma_\theta^2}{2\pi} \int_{-\pi}^{\pi} |F(e^{j\omega})|^2 d\omega \qquad (21)$$

where σ_θ^2 is given by (5).
A Three-dimensional profile for the jitter in dB based on (21) with σ_θ^2=0dB, for different complex pole location is shown in figure 6a. The acquisition time is shown in figure 6b, and is obtained by simulation. It is seen that it is impossible to get a short acquisition time and at the same time have small jitter. A good location of the poles is on the positive part of the x-axis.

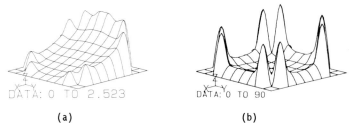

(a) (b)

Figure 6. Jitter in dB and acquisition time for different complex pole locations.

6. SIMULATION WITH NON-LINEARITIES

Figure 7 shows the step response for the transfer function given in (16) and a simulation with the estimate in (6). The following parameters are used: L=14, N=36, Pe=0.01, KK_1=0.19, K_2=1.05 corresponding to a double pole in 0.9 and a phase offset on δ=4. The system is simulated for QPSK where two bits are destroyed by the truncation, number 5 and 6 for this simulation. ψ is there only determined to a precision of 1 symbol. The negative bias for the estimate manifests itself by a trend to approaching the lower limit instead of the upper limit.

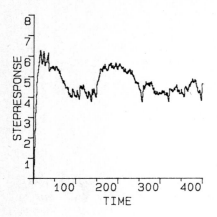 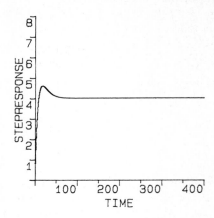

Figure 7. Step response and simulation of step response.

SUMMARY

The proposed frame acquisition, based on a metric truncation method, can be described with a second order digital PLL. Two methods for estimation of the location of the truncation have been described. The first method gives a bias free estimate of the location but requires knowledge of the bit error rate. The second method does not require this knowledge, but has a small bias. Due to the random bits in the truncated part of the metric some jitter will occur, but this jitter has negligible importance, if the parameters are correctly selected.

REFERENCES
[1] S. Joseph Campanella and Thomas Inukai, SS/TDMA FRAME SYNCHRONISATION, IEEE International Conference on Communications June 1981 pp. 5.5.1-7
[2] Alan V. Oppenheim and Ronald W. Schafer, Digital Signal Processing, Prentice-Hall 1975

FODA-TDMA satellite access scheme: description, implementation and environment simulation.

Nedo Celandroni, Erina Ferro
CNUCE Institute
Via S.Maria 36 - 56100 Pisa (I)
Phone: + 50-593207/593246
Telex: 500371 CNUCE I

A Fifo Ordered Demand Assignment-TDMA (FODA) satellite access scheme is briefly presented and few notes about the current implementation are given. The scheme is designed to handle together packetized data, voice and immages traffic in a multiple access satellite broadcast channel of Mbits band.
The channel is shared by as many as 64 simultaneously active earth stations capable to optimally share the channel capacity and also to multiplex the different traffic types, incoming from satellite users attached to Local Area Networks (LANs), taking into account their different requirements in terms of quality of service.
As the system needs to be tested in a sophisticated user environment, the performance extimation has been carried on in a double way.
The simulation of the whole system at 2 and 8 Mbit/sec produced an exaustive set of results and few of the most significative are here presented. A measurement system has been tuned up, using a suitable traffic generator, simulating the incoming of real data from users connected to the LAN and whishing to send data via satellite.
The comparison between the results of the simulation and the measurement tests has been made possible.

1. THE FODA-TDMA ACCESS SCHEME

1.1 Short description
The aim of the present work is the realization of a reliable trasmission device suited at optimizing the use of a Megabits satellite channel in presence of mixed digital traffic (stream + datagram) coming from an eterogeneous network environment.
The system is not restricted to fixed point-to-point links, but supports the special requirements of distributed computing and information dissemination.
A group of users (up to 64 active stations in a range of 255 addressable ones) shares a satellite channel in a time division multiple access (TDMA) mode on a demand basis: it means that only when packets are to be sent, transmission time slots are actually allocated to a station.
The access to the satellite channel is given by the satellite TDMA controller running the FODA software. It receives packets from individual users via a LAN and transfers them over the satellite channel to the addressed user or host.

The system not only provides computer data trasmission facilities, but also voice or slow scan or compressed video communications. These different services have their own "quality of service" parameters: required bit rate, maximum tolerable bit error rate, priority of the service and burstiness of the data.

The FODA access scheme is essentially based on reservation of bandwidth. The time is divided into slots in which the various stations alternate in their use of the entire capacity of the channel. The assignment of the time slots is made dynamically upon demand of the users (earth stations). Requests for datagram and for stream slots are managed by the system in different ways.

One of the active stations plays the role of master (channel dispatcher) other than its normal slave role. Techniques to replace the master station, in case of fault, are provided, in order to create the minimum trouble to the other users. The design is based on the following assumptions:
- the satellite network offers only its best efforts to satisfy the quality of service needed by higher level protocols to match the requirements of the different applications;
- operations such as opening/closing of virtual circuits, and retransmission of corrupted packets for applications requiring error free delivery, must be performed by higher level protocols;
- the data packets must be organized at the stations in different FIFO queues, according to the quality of service (priority). There will be a stream queue and two different queues for datagram traffic, since higher priority is given by the station to sending interactive as opposed to bulk traffic;
- in order to make the time-slot length independent by the packet length, fragmentation/reassembly techniques are foreseen. Channel saturation control techniques are also foreseen.

The transmission time is divided into frames 31.25 ms long, in order to synchronize events like the stream slot repetition and the datagram slot assignments. The master station sends a reference burst at the beginning of each frame for synchronization purposes and for distributing control information to all the stations.

A bigger period of 2 seconds is chosen to schedule events varying with lower frequency, e.g. the stream channels setting. This super-frame is compound by 64 frames, numbered from 0 to 63.

Each frame is divided into 3 sub-frames:
- The Control sub-frame, containing 4 small slots assigned cyclically to all the active stations. They are used by the stations to send the requests for stream and datagram capacity, as well as to send control information. Datagram and stream requests can be also piggy-backed with data.
- The Stream sub-frame, containing fixed sized stream slots, assigned regularly in time to the requestors, up to a limit (tipically half frame), to be tuned up accordingly to traffic experiences.
- The Datagram sub-frame, containing datagram slots, variable in size with small granularity, for interactive and bulk data traffic.

Fig 1.1 shows the time frame structure (the relative times are not to scale).

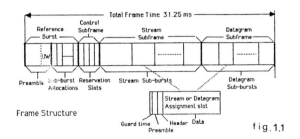

fig.1.1

1.2 Definitions

Some definitions are here needed to understand the following.

Elementary Slot (ES) is the interval of time which allows 16 octects of information, after eventual coding, to be transferred.

Number of Consecutive elementary Slots (NCS) is any number of consecutive elementary slots, relative to an allocation.

Stream Assignment Slot (SAS) is the smallest portion of time assigned for stream. It allows 256 octects of information (16 elementary slots), after eventual coding, to be transferred. The guard time, the preamble and the header are also included in the SAS size. The SAS is used to transmit stream traffic but the station can use it also for datagram, not to waste already allocated time, in absence of stream packets, such as during the silent periods of the voice transmission.

Datagram Assignment Slot (DAS) is the portion of time assigned to a station allowing to transmit a variable amount of data contained in a NCS. The guard time, the preamble and the header are also included inthe DAS size. The DAS is used for transmission of datagram only.

1.3 Stream requests and assignment

The request for the stream channels is made only once and, if accepted, it will be considered valid until the station sends a relinquish indication or it is declared dead. The request is made on the basis of a multiple of the "stream channel", whose throughput is 16 kbit/sec, corresponding to the allocation of one slot every fourth frame.

The requests of the stream are maintained by the master in a FIFO queue and are granted in the same order.

The stream assignments are made on a super-frame basis; i.e. they may vary only at the beginning of the super-frame.

The stream assignment distribution is repetitive by a period of 4 frames (125 msec), whatever the bit rate is. The uncompressed voice, for example, having a speed of 64 kbit/sec, correspond to 4 channels: 1 slot per frame.

The distribution of the SASs among the stations is done in such a way to optimise the transmission duty cycle of all the stations. If more than one SAS per frame is assigned to the same station, they are allocated in a contiguous way.

1.4 Datagram requests and assignments

The DASs are assigned in the remaining space of the time frame, after the stream allocations. The datagram allocations generally change in each frame.

A datagram request contains the number of elementary slots needed by the station at the moment. After a complex study of the system (4), a suitable expression has been chosen for the datagram request (UR), in which a term proportional to the incoming traffic has been added to the backlog (i.e. the volume of data waiting to be sent on satellite).

$$UR = BACKLOG + H * TRAFFIC \qquad (1.1)$$

Where H is a suitable constant of proportionality.
The request must keep into account the increasing in size of the data, due to the eventual FEC coding.
Each request is put in a FIFO queue by the master station. Any further request, other than the first, is considered an updating and replaces the previous value. A request indicating zero elementary slots is removed in the last position of the queue, without being cleared. This entry will be cleared when found at the first position of the queue if the request is still null.
To accomplish the allocation duty, the channel dispatcher must perform the following operations:

- takes the first entry in the FIFO queue of the requests;
- removes from the queue a request indicating zero slots when found as the first entry of the queue;
- assigns a NCS proportional to the request (inside the interval between a minimum and a maximum established threshold values);
- subtracts the assigned NCS from the number of the requested ones;
- puts the resulting request, containing a positive or null value, in the last position of the queue;
- repeats the cycle till the datagram time available in the frame is exhausted.

This scheduling technique has a basic fairness in case of heavy global traffic and allows a heavy loaded station to own almost the whole datagram subframe in case of light traffic from the remaining stations.

2. THE IMPLEMENTATION

A satellite TDMA controller has been designed and developed which consists of 3 maior elements:
a) The system bus, processor, memory and communication support hardware.
   The actual transmission and reception of bursts is handled by two separated microprocessors Motorola 68000, having specific transmitting and receiving functions and each running the part of the FODA software relative to its function (Up and Down processors respectively). The two boxes constituting the TDMA controller communicate between them via PIO interface.
   The interface with the attached LAN is also provided for both the processors.
b) A variable rate digital burst modem which is capable of operating at 1, 2, 4 or 8 Mbit/sec inside a 5 Mhz frequency band.
c) An encoder/decoder implemented in VLSI and which is specifically adapted to the modulation scheme. The encoding rate may be changed on each

sub-burst within a burst allowing different BER services for voice, video or data.
The combination of the modem and the encoder will deal with most fade conditions and will also allow a wide variety of BERs, between $10^{**}-4$ and $10^{**}-9$, for user services.

2.1 Software architecture on Motorola 68000

The FODA access scheme software has been written mostly in C language and only small parts in assembler. The operating system running on both the Up and Down machines, called C-Exec, is a Unix like multitasking operating system.
In order to speed up the operations requiring very quick responses, they are performed at interrupt level, so the CPU works in system mode, at rather high priority level. It allows to avoid the overhead due to the operating system and the use of the entire set of machine istructions other than a dynamical management of the CPU priority, according to the operation under execution.
More precisely, all the operations relative to the satellite access method are performed by several processors entered by either hardware or software interrupts. The other processors called at interrupt level are the line drivers (terminal and LAN interface) supported by the operating system.
The interrupt level processes are:
On the Up machine:
- DMA end of block: the DMA board has finished the transfer of a burst on satellite.
- PIO read: the receiving of the first word from the Down machine is completed.
- Timer: this interrupt occurs once per frame, in a fixed position.
- LAN end of block: the LAN handler has finished to read a packet which is ready to be sent on satellite.

On the Down machine:
- PIO read: the receiving of the first word from the Up machine is completed.
- Timer: it occurs once per frame in a fixed position.
- DMA end of block: the DMA board has finished the read of a burst from the satellite.
- LAN end of block: the DMA interface has finished the transfer to the LAN of a block of data.

Communications between interrupt level processes and C-Exec tasks are needed in both ways. They are made possible according to the following procedures:
- Interrupt level process to C-Exec task:
  the process uses the C-Exec routine "wint" to put a byte in a pipe exactly like the C-Exec drivers do. After the return from exception (RTE) instruction the system will alert the receiving task by its priority.
- C-Exec task to interrupt level process:
  the task issues a software interrupt provoking the CPU jumping to the address specified in that software interrupt vector, entry-point of the called process.

The PIO write is performed by a routine called at task or at system level. No interrupt is generated after the transfer of each word at the writing machine. As the writing provokes an interrupt at the reading machine, that is kept to high priority during the whole transfer, it has been taken care of performing also the write operation at rather high priority level, making fast enough the message transfer.

3. STUDY OF THE SYSTEM PERFORMANCES

In order to tune up the system design and find out the performances, a theoretic study has been carried out (4). The analitic expression of the data delay due to the satellite network crossing versus the different channel loading conditions has been obtained for the stationary case. Moreover two more tools have been developed: a complete system simulation and a measurement system, both herein shortly described. The comparison among theory, simulation and measurement results has been made possible in some simple cases.

3.1 The simulation

The simulation program is written in SIMULA67 and uses discrete simulation techniques. The various processes constituting the system under investigation are modelled and a scenario of the system performances is produced at regular intervals of time.

The program is driven by an input file describing the general characteristics of the system and giving the detailed description of all the active stations and of the traffic generators connected to each station.

Several not correlated traffic generators can be defined for each station, contributing to produce the total traffic.

The generators produce packets following an inter-arrival time distribution according to one of the following types:

FIXED RATE   the data packets are generated at constant intervals of time.
POISSON   the intervals of time in between the packets generation follow a negative exponential distribution.
VOICE   the information flow of a digitalised speech is thought to generate data packets. No packet is generated if the "silent" period is long enough. The talk-silent periods are sorted in a well experienced distribution, relative to the English speech.

Herein a set of diagrams is reported, taken from ref. (4), showing the results of few significative simulation tests.

The figures 3.1, 3.2, and 3.3 show the single station delay versus its outgoing datagram traffic (expressed in channel capacity percentage). They are obtained making 7 runs in which the traffic of each station is multiplied by a loading factor varying between 0.3 and 1.0. The product of each run is a set of points, coherent approximately to a horizontal (dashed) line. The delay is relative to

FODA—TDMA Satellite Access Scheme

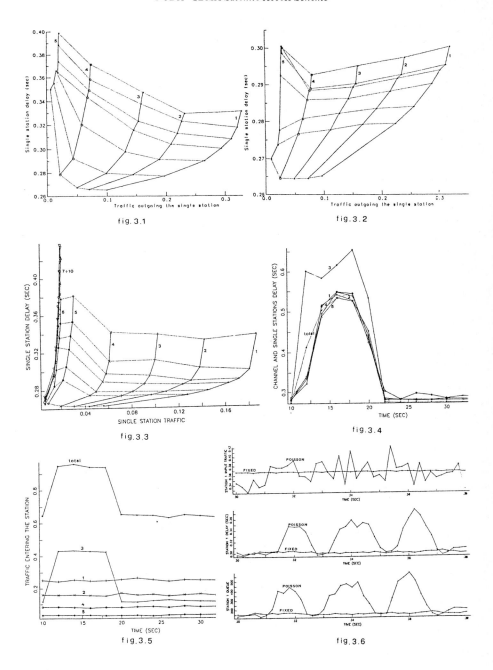

fig. 3.1

fig. 3.2

fig. 3.3

fig. 3.4

fig. 3.5

fig. 3.6

the satellite network crossing. It is the sum of the round trip time (assumed as 255 msec) and the queuing time.
The diagrams show the steady state behaviour of the system. The delay values are averaged over a period of 20 sec (50 sec for the diagram of fig 3.3) starting 10 sec after the beginning of the simulation, to exclude the effect due to initial transitory.
Fig. 3.1 is a 5 stations simulation at 2 Mbit/sec using Poisson traffic generators. In all the diagrams the values of the traffic represent user data only, without headers and preambles.
Fig 3.2 is the same as fig 3.1 but a fixed rate traffic generator is used.
Fig 3.3 is a 10 stations simulation at 8 Mbit/sec, using Poisson datagram traffic generators. 30 stream channels are totally assigned to the first 6 stations and used for voice traffic. It is evident the lower delay with respect to the other stations, due to the fact that the first 6 stations use also the silent periods of the voice to send datagram other than the normal datagram assignment slots.
Fig. 3.4 shows a transitory. The various stations delays are shown, at 8 Mbit/sec, as a temporal function, when the input traffic is the one of fig. 3.5.
In fig. 3.6 a comparison is made between the behaviours of the system when excited with fixed rate or Poisson traffic generators. The simulation is made at 8 Mbit/sec. Both the delay and the datagram queue lengths are shown, relative to one of 5 stations.

3.2 The measurement system

The problem was to create a tool able to measure the performances of the FODA system without taking into account the LAN crossing. For that purpose a traffic generator was implemented (5) internally to the satellite controller.
The generator is able to produce a calibrate amount of different types of traffic (i.e. stream, interactive and bulk). It is realized in such a way that, at each timer interrupt on the Up machine (once per frame), the selected quantity of data is generated and enqueued in the relative queue, simulating their incoming from the attached LAN.
The software generator has been written in C language and it is part of the FODA software running on the Up machine.
From the operative point of view, the generator can be managed by the system console. At the initialization phase, the operator can choose to start the traffic generator facility. In this case, he must specify the required stream, interactive and bulk traffic amounts, expressed in percentages of the channel capacity, together with their relative coding rates. The starting, stopping and restarting with new parameters are driven by commands as well.
At present the limitation of this tool is that the generator is designed only for fixed rate traffic. A more sophisticated system, including the Poisson and the voice models, like the ones used in the simulation, will be possible when more CPU power will be added to the satellite controller. This problem is in course of study.
In order to measure the delay of the satellite network crossing, a Delay Measurement Packet (DMP) is sent, properly flagged inside the header. At regular intervals of time (synchronously) or asynchronously (by command) the DMP is created by the Up machine and properly enqueued. The sending time is recorded in the form of frame number and time inside the frame. On

receiving of the DMP, the Down process notifies the Up process about it. The delay time is so computed by difference of times and the relative value is stored in memory and displayed at the console. The value of the relative queue length is recorded as well.

Making a set of tests with different values of the traffic rates and gathering the values of the delays, it is possible to produce diagrams like the ones of fig. 3.2 or fig 3.4, for fixed rate generators, other than diagrams showing the length of the queues versus the stations traffics.

The whole system has already been tested in base band and it will run on satellite with 3 stations, hopefully in a short time.

References

1. R.Beltrame, A.B.Bonito, N.Celandroni, E.Ferro, "FODA-TDMA: final report on the new protocol for mixed traffic. Theoretical study and first simulation results", CNUCE Report C85-03.
2. C.J.Adams, J.W.Burren, "A Proposal For Experiments In Satellite Access Methods", Rutherford Appleton Laboratories, Chilton, Nr Didcot, Oxon, November 1984.
3. E.Ferro, "The implementation of the FODA-TDMA satellite access scheme on the New Satellite Bridge TDMA controller", CNUCE Report C86-15.
4. R. Beltrame, N. Celandroni, "The performances of the FODA access scheme: theory and simulation results", CNUCE Report C86-19.
5. E. Ferro, "A fixed rate traffic generator to test the FODA system", CNUCE Report C87-10.

COMPARISON OF DIGITAL MODULATION TECHNIQUES FOR FDMA/SCPC SATELLITE
COMMUNICATION WITH SMALL EARTH STATIONS

Philip SANDERS and Marc MOENECLAEY

University of Ghent
Communication Engineering Lab.
Sint-Pietersnieuwstraat 41
B-9000 GENT, BELGIUM

A comparative study of digital modulation techniques for satellite communication with small earth stations in a FDMA/SCPC environement is presented. Operation of the earth station's HPA at several input backoff values well into the nonlinear region is assumed. A wide class of modulation schemes are investigated and for the fluctuating envelope schemes computer simulations are performed. Not only the spectral and BER performances are evaluated, but also the receiver complexity. Within each class of complexity, the most cost-effective schemes are identified.

1. INTRODUCTION

This paper presents a comparative study of several digital modulation techniques for satellite communication with low-cost small earth stations. Frequency division multiple access (FDMA) and single channel per carrier (SCPC) operation is assumed. We restrict our attention to small earth stations, which transmit only one carrier. Intended applications include maritime, rural and mobile satellite communications.

As the earth station's HPA amplifies only one carrier, it can be operated in its nonlinear region in order to keep costs as low as possible. Because FDMA is employed, the satellite's TWTA must be operated within its linear region so that severe intermodulation is avoided. Hence, we have neglected the effect of the TWTA nonlinearity in this study.

In this paper we consider not only conventional modulation schemes, but also more advanced formats, such as coded 8PSK (C8PSK) and continous phase modulation (CPM). These advanced modulation formats have received much attention during the past decade, because of their promising spectral and bit error rate (BER) properties.

This research has been supported by the Flemish Government (IWONL VL 1/4-8994/075), by the Belgian National Fund for Scientific Research (NFWO) and by NEWTEC Cy.,B-2018 ANTWERP, BELGIUM.

The comparison of the modulation schemes is performed on the basis of:
- spectral efficiency at the HPA output
- BER performances
- receiver complexity

In this study degradations due to synchronization imperfections are neglected; that is, perfect recovered carrier and clock are assumed to be available at the receiver. Furthermore, the hardware required for synchronization is not included in the receiver complexity comparisons.

The organization of the paper is as follows. In the second section, the modulation schemes considered will be enumerated. The third section presents the simulation model that is used for evaluation of the non-constant envelope schemes on the nonlinear channel, along with the results obtained. In the fourth section, the constant envelope schemes are considered. In the final section, the actual comparison is made and conclusions are drawn.

2. MODULATION SCHEMES

The following modulation techniques are investigated:

(a) <u>Conventional</u> :
- Filtered QPSK schemes with either a 40% cosine rolloff (CRO) or 100% CRO power spectrum at the HPA input. Both no coding and convolutional encoding are examined. The coded is a rate 1/2 constraint length 7 code as defined by the polynomials $1+D+D^2+D^3+D^6$ and $1+D^2+D^3+D^5+D^6$.
- Same schemes as above, but using filtered offset QPSK (OQPSK) instead of filtered QPSK.

(b) <u>C8PSK</u> :
Filtered coded 8PSK with either a 40% CRO or 100% CRO power spectrum at the HPA input. Considered codes are rate 2/3 Ungerboeck codes with 8 states and 16 states [1].

(c) <u>CPM</u> :
- Tamed frequency modulation (TFM) [2].
- Generalized TFM (GTFM) with B=0.62 and r=0.36 [2].
- Binary CPM schemes with h=0.5, using raised cosine frequency pulses with durations equal to 2 symbol intervals (2RC) and 3 symbol intervals (3RC).
- Quaternary CPM scheme with h=0.5, using a raised cosine frequency pulse with duration equal to 3 symbol intervals (Q3RC).
- CORPSK(4-5) and CORPSK(4-7,1+D) schemes [3].

For comparison purposes, the BER is expressed as a function of E_{sat}/N_o, E_{sat} being defined as $E_{sat} = P_{sat}/f_b$, where P_{sat} and f_b denote the maximum HPA output power and the information bitrate, respectively. The HPA cost is directly reflected by using E_{sat} instead of E_b, the energy per bit at the HPA output.

The CPM formats generate a constant envelope signal; the conventional and C8PSK schemes, on the other hand, all include a transmit filter after the PSK modulator and thus generate a signal with fluctuating envelope. The performances of these filtered schemes will be evaluated first.

3. EVALUATION OF FILTERED SCHEMES.

The spectral and BER performances of the fluctuating envelope schemes on a nonlinear channel are difficult to evaluate analytically. Therefore, the performances are obtained through extensive computer simulations. For each modulation scheme, simulation runs are performed for several backoff values as well as for a linear channel model.

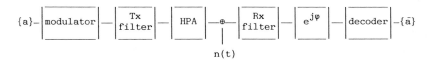

Fig.1. Model used for simulations.
n(t): White Gaussian noise

The model used for the simulations is shown in fig.1; complex envelope notation is assumed for the bandpass signals. The AM/AM and AM/PM conversions of the memoryless HPA nonlinearity are shown in fig.2. The transmit and receive filters are square root RCO filters with equal rolloff factors, namely 40% or 100%. Hence, the receive filter is matched to the transmit filter, which is optimum for a linear AWGN channel. In this model, the receiver uses both the transmitter clock and carrier for the synchronization. The optimum value for the receiver clock is obtained analytically and it is found that it equals the transmitter clock except for a delay equal to the group delay of the transmit and receive filter combination. The carrier sychronization is implied in the complex envelope notation and the receiver phase shifter. The optimum value for this phase shift is determined using an estimation from a simulation run prior to the BER simulation run. During the latter, the phase shift is kept constant. It was found that the required phase shift can be approximated within a fraction of a degree by the AM/PM conversion caused by the HPA nonlinearity on a constant envelope signal with equal input backoff as the modulated signal, at least for drive levels up to 0 dB. The output backoff values, however, cannot be well approximated using the AM/AM conversion alone.

The spectra of the NL's output signal, i.e. "the transmitted signal", are calculated using the method of averaging of modified periodograms. A total of

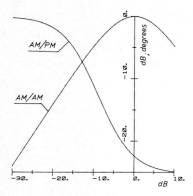

Fig.2 AM/AM and AM/PM conversions

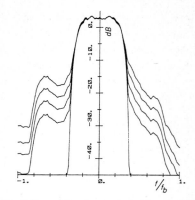

Fig.3 Normalized spectra of unc.QPSK (left) and unc.OQPSK (right) with 40% ro. for (top to bottom curves) 0,3,6,10dB IBO and linear channel

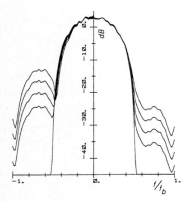

Fig.4 Normalized spectra of unc.QPSK (left) and unc.OQPSK (right) with 100% ro. for (top to bottom curves) 0,3,6,10dB IBO and linear channel.

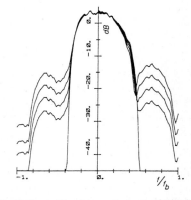

Fig.5 Normalized spectra of C8PSK 40% (left) and C8PSK 100% (right) for (top to bottom curves) 0,3,6,10dB IBO and linear channel.

249 overlapping sections of 32 channel symbols is used. Each section is windowed with a $\cos^4$ window and has a 50% time overlap with the previous section. This results in a 95% confidence interval in the order of +/- 1 dB.

The spectra for the uncoded QPSK and OQPSK schemes and for the C8PSK schemes are presented in fig.3 to 5. All spectra are normalized to power equal to one in order to facilate comparison; this corresponds with a center density of +3 dB. The spectra for the coded (O)QPSK schemes are found to be identical to the ones for the corresponding uncoded schemes operated at identical channel symbol rate, i.e. at twice the information bitrate. This result is expected for a linear channel because it can be proved that the encoder generates channel symbols that are uncorrelated.

The BER performances for the uncoded schemes are obtained with semi-analytically methods; a bit per bit threshold decision logic is assumed. Fig.6 presents some BER performances as a function of E_b/N_o. Hence, the

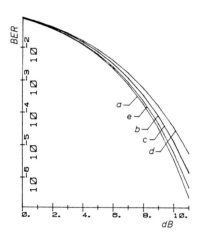

Fig.6 BER versus E_b/N_o
1) lin.chan. unc.(O)QPSK(a)
2) 0dB IBO: unc.QPSK 40%(b) and 100% (c); unc.OQPSK 40%(d) and 100%(e)

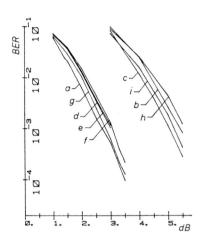

Fig.7 BER versus E_b/N_o
1) lin.chan.:cod.(O)QPSK(a),8state C8PSK(b),16state C8PSK(c)
2) 0dB IBO: cod.QPSK 40%(d) and 100% (e); cod.OQPSK 40%(f) and 100%(g); 8state C8PSK 40%(h);16state C8PSK 40% (i)

degradation due to the distortions can be evaluated. By taking the output backoff value into consideration, the BER curves as a function of E_{sat}/N_o are obtained. The BER performances for the coded schemes are evaluated using Monte-Carlo simulations. A Viterbi decoder with a decision delay equal to 32 symbol intervals is simulated. Some results are presented in fig.7. It is found that the degradation of a coded scheme at E_s/N_o approximates the degradation of the corresponding uncoded scheme at the same E_s/N_o; E_s being the transmitted energy per channel symbol. The degradation at a given SNR is defined as the decrease in noise level required on the nonlinear channel in order to maintain the BER obtained on a linear channel with the given SNR. This holds for the (O)QSPK schemes well as for the 8PSK schemes. The latter required the BER evaluation of uncoded 8PSK schemes; these schemes are not included in the set of schemes selected for low-cost applications because their power efficiency is too low. Fig.8 shows the E_{sat}/N_o required for BER=10^{-5} versus the -30 dB level single-sided bandwidth for all schemes considered. No data for input backoff values greater than 3 dB are shown because they result in significantly worse BER performance with only minor decrease in -30 dB level bandwidth.

As to the receiver complexity, we note that all filtered schemes considered require only 2 baseband filters. The uncoded schemes do not require a Viterbi decoder, the C8PSK schemes require a 8 state or a 16 state decoder and the coded (O)QPSK a 64 state decoder.

4. CONSTANT ENVELOPE SCHEMES.

The CPM formats generate a constant envelope signal so that no spectral or BER performance degradation results from the HPA nonlinearity. Hence, a HPA operated at maximum output power (0 dB backoff) is most cost-effective. The presented figures for CPM are simply taken from the literature.

In principle, the optimum MLSE receiver contains a bank of matched filters and a Viterbi decoder. For M-ary schemes with rational $h=k/p$ values and frequency pulses that are time-limited to length L, the number of matched baseband filters and the number of states required in the Viterbi decoder are respectively [4]:

$$F=2M^L$$
$$S=pM^{L-1}$$

and hence, receiver complexity increases exponentionally with the frequency pulse duration. It should be noted that certain schemes require some form of precoding (rate 1/1 coding) in order to suppress error propagation. This will not be taken into account in the receiver complexity comparisons because the presence of precoding does not significantly increase the complexity. Hence, the required number of baseband filters for 2RC, 3RC and Q3RC are 8, 16 and 64 respectively, and the number of decoder states are 8, 16 and 32 respectively. A suboptimum receiver for Q3RC based on a truncated frequency pulse with L=2 requires only 16 baseband filters and 32 decoder states [5]; its performances are comparable to the optimum receiver. The (G)TFM and CORPSK schemes all have infinite frequency pulses and thus in principle only suboptimum receivers are feasible. For the TFM and GTFM schemes, it has been found that near optimum performances can be obtained with a MSK-like receiver, thus requiring only 2 baseband filters and no Viterbi decoder [2]. For the CORPSK(4-5), a suboptimum receiver based on a truncated frequency pulse with L=1 and thus yielding 8 baseband filters and 4 decoder states is considered, whereas for a suboptimum receiver for CORPSK(4-7,1+D) these figures are L=2, 32 baseband filters and 16 decoder states [5].

The performances of the CPM formats are included in fig.8.

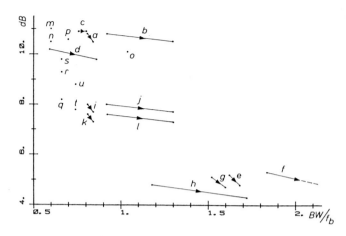

Fig.8 E_{sat}/N_o for BER=10^{-5} versus -30 dB bandwidth
3 to 0 dB IBO (→) for (a) to (l) ; 0 dB IBO for (m) to (u)

(a) unc.QPSK 40%
(b) unc.QPSK 100%
(c) unc.OQPSK 40%
(d) unc.OQPSK 100%
(e) cod.QPSK 40%
(f) cod.QPSK 100%
(g) cod.OQPSK 40%
(h) cod.OQPSK 100%
(i) 8 state C8PSK 40%
(j) 8 state C8PSK 100%
(k) 16 state C8PSK 40%
(l) 16 state C8SPK 100%
(m) TFM
(n) GTFM
(o) 2RC
(p) 3RC
(q) Q3RC
(r) CORPSK(4-5)
(s) CORPSK(4-5) subopt.Rx
(t) CORPSK(4-7,1+D)
(u) CORPSK(4-7,1+D) sub-
    opt.Rx

5. COMPARISON AND CONCLUSIONS

Within the classes of conventional schemes, either uncoded or coded, simulation results show that OQPSK with 100% rolloff and 0 to 3 dB input backoff yields the best BER and spectral performances. Inclusion of the rate 1/2 code considered leads to an increase in power efficiency of more than 5 dB but also a doubling of the bandwidth. Within the classes of C8PSK schemes, best performances are obtained with 40% rolloff and 0 to 3 dB input backoff; the number of states has no influence on the spectra at the HPA output, but going from 8 states to 16 states increases the power efficiency at BER=10^{-5} by 0.4 dB.

When all schemes considered are compared and receiver complexity is taken into account, than the following results are obtained:
- Low complexity (MSK-like) receiver:

Uncoded 100% CRO filtered offset QPSK schemes with 0 to 3 dB input backoff values do not only outperform other uncoded conventional schemes, but also the TFM and GTFM schemes. This yields BER=10^{-5} for E_{sat}/N_o about 10 dB and

a -30 dB level bandwidth from 0.6 to 0.9 f_b.
- Medium complexity receiver (Viterbi decoder with up to 16 states):
  40% CRO filtered C8PSK with 16 states and 0 to 3 dB input backoff outperform other filtered C8PSK schemes and CPM schemes with similar complexity. This yields BER=10^{-5} for E_{sat}/N_o about 7.5 dB and a -30 dB bandwidth from 0.8 to 0.85 f_b.
- High complexity receiver (Viterbi decoder with 64 states):
  Coded 100% CRO filtered offset QPSK schemes with 0 to 3 dB input backoff outperform other coded conventional modulation schemes. This yields BER=10^{-5} for E_{sat}/N_o about 4.5 dB and a -30 dB level bandwidth from 1.2 to 1.8 f_b.

Although spectral efficiency was so far expressed as the -30 dB level bandwidth, our study confirmed similar results if the spectral efficiency is expressed as the 99.9% in-band power bandwidth.

The following conclusions can be drawn:
- For a -30 dB level bandwidth in the order of 0.8 f_b, a trade-off exists between BER performance and complexity.
- A still better performance can be obtained at the expense of both complexity and bandwidth.
- Although constant envelope CPM schemes do not exhibit any performance degradation when using a nonlinear HPA, their resulting BER performance is still worse than when using some specific fluctuating envelope schemes with 0 to 3 dB input backoff and comparable complexity.

REFERENCES

[1] G.Ungerboeck,"Channel coding with multilevel/phase signals," IEEE Trans. Inform. Theory, vol. IT-28, pp. 55-67, Jan. 1982.
[2] K.-S.Chung,"Generalized tamed frequency modulation and its application for mobile radio communications," IEEE Journ. Sel. Areas, vol. SAC-2, pp. 487-497, July 1984.
[3] D. Muilwijk,"Correlative phase shift keying - a class of constant envelope modulation techniques," IEEE Trans. Commun., vol. COM-29, pp. 226-236, March 1981.
[4] T.Aulin, B.Persson and N.Rydbeck, "Spectrally-efficient constant amplitude modulation schemes for communication satellite apllications", Estec contr.nr. 4765/81/NL/MD, May 1982.
[5] D.Muilwijk and J.H.Schadé,"Correlative phase modulation for fixed satellite services," part 1 and supll.report,Estec contr.nr. 4485/80/NL/MS(SC), Jan.83.

ADAPTIVE CHANNEL CODING AS A FADE COUNTERMEASURE IN MILLIMETER
WAVE SATELLITE COMMUNICATIONS

G. TARTARA, R. CARENA

Politecnico di Milano, Dipartimento di Elettronica and
Centro Telecomunicazioni Spaziali, CNR

1. INTRODUCTION

Future satellite communication systems will use millimeter waves (Ka-band, 20/30 GHz) to meet the demand for satellite communications. Lower bands such as C-band and Ku-band are becoming saturated and are more affected by interference from terrestrial radio links. At Ka-band frequencies, narrow beams can be generated using satellite antennas of moderate size, providing multiple spot beams and thus a certain degree of frequency reuse. This results in a substantial capacity increase.

But when using these high frequencies a major problem has to be faced, namely the high attenuation caused by rain. This may require very high fade margins if a high availability is needed in rainy areas. These margins may be too large to be overcome only through the brute force approach of increasing the available peak power, particularly if we consider the fact that the reserve power for counteracting rain effects needs to be used only very unfrequently. For this reason, fade countermeasures are needed that can be adapted to the characteristics of the rain attenuation and also to the type of communication service.

One concept is based on classical site diversity or frequency diversity. These are costly measures since the former implies another station located at a distant site, sufficiently far apart to get a low probability of joint rain occurrence at the two sites, while the latter uses a lower frequency band less affected by rain as a backup.

Another concept is based on providing a common resource to be shared among the stations that at a certain time need protection against fades: this additional reserve resource to be assigned on demand to the links affected by rain, could be a fraction of the bandwidth of the satellite transponder in FDMA systems, or a fraction of the frame interval in a TDMA system [1][2]. This is an efficient solution since it takes advantage of the fact that usually the number of links requiring protection simultaneously is small. Thus a modest amount of reserve capacity is needed to satisfy requests for protection.

Finally the simplest type of countermeasure is to adapt the transmission (modulation and coding) on the affected link to the link attenuation values. One way of doing this without changing modulation and symbol rate (i.e., bandwidth) consists of reducing the source information rate while increasing, correspondingly, the redundancy for error control.

Obviously the above mentioned countermeasures may be used in combination.

2. USE OF FORWARD ERROR CORRECTION

If we consider the various forms in which redundancy for channel coding may be introduced, namely by reserving a fraction of the transponder bandwidth or of the frame interval, or by reducing the information transmission rate from the source without changing the occupied bandwidth or time, it turns out that the power gain achieved with FEC is given by

$$G = G_c + 10\log(1/r) \quad \text{dB} ;$$

G_c is the coding gain of the particular code used, that depends on the ratio (energy per information bit/ noise power spectral density); r is the code rate (information bits/ encoded bits). The available signal power per bandwidth unit is assumed constant.

Practical values for G_c using efficient coding schemes of reasonable complexity may be on the order of 5 to 7 dB: typically we may consider a convolutional code with max. likelihood (Viterbi) decoding or some concatenated scheme for a higher performance. A reasonable value for the code rate r may be 1/2 since lower values for r give marginal improvements [3]. This sets the range of application of coding: FEC can be used for compensating relatively low fades, up to a value, say, of 8 to 10 dB, for reasonable values of the size of the common resource pool, when the information rate is kept constant. When reducing the source information rate, gain improvements that are proportional to rate reduction can be obtained. A convolutional code with r=1/2 (64 states) with soft quantized Viterbi decoding provides about 5.5 dB for G_c. Further increase in gain (without increasing too much the redundancy introduced) can be obtained by concatenating some simple block code (outer code) on the inner convolutional code previously considered [3]. A natural solution for the outer code is given by a Reed-Solomon (RS) code that combines high efficiency with reduced redundancy and complexity.

Figure 1 shows the performance (error probability P) respectively when transmitting without coding (P_o), when using the convolutional code r=1/2 (P_1), and when using the same convolutional code plus a concatenated RS code (P_2). The outer code considered is a (63,57) or a (127,121) RS code. It provides a supplementary coding gain with respect to the inner code of about 2 dB (at $P=10^{-6}$), so that the total code gain is about 7.5 dB.

All this concerns baseband processing and does not affect modulation. Rate reduction and power gain can also be obtained in multilevel phase or amplitude modulation by reducing the number of modulation levels. Note that by using punctured codes [4] it is possible to introduce a variable degree of redundancy without changing the codec basic structure.

For the operation of the rain fade protection schemes, it is required to monitor accurately the channel conditions. The attenuation should be estimated in a precise and rapid way in order to detect the occurrence of rain fades; measurements can be carried out either directly or through the estimation of the bit error probability.

The allocation of a fade countermeasure requires a number of operations for fade detection, signalling, resource assignment and network reconfiguration. A major problem is given by the rate of change of attenuation, and fast fluctuations during rain events.

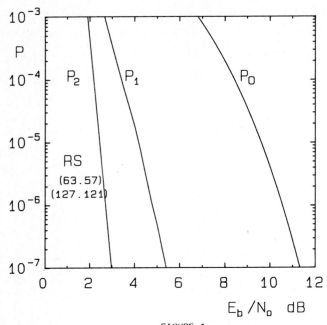

FIGURE 1
BER performance without coding (P_o); with a convolutional code r=1/2(P_1); with the same convolutional code plus a concatenated RS code (P_2).

3. ESTIMATION OF THE BIT ERROR PROBABILITY

A reliable estimate of the bit error rate (BER) requires counting a sufficient number of errors. Therefore, at low values of BER, a too long observation time may be required.

A technique to reduce substantially the observation time consists in increasing artificially the errors in a parallel secondary receiver, by generating pseudoerrors [5]. Then the BER is calculated through extrapolation from the pseudoerror rate. A simple example of secondary receiver for pseudoerror observation is obtained by displacing the data detector thresholds with respect to the optimal thresholds in the main receiver. The error rate in the secondary receiver is increased in this way by a multiplication factor F. The multiplication factor between actual errors and pseudoerrors depends on the actual BER. A value of F=10 is reasonable for BER around 10^{-4} and may grow to the value of 100 for BER around 10^{-6}.

The required accuracy sets the minimum number of pseudoerrors to be observed. For low transmission rates, even the pseudoerror technique may give a too large BER estimation time, if we consider the rate of change of the attenuation.

4. ACTIVATION OF RAIN FADE COUNTERMEASURES

Activation of the adopted fade countermeasure presents some problems due to the dynamic characteristics of the rain attenuation. To obtain a protection against fades perfectly matched to the time-varying channel conditions, the response time of the network control system should be fast enough with respect to the rate of change and fluctuation period of the attenuation. In an ideal system, when the quality of the transmission decays below a certain threshold as measured by a precise and sufficiently fast monitoring system, the adopted fade countermeasure becomes effective with a negligeable delay. Depending upon the type of countermeasure, a certain degree of network reconfiguration and traffic rerouting could be necessary. In practice because of delays and inaccuracies, the crossing of the attenuation threshold S corresponding to the minimum acceptable quality, should be foreseen in advance in order to allow to set the protection against fades before the quality threshold is reached. The anticipation margin M (the protection procedure is initiated when the level S-M is reached for the attenuation) depends on:
- inaccuracy in estimating the transmission quality
- time required to reconfigure the system from the normal condition to the protected one
- rate of change of attenuation

With a simple BER estimator based on pseudoerrors counting, a multiplication factor of the error rate on the order of 10 to 100 may be obtained depending on the value of the error probability. With reference to a QPSK transmission at a rate of 1 Mbaud, the time required to estimate the error probability may be on the order of 10 msec when the error probability is 10^{-5}.

The other relevant source of delay is the time required to reconfigure the network, and this is basically related to the round-trip delay of the transmission through the satellite (about 0.5 sec) and to the type of protocol used. The total delay and the rate of change of attenuation determine the anticipation margin M. This should also take into account the inaccuracy, if any, in measuring the attenuation.

The rate of change R of the attenuation depends on the level of attenuation considered. From the SIRIO satellite experimental data, measured at 11.6 GHz, a statistical distribution of R can be derived [6]. This can be extrapolated at higher frequencies: at 11 Ghz the value of R turns out to be below 1-1.5 dB/sec (depending on the threshold) in the 90% of the cases, so that using extrapolation formulas, we can assume at 20 GHz a value of 2-3 dB/sec as the maximum of the rate of change of attenuation. If we assume a too high value for the maximum R, the resulting margin M may be too large: this implies a low efficiency and the occurrence of frequent false alarms.

Another problem arises because of fluctuating attenuation: during a rain event, attenuation fluctuates and, given a threshold S, several threshold overcoming intervals ("fades") of various different durations may occur. If we examine the statistics of the fade duration, we find that the total time in which attenuation exceeds the threshold is basically determined by the fade intervals of long duration, but the large majority of the fade intervals is of short duration, say less than 10 sec. This fact should be taken into account in designing the protocol for applying the fade countermeasure.

A possible way to cope with this aspect is to introduce a hysteresis in the process of activating the countermeasure when a fade arises and then restoring the normal configuration when the fade ceases. If T is the threshold level at which the procedure of inserting the countermeasure is initiated, the procedure for excluding it restoring the normal configuration is initiated when the attenuation falls below the level T-H, H being the hysteresis value. This in practice reduces drastically the number of fades of short duration, but obviously decreases the efficiency with respect to an ideal system.

Some results based on experimental data derived from SIRIO satellite measurements at the frequency of 11.6 GHz over a three years period of observation time, may clarify the issue (Table 1).

TABLE 1

T (dB)	H=0		H=0.5 dB		DD
	N	D	N	D	
2	79%	94%	22%	99.6%	25%
4	70%	96%	19%	99.6%	8%
6	66%	97.6%	22%	99.4%	6%

frequency: 11.6 GHz

N represents the relative frequency of fades of duration less than 16 sec; D represents the fraction of time in which the threshold T is exceeded because of fades longer than 16 sec; DD represents the increase of the total cumulative duration of fades due to a hysteresis H=0.5 (with respect to the no hysteresis case).

These data show that the large majority of fades are short, even if they contribute not much to the total cumulative fading time. Note that data of Table 1 may be used also at higher frequencies after a scaling of the values of T and H; for instance, the range 2-6 dB at 11.6 GHz may correspond approximately to the range 5-14 dB at 20 GHz.

Other strategies may be used to cope with fast fluctuations of the attenuation, depending also on the type of countermeasure and of the information signal transmitted.

5. CONCLUSIONS

Various types of countermeasures may be considered for protecting the information signal to be transmitted over channels affected by rain attenuation. The most suitable countermeasure depends on the type of signal and on the level of availability and quality required. In some cases hybrid combinations of different types of countermeasures could be convenient.

Use of countermeasures implies switching between different transmission conditions, and this requires a careful definition of the switching protocol because of the dynamic characteristics of the rain attenuation process, in particular because of the fluctuations of the attenuation during a rain event. Introducing some degree of hysteresis in the switching process reduces substantially the number of short fades, without penalizing too much the transmission efficiency.

REFERENCES

[1] A. S. Acampora "The use of resource sharing and coding to increase the capacity of digital satellites" IEEE Journal on Selected Areas in Communications, vol. SAC-1, n.7, January 1983, pp. 133-142 .
[2] F. Carassa "Adaptive methods to counteract rain attenuation effects in the 20/30 GHz band" Space Communications and Broadcasting, vol.2, n.3, Sept. 1984, pp. 253-269.
[3] G. C. Clark and J. B. Cain "Error correction coding for digital communication" 1981 Plenum Press, New York
[4] Y. Yasuda, Y. Hirata, K. Nakamura, S. Otani "Development of variable rate Viterbi decoder and its performance characteristics" 6th Int. Conf. Digital Satellite Commun., Phoenix, Sept. 1983.
[5] E. A. Newcombe, S. Pasupathy "Error rate monitoring for digital communications" Proc. of IEEE, vol.70, n.8, August 1982, pp. 805-828.
[6] E. Matricciani, M. Mauri, A. Paraboni "Dynamic characteristics of rain attenuation: duration and rate of change" Alta Frequenza, vol. 56, Jan. 87, pp. 33-45.

ANALYTICAL PERFORMANCE COMPARISON OF TRELLIS CODED 8-PSK AND QPSK
OVER HARD-LIMITED SATELLITE CHANNELS IN THE PRESENCE OF INTERFERENCE

Nympha Jayamanne[+], Ikuo Oka[++], and Shinsaku Mori[+]

[+]Dept. of Elect. Engineering
KEIO University,
Hiyoshi, Yokohama,
223 JAPAN.

[++]University of Electro-Communications
Chofu, Tokyo,
182 JAPAN.

Abstract

In this paper we present an analytical performance comparison of Trellis coded 8-PSK and BW expanded coded QPSK with soft decision Viterbi decoding, over Hard-Limited (HL) satellite channels in the presence of up-link and down-link interferences in addition to the up and down link noise, with the aid of the linear metric approach and the bounding technique. Accordingly, numerical results are presented for hard limited satellite channels, linear satellite channels, as well as for linear terrestrial channels with the pertinent comments, for 8-PSK coded modulation and the performances are compared with the counterpart uncoded QPSK, and also with the conventional bandwidth (BW) expanding coded QPSK of 8-level soft decision, for various system parameters. We note that in coded 8-PSK and in the interference environment, hard limited satellite channels outperforms the linear satellite channels. It is interesting to note that, substantial larger coding gains could be achieved at higher interference levels than the lower levels and, in particularly than the Gaussian noise alone. For example, in HL channels at a BEP of 10^{-5} and up-link interference (CIU) level of 30 dB, the coding gain is 1.9 dB while for CIU = 20 dB, it is 2.2 dB for coded 8-PSK where as in BW expanded coded QPSK it is 5.9 dB and 7.1 dB respectively.

1. INTRODUCTION

The Trellis Coded Modulation (TCM) has evolved over the past decade as a combined coding and modulation technique for bandwidth and power efficient digital transmission [1]. On the other hand, the modulation system design can be particularly complicated when the channels are severely bandwidth limited, power limited, and subject to impairments caused by interferences. Further, in order to compensate for both AM-AM and AM-PM conversions, a limiter ahead of the TWT amplifier in the satellite, is commonly adopted.

Promising results were obtained by several authors in previous work done in this area of coded modulation, mainly in additive white Gaussian noise (AWGN), which employ convolutional and Viterbi decoding, and further, with related to nonlinear channels ([2] and references there on). Further Miyake et. al. consider coded 8-PSK with simplified metric approach [3]. Almost all the published works on TCM schemes assume an ideal Gaussian channel accept [4] (for QAM with different approach) and with some computer simulation works with interferences [5].

The primary purpose of this paper is to present an analytical approach, to include, hard limited satellite repeater channels of coded MPSK with soft decision Viterbi decoding, subject to degradations, of up-link and down-link Continuous Wave Interferences (CWI), apart from up and down link noise. The proposed analytical method is valid in the bit error probability analysis for band-pass hard limiter (BPHL) type nonlinearity and when the CWIs have constant amplitude and random phase.

Here we employ linear symbol metric quantization and, using the probability density function (pdf) of the difference of the correct and incorrect symbol metrics and integrating over the error region, we get the first event error probability and, the necessary tight upper bound of the bit error probability of soft decision Viterbi decoding, is obtained with the aid of the generating function of the employed code. In addition we have considered the BW expanded coded QPSK (BECQ) with independent coding and modulation, with 8-level soft decision Viterbi decoding and in the analysis of this case we have used the moment technique, in deriving the soft decision probability.

2. SYSTEM MODEL

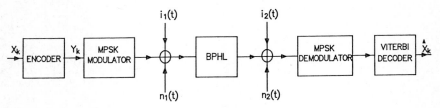

Fig.1. System Model

The overall system model is shown in Fig.1. At the transmit side, the input data sequence X_k is encoded by the convolutional encoder of rate 2/3. The encoded sequence Y_k is then mapped into the channel symbol sequence and modulated to the MPSK signal by the MPSK modulator. The modulated signal $s_0(t)$ is then disturbed by the up-link CWI $i_1(t)$ and up-link noise $n_1(t)$ and then passed through the bandpass hard-limiter (BPHL) on board the satellite. The output of the BPHL is then disturbed by the down-link CWI $i_2(t)$ and down-link noise $n_2(t)$. Finally the received signal is demodulated and Viterbi decoded. The vector diagram of the considered system model is shown in Fig.2 and the phasor diagram is shown in Fig.3. The received signal at the BPHL, u(t), can be written as

$$u(t) = s_0(t) + i_1(t) + n_1(t) \qquad (1)$$

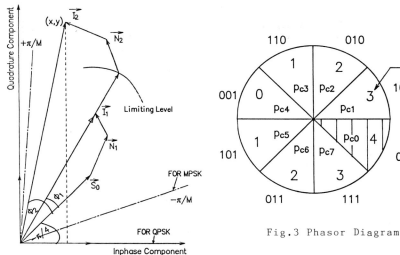

Fig.2. Vector Diagram

Fig.3 Phasor Diagram

The transmitted MPSK signal of unit power is given by
$$s_i = 1 \cdot \exp(j\theta_i) \quad (2)$$
where $\theta_i = \pi(2i+1)/M$ and $i=0,1,2,\ldots,M-1$.
For the considered channel cases, the probability, p_{ci}, of the desired signal to be in the region i could be proved by using the pdf of the received phase [8] as

$$p_{ci} = \frac{1}{M} + \sum_{\nu=1}^{\infty} \frac{\nu}{2\pi\Gamma^2(\nu+1)} \operatorname{Sin}\frac{\nu\pi}{M} \operatorname{Cos}\frac{2\nu\pi}{M}(i+1) \cdot f_1(\nu) \cdot f_2(\nu) \quad (3)$$

where
$$f_k(\nu) = \sum_{n=0}^{\infty} \alpha_k^{\nu/2+n} \frac{(-1)^n}{(n!)^2} \gamma^{-n} \Gamma(\nu/2+n) \cdot {}_1F_1(\nu/2+n; \nu+1; -\alpha_k) \quad (4)$$

α_k = Effective carrier to noise power ratio which depends on the channel case

γ_k = Effective carrier to interference power ratio which depends on the channel case
(k=1 up-link and k=2 down-link)

For HL $\alpha_k = 1/2\sigma_k^2$ and $\gamma_k = 1/|i_k|^2$ where $2\sigma^2$ is the noise power and $|i_k|$ the amplitude of interference.

$\Gamma(.)$ = Gamma Function

${}_1F_1(.;.;.)$ = Confluent Hypergeometric Function

3. BIT ERROR PROBABILITY EVALUATION
(a). For Coded Modulation:-

Without loss of generality we analyze the bit error probability (BEP) for the correct path of all zero sequence. The probability of bit error P_b is bounded by [6]

$$P_b < 1/L_0 \sum_{K=d}^{\infty} C_K P_K \qquad (5)$$

where C_K is the total number of bit error contributed by all the incorrect paths which are at distance K from the correct path, P_K is the first event error probability, L_0 is the number of input bits enters to the encoder at a time, and d is the minimum free distance of the code. The coefficient C_K is uniquely determined by the generation function of the employed code and the corresponding generating function can be represented as

$$T(D,N) = ND^4 + \sum_{h=0}^{\infty} \sum_{a=0}^{[h/2]} \sum_{b=0}^{1+h-a} \sum_{c=0}^{a} 4 \binom{h-a}{a}$$
$$\cdot \binom{1+h-a}{b} \binom{a}{c}$$
$$\cdot N^{4+2h-a} D_1^b D_2^2 D_3^{1+h-a-b} D_4^c \qquad (6)$$

where N is the incorrect input bits, D_k is the distances from the correct path and [h/2] is the nearest integer to h/2 and $\binom{h}{a}=h!/(h-a)!a!$. In derivation of eqn.(6) we have set the path length to (3+h) and L=1, assuming same number of incorrect input bits for large N as well as for small N and have used the Binomial expansion. The coefficient C_K is given by

$$\left.\frac{dT(D,N)}{dN}\right|_{N=1} = \sum_{K=d}^{\infty} C_K D^K \qquad (7)$$

Let J be the metric of the correct path, and J^+ as that of the incorrect path of distance $[d_1,d_2,d_3,d_4]$. Then the first event error probability P_K is given by

$$P_K = \text{Prob.}[J^+ - J \geqslant 0]$$
$$= \text{Prob.}[\sum_{j=1}^{d_1} \{m_1^+(j)-m_1(j)\} + \sum_{j=1}^{d_2} \{m_2^+(j)-m_2(j)\}$$
$$+ \sum_{j=1}^{d_3} \{m_3^+(j)-m_3(j)\} + \sum_{j=1}^{d_4} \{m_4^+(j)-m_4(j)\} \geqslant 0] \qquad (8)$$

where $m_i^+(j)$ and $m_i(j)$, (i=1,2,3,4) are the branch metrics of the incorrect and correct paths respectively and the branches of distance d_i is denoted by subscript i. Let the pdf of $m_i^+(j)-m_i(j)$ be $p_i(m)$, and then the pdf $p_J(J)$

becomes

$$p_J(J) = p_1(m)*\ldots*p_1(m)*p_2(m)*\ldots*p_2(m)$$
$$*p_3(m)*\ldots*p_3(m)*p_4(m)*\ldots*p_4(m) \quad (9)$$

where * denotes the convolution integral. For 8-PSK and for integer metric 0,1,2,3,4 $m_i^+(j)-m_i(j)$ takes the integer values of $-4,\ldots,+4$. The probability of $(m_i^+ - m_i = n)$ is given by,

$$p_i(n) = \sum_{[m_i^+(j)-m_i(j)=n]} p_c \quad (10)$$

where $n=-4,-3,\ldots,0,\ldots,3,4$ and $i=1,2,3,4$ and p_c is given by eqn.(3). Using $p_i(n)$ in eqn.(9) we get $p_J(J)$.

Integration of (8) over the error region $(0,2\pi)$ yields P_K of (5) and can be expressed as

$$P_K = \int_\Delta^\infty p_J(J)dJ + 1/2 \int_{-\Delta}^\Delta p_J(J)dJ \quad 0<\Delta\ll 1 \quad (11)$$

where equality implies that half of the events are in error in the case of, correct and incorrect paths have same path metric.

(b). For BW expanded coded QPSK:-

In BW expanded coded QPSK we have considered 8-level soft decision Viterbi decoding [7]. It is necessary to determine the pdf of the up-link received phase of $(\phi_1-\pi/4)$ with respect to the inphase component (see Fig.2). The corresponding pdf is given by

$$p(\phi_1-\pi/4) = \frac{1}{2\pi} + \sum_{\nu=1}^\infty \frac{\nu}{2\pi\Gamma(\nu+1)} \cos\nu(\phi_1-\pi/4) \cdot f_1(\nu) \quad (12)$$

where $f_1(\nu)$ is given by eqn.(4).

In determining the necessary soft decision probability for 8-level soft decision Viterbi decoding, the moment technique and the Gram-Charlier expansion is used [9].

4. NUMERICAL RESULTS

In numerical calculations, 8-PSK coded modulation is considered for three different channel cases. In linear terrestrial channel case, interference free down-link is taken into account. The number h in (6) is set to 10, which was fully sufficient for the employed code. In BECQ with 8-level soft decision, decision threshold of equal spacing is assumed and the optimum spacing value which minimize BEP was obtained as 0.2. In this case, to compensate the BW doubling we have reduced carrier-to-noise levels of both up and down-link by 3 dB for comparison purposes.

Fig.4-6 show the bit error probability versus up-link carrier-to-noise ratio (CNU), with up-link carrier-to-interference ratio (CIU) as parameter, for hard limited satellite channel, linear satellite channel and the linear terrestrial channel cases and are compared with the counterpart uncoded QPSK of each case. Further analytically verified bit error performance for BECQ with 8-level soft decision are also shown in the same Fig.4-5, for hard limiter and linear satellite channel cases. Here the down-link carrier-to-noise ratio (CND) and down-link carrier-to-interference (CID) is kept constant at 15 dB and 20 dB

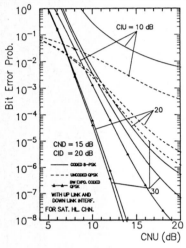

Fig.4 BEP for Hard-Limited Satellite Channel

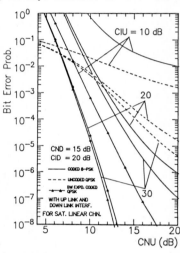

Fig.5 BEP for Linear Satellite Channel

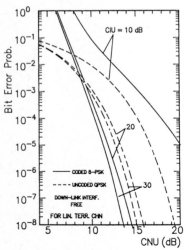

Fig.6 BEP for Linear Terrestrial Channel

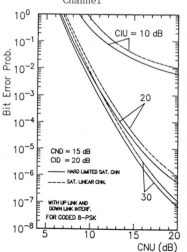

Fig.7 Comparison of BEP for HL and Linear channel for coded 8-PSK

respectively in first two cases. It is interesting to note that, for 8-PSK coded modulation as well as for BECQ, substantially larger relative coding gains could be achieved with higher interference levels than in low interference levels and definitely than Gaussian noise alone.

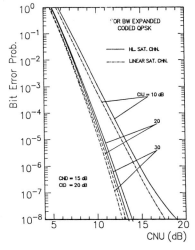

Fig.8 Comparison of BEP for HL and Linear Channel for BECQ

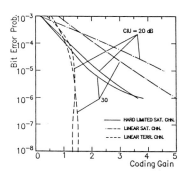

Fig.9 Coding Gain Performance of Coded 8-PSK

Table-1 Obtainable Coding Gains

	CODING GAIN at BEP of 10^{-5} (in dB with respect to Uncoded QPSK)			
	Coded 8-PSK		BW Expanded Coded QPSK	
	CIU=20 dB	CIU=30 dB	CIU=20 dB	CIU=30 dB
Hard-Limited Satellite Channel	2.2	1.9	7.1	5.9
Linear Satellite Channel	4.5	3.5	10.5	8.3
Terrestrial Channel (Down-Link Inter. Free)	1.4	1.3	-	-

In Fig.7 the coded 8-PSK performance for hard limited satellite channel is compared with the linear satellite channel for the above same parameters. Here we note that the hard limited satellite channel case has better performance than the linear satellite channel case in the interference environment even in coded modulation. Fig.8 shows the corresponding performance for BECQ. Here the linear channel outperforms the HL channel.

Fig.9 shows the bit error probability, versus coding gain for coded 8-PSK, with respect to the counterpart uncoded QPSK, with CIU as parameter, for the above considered three channel cases. The coding gains achievable at a BEP of 10^{-5} is summarized in Table.1.

5. CONCLUSIONS

An analytical method has been presented to investigate the performance of coded 8-PSK and BW expanded QPSK, over hard limited satellite channels in the presence of up-link and down-link interferences in addition to the link noise. Using the generating function of the code and the pdf of the metric, we have obtained the bit error probability with the aid of bounding technique. It is found that the hard limited satellite channel has better performance than the linear satellite channel in the interference environment. It is interesting to note that substantial larger coding gains could be obtained at higher interference levels than the lower interference levels and definitely than the noise alone. Further, if we stick in terms of coding gains, for coded 8-PSK, at a BEP of 10^{-6}, HL channel can give rise to coding gain of 3.5 dB at a CIU=30 dB where as for linear channel it is about 5 dB. But the overall performance is better in the case of HL. In BW expanded coded QPSK although the penalty is doubled BW, upto a about 5.9 dB coding gain is achievable for HL channel at a CIU=30 dB where as for CIU=20 dB upto about 7.1 dB is obtainable.

REFERENCES

[1] G. Ungerboeck, "Channel Coding with Multilevel/Phase Signals," IEEE Trans. Inform. Theory, vol.IT-28, pp.55-67, January 1982.
[2] E. Biglieri, "High-Level modulation and coding for nonlinear satellite channels," IEEE Trans. Commun., vol.COM-32, pp. 616-626, May 1984.
[3] M. Miyake, T. Fujino, Y. Umeda and E. Yamazaki, "A study of coded 8PSK modem with Viterbi decoder using simplified metric calculation," in Conf. Rec., PTC, Hawaii, pp.279-286, January 1985.
[4] M. Kaveharad and C. W. Sundberg, "Bit Error Probability of Trellis-Coded Quadrature Amplitude Modulation Over Cross-Coupled Multidimensional Channels," IEEE Trans. Commun.,vol.COM-35,no.4, pp. 369-381, April 1987.
[5] J. Hui and R. J. F. Fang, "Convolutional Code and signal waveform design for band-limited satellite channels," in Conf. Rec., Int. Conf. Commun., Denver, CO, pp.47.5.1-47.5.10, June 1981.
[6] A. J. Viterbi, "Convolutional Codes and Their Performance in Communication Systems," IEEE Trans. Commun., vol.COM-19, pp.751-772, October 1971.
[7] Y. Yasuda, Y. Hirata and A. Ogawa, "Bit Error Rate Performance of Soft Decision Viterbi Decoding," Trans. IECE Japan, vol.E64, no.11,pp.700-707, November 1981.
[8] I. Oka et al, "Interference Immunity Effects in CPSK Systems with Hard-Limiting Transponders," IEEE Trans. vol.AES-17, pp.93-100, January 1981.
[9] N. Jayamanne, I. Oka and S. Mori, "Combined Effects of Adjacent Channel, Intersymbol and CW interference in MSK and OQPSK Hard-Limited Satellite Systems," in Global Telecommun. Conf., Rec., GLOBECOM'86, pp. 29.6.1-29.6.5, December 1986.

Satellite Integrated Communications Networks
E. Del Re, P. Barthelomé and P.P. Nuspl (eds.)
© Elsevier Science Publishers B.V., 1988

SYNCHRONIZATION ALGORITHM FOR CONTINUOUS
PHASE MODULATED SIGNALS

Joachim Habermann
Brown Boveri Research Center,
CH-5405 Baden-Dättwil, Switzerland

ABSTRACT This paper presents an open loop synchronization algorithm to be used for carrier phase and symbol timing extraction in transmission systems which utilize continuous phase modulation. Performance is expressed in terms of jitter variances which are obtained both with analytic calculations and by simulations. The performance of a complete continuous phase modulated transmission scheme, including synchronization, is given for Gaussian and Rayleigh fading channels with the aid of a computer simulation.

1. INTRODUCTION

Continuous phase modulation (CPM) is known to be a bandwidth and power efficient modulation scheme. In transmission channels with nonlinear distortions - such as the satellite or the land mobile radio channel - CPM is even more useful due to its constant envelope. To preserve the power efficiency, it is required to detect CPM signals coherently.

Several synchronization algorithms for CPM signals have so far been investigated in the literature. All these algorithms have in common that they use closed loop structures, either on a data aided or on a conventional phase locked loop basis (see [1], [2]).

To achieve fast acquisition, which is of major importance in fading channels, and strong noise suppression we propose an alternative synchronization scheme, namely an open loop structure.

In section 2 the algorithm is described in detail. Section 3 introduces a procedure which gives optimum CPM pulses in terms of synchronization performance. In section 4 we develop the jitter variance of the algorithm, and in section 5 we present the overall system performance of one CPM scheme in two different transmission channels.

2. PRINCIPLE OF OPEN LOOP SYNCHRONIZATION ALGORITHM

Because CPM signals have been widely investigated we don't want to explain the structure of CPM signals in this paper and therefore refer to the standard literature [3] where the definition of a CPM signal is given and modulator and demodulator structures are treated.

In order to estimate the carrier phase introduced by the channel one has to separate the channel phase from the data dependent phase, which carries per definition the information in a CPM signal. In an AWGN channel the channel phase is a random variable and in a fading channel the channel phase varies with time, i.e. the phase is a stochastic process. In the latter case, however, the bandwidth of this process is limited by the maximum doppler frequency f_d, and therefore is in most cases small compared to the inverse of the symbol duration T. This property allows a reliable estimation of the channel phase.

An idea of accomplishing a separation of channel and data dependent phase can be obtained by investigation of Figure 1 which shows the phase eye diagram of a 4REC CPM (rectangular frequency pulse of length 4T) signal without channel disturbance. It can be seen that the phase takes on one of a small number of values at a special time instant (in most cases the sampling instant) and shows a more scattering behaviour for all other time instants. This behaviour can now be advantageously used by a synchronization algorithm. If the P phase values (in this example P = 8) at the sampling time instant are reduced to one phase value, then for all other time instants the phase values are not reduced to a single point but to a region. The size of this region is increasing with increasing distance from the sampling instant (maximum for an offset of half a symbol duration).

An open loop estimator, which uses this fact, can now work as follows: Similar to the implementation of a maximum likelihood estimator, a symbol interval of length T is subdivided into M subintervals of length ΔT. After reducing the phase as mentioned above, the property of the phase eye diagram is utilized. This is done by observing the received signal over 2N+1 symbols (N ε $\mathbb{N}$) with the aid of a sliding window.

Figure 2 gives the block diagram of the combined carrier phase and sampling instant estimator. After splitting the received signal into quadrature components we introduce analog low pass filtering. Then we sample the incoming signal with an oversampling factor of M - which gives the time resolution of the sampling instant - and do analog to digital conversion. In order to reduce the phase at the sampling instant to one phase value or (for other pulse shapes) to a small phase region, we first multiply the phase of the signal by P. Observing the CPM signal with a sliding window estimator of length KT = (2N+1)T, we then combine K samples which are spaced one symbol interval. We do this for every subinterval (out of M) of the symbol, located in the middle of the sliding window estimator. Of these combined values we calculate M mean values and M variances. Adding the variances of the I- and Q- channel we search for the minimum total variance. The corresponding time value of this variance becomes our estimated "optimum" sampling instant. The mean phase at this time instant selected is calculated via the arctan-function from both quadrature mean values and gives an estimate of the channel phase.

As will be shown in section 4, and as it has been shown for other synchronizers [4], the nonlinearity which multiplies the phase of the signal by P leads to a significant increase in jitter variance at low signal to noise ratios when P is large (P > 4). It is therefore convenient to have small values of P. How this can be accomplished is shown in the next section.

3. TIME DEPENDENT TRELLIS AND OPTIMIZATION OF CPM SIGNALS

In this section we want to extend the above algorithm in such a way that the multiplication factor P in Figure 2 can be reduced and we then present an optimization procedure which yields easily synchronizable CPM signals.

In Figure 3 the phase trellis of tamed frequency modulation (TFM) is shown, which is one special CPM scheme. We observe that the phase values are concentrated at $\phi = 0$ and $\phi = \pi$ for time instant $t = 4T$. Furthermore we notice, that no transitions to $\phi = \frac{\pi}{2}$ and $\phi = 3\pi/2$ occur. A similar behaviour can be observed if $t = 5T$, however all phase values are shifted by $\pi/2$. Further investigations show, that the phase trellis is time variant, having a period of 2T. Figure 4a shows the phase plot of TFM at the sampling instant. These values are mean values - averaged over every symbol interval - and all occur with the same probability. In this case a multiplication factor of P = 8 is necessary to reduce these phase values to one point. If, however, only every second symbol interval is used for averaging - according to our observation in Figure 3 - we obtain the phase plots at the sampling instant given in Figures 4b and 4c for even and odd indices of the symbols. It is easy to verify that the phase values which are located on the coordinate axes have a higher probability of occurrence (numbers in Figure 4) than other values.

This fact can be used by the synchronization algorithm to reduce the factor P. In the above example of TFM we can reduce the multiplication factor from P = 8 to P = 2 by following the time variant trellis of the phase with a phase shift of $\pi/2$ every other symbol. Due to the phase values which are not equal to 0, $\frac{\pi}{2}$, π, and $\frac{3\pi}{2}$ we have introduced a data dependent variance of the phase even at the sampling instant. The effect on the estimator performance is studied in section 4.

We next want to give a system of linear equations which allows us to generate CPM waveforms which concentrate all phase values at the sampling instant at few regions, hence are optimum in terms of synchronization. Because we want to optimize the CPM phase pulse at multiples of the symbol duration (sampling instant), we only have to define the CPM phase pulse $q(t)$ of length L·T [3] in the following way:

$$q(n \cdot T) = \begin{cases} 0 & ; \quad n \leq 0 \\ \frac{1}{2J} \sum_{i=1}^{n} b_i & ; \quad 0 < n \leq L \\ 0.5 & ; \quad n > L \end{cases} \quad (1)$$

where

$$J = \sum_{i=1}^{L} b_i \quad , \quad n \in \mathbb{N}$$

If we now claim that the phase plot (see Fig. 4b,c) is time variant, that the phase values are concentrated around few mean values (given by the modulation index h, and the alphabet size via the accumulated phase [3]), and that the variances to these mean values have to be minimum, we find:

$$F_{b_1} + \lambda_1 + \lambda_2 \cdot \Upsilon_{2,b_1}(b_1\ldots b_L) + \ldots + \lambda_m \cdot \Upsilon_{m,b_1}(b_1\ldots b_L) = 0$$
$$\vdots \qquad \qquad \qquad \qquad \qquad \qquad \qquad \qquad \vdots \qquad (2)$$
$$F_{b_L} + \lambda_1 + \lambda_2 \cdot \Upsilon_{2,b_L}(b_1\ldots b_L) + \ldots + \lambda_m \cdot \Upsilon_{m,b_L}(b_1\ldots b_L) = 0$$

where

$$\Upsilon_{i,b_j} = \frac{\partial \Upsilon_i(b_1\ldots b_L)}{\partial b_j} \quad , \quad \text{and } \Upsilon_i(b_1\ldots b_L) = 0$$

and where $\Upsilon_i(b_1\ldots b_L)$ are the constraints, which can, for instance, influence the spectral shape of the signal. Further we have

$$F_{b_k} = \sum_{i=1}^{Q^{L-1}} 2 \cdot (\underline{\alpha}_i^T - \underline{I}^T \cdot C_i) \cdot (\underline{\alpha}_i^T \cdot \underline{I}_k - C_i) \cdot \underline{b} \qquad (3)$$

$$\underline{\alpha}_i = \begin{bmatrix} \alpha_n^i + \alpha_{n-1}^i + \ldots \alpha_{n-L+2}^i \\ \alpha_{n-1}^i + \alpha_{n-2}^i + \ldots \alpha_{n-L+2}^i \\ \vdots \\ \vdots \\ \alpha_{n-L+2}^i \\ 0 \end{bmatrix} \quad ; \quad \underline{b} = \begin{bmatrix} b_1 \\ \vdots \\ \vdots \\ \vdots \\ b_L \end{bmatrix} \quad \underline{I}_k = \begin{bmatrix} 0 \\ 0 \\ \vdots \\ 0 \\ 1 \\ 0 \\ \vdots \\ 0 \end{bmatrix} \quad ; \text{1 at position k}$$

In Equation 3 we have to insert that vector $\underline{C}$ which leads to a minimum phase variance, where

$$\underline{C} = \begin{bmatrix} C_1 \\ \vdots \\ \vdots \\ C_{Q^{L-1}} \end{bmatrix} \quad \text{and} \quad C_i = \begin{cases} \pm 1, \pm 3, \pm 5 \ldots \\ \text{or} \\ 0, \pm 2, \pm 4 \ldots \end{cases} \qquad (4)$$

α_n^i : value of data at $t = nT$, when sequence i is transmitted

b is the desired vector which gives the phase pulse at the sampling instants. α is the value of the Q-ary input data, and with the modulation index h (which is not included in Equations 2 and 3) we can select the number of mean values. h = 1/2 gives two mean values. It can easily be shown that TFM is optimum in terms of synchronization defined in Equation 2 (L = 3). Other pulse shapes can be found. For a pulse length of L = 5, h = 1/2 we can have both very good bandwidth efficiency - introduced via suitable constraints - and synchronization performance. The above investigations show that if h = 1/n (n ε ℕ) it is possible to use the time variant phase trellis advantageously for synchronization in such a way that the multiplication factor is equal to P = n.

4. JITTER VARIANCE IN AWGN CHANNEL

For comparison of the algorithm given in sections 2 and 3 with other algorithms and bounds we have to define the single sided bandwidth B_s of the algorithm. Because the algorithm observes the signal over a time interval of (2N+1)T, we find the overall pulse response as:

$$h_s(t) = \begin{cases} \dfrac{1}{(2N+1) \cdot T} & ; \quad |t| \leq NT \\ 0 & ; \quad \text{elsewhere} \end{cases} \qquad (5)$$

This gives immediately the equivalent bandwidth B_s of the system:

$$B_s = \frac{1}{(4N+2)T} \qquad (6)$$

For further investigations we assume that carrier phase and symbol timing estimation are independent.

4.1. Carrier phase jitter variance

In this subsection we assume that the symbol timing is ideal, i.e. the sampling instant t_{opt} is known.

We have to distinguish between two cases. First, if the phase reduction is accomplished by a high multiplication factor P in such a way that all phase values coincide in one point, then no data dependent term exists, hence the phase jitter variance is only a function of the noise. The noise dependent variance, however, has been calculated in [4]. Second, if we use the modification of the algorithm described in section 3, we introduce a data dependent variance. Considering that noise and data are independent, we get:

$$\sigma_\phi^2 = \sigma_{\phi,noise}^2 + \sigma_{\phi,data}^2 \quad , \tag{7}$$

which is only an approximation due to the modulo function of the phase. The data dependent variance can be calculated with the aid of the known phase pulse q(t) as:

$$\sigma_{\phi,data}^2 = E_\alpha \left\{ \left[\frac{1}{P(2N+1)} \cdot \sum_{n=-N}^{+N} \mathrm{mod}_{2\pi} \left\{ P2\pi h \right. \right. \right.$$

$$\left. \cdot \sum_{i=n-L+1}^{n} \alpha_i \cdot q(t - iT + nT) \right. \tag{8}$$

$$\left. \left. + P\pi h \cdot \sum_{i=-\infty}^{n-L} \alpha_i \right\} \right]^2 \right\} \quad ; \quad t = t_{opt}$$

where we have assumed that the mean value of the phase is zero.

In Figure 5 the results are compared with the Cramer Rao bound [5] and an algorithm investigated in [2]. TFM is used as modulation format. Estimator bandwidth is $B_s \cdot T = 5 \cdot 10^{-3}$, and input filter bandwidth is $B_{LP} \cdot T = 0.6$. Figure 5 shows that utilizing the modification of section 3 (P=2), gives good performance at low signal to noise ratios, which is of interest for mobile radio applications. If the signal to noise ratio is guaranteed to be high, a high multiplication factor (P=8) leads to nearly optimum results.

4.2. Symbol timing jitter variance

An analytic approximation of the jitter variance can be obtained, which, however, is rather lengthy. Results are comparable to those of the carrier phase in section 4.1.

5. PERFORMANCE OF SYSTEM INCLUDING SYNCHRONIZATION

Here we give the performance of a transmission system which utilizes TFM to modulate the incoming data. A coherent receiver is used, which incorporates optimum detection. Synchronization of carrier phase and symbol timing is accomplished by the above described algorithm. Figure 6 shows the performance in the AWGN and Figure 7 in the Rayleigh fading channel. Parameters of the simulation are: $B_s \cdot T = 10^{-2}$, $B_{LP} \cdot T = 0.6$ and $f_d \cdot T = 2 \cdot 10^{-3}$ (Rayleigh channel).

It can be seen that the loss due to the synchronization algorithm is relatively small for low signal to noise ratios. An improvement in performance compared to a noncoherent systems can be obtained [6]. In the fading channel we observe an irreducible error rate of 10^{-3}, which is caused by the synchronization algorithm. It should however be pointed out, that in most cases error rates of less than 10^{-3} are of no relevance, since an additional error correcting code could cope with such error rates.

6. CONCLUSIONS

An open loop estimator for carrier phase and symbol timing of CPM signals has been presented. The algorithm is, in principle, applicable to all CPM schemes, but if the time dependence of the phase trellis is used for enhancing the system performance, specific modulation indices have to be used (h = 1/n; n ε **N**). It is shown that synchronization losses are quite small and that the algorithm can be used even in fading channels with reasonable Doppler frequencies.

REFERENCES

[1] G. Ascheid, "Optimale Detektion und Synchronisation mit digitaler Signalverarbeitung", Dissertation, Technical University Aachen, FRG, 1984.
[2] Andrea, Mengali, Reggiannini, "Synchronization of CPM signals", Proceedings second Tirrenia workshop, Italy, 1985.
[3] Anderson, Aulin, Sundberg, "Digital phase modulation", Plenum Press, 1986.
[4] A.J. Viterbi, A.M. Viterbi, "Non-linear estimation of PSK modulated carrier phase with application to burst digital transmission", IEEE Trans. on Inf. Theory, Vol. IT-29, July 1983.
[5] I. Bruyland, M. Moeneclaey, "The joint carrier and symbol synchronizability of CPM-waveforms", Proceedings ICC, p. 31.5.1-31.5.5, 1986.
[6] M. Aldinger, H.P. Kuchenbecker, "Schmalbandmodulationsverfahren mit konstanter Hüllkurve für die digitale Signalübertragung, NTG Fachtagung: Bewegliche Funkdienste, 1985.

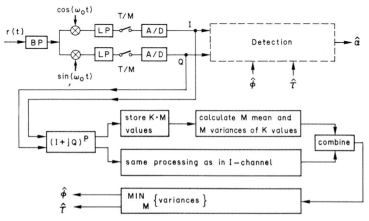

Fig. 2: Block diagram of estimator.

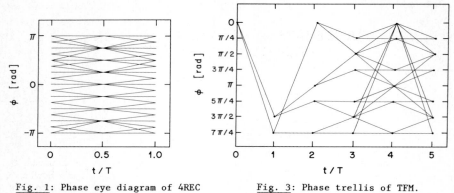

Fig. 1: Phase eye diagram of 4REC CPM signal.

Fig. 3: Phase trellis of TFM.

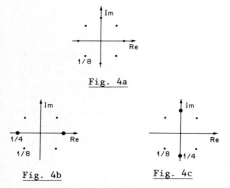

Phase plots of TFM at sampling instant.

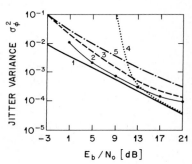

Fig. 5: 1: Cramer Rao bound
2: analytic, $P = 2$
3: simulation, $P = 2$
4: analytic, $P = 8$
5: Other algorithm [2]

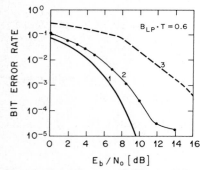

Fig. 6: Performance of TFM in AWGN channel
1: ideal PSK (reference)
2: coherent, $P = 2$
3: noncoherent 3RC CPM, [6]

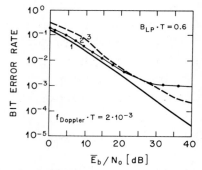

Fig. 7: Performance of TFM in Rayleigh channel
1: ideal PSK (reference)
2: coherent, $P = 2$
3: noncoherent 3RC CPM, [6]

PULSE SHAPE OPTIMIZATION IN PARTIAL RESPONSE CPM SYSTEMS

M. Campanella, U. Lo Faso, G. Mamola

Dipartimento di Ingegneria Elettrica, Università di Palermo.
Viale delle Scienze 90138 Palermo, Italy (*).

Abstract. A procedure for the determination of the baseband pulse shape able to yield an optimum spectral performance with a prescribed minimum Euclidean distance is illustrated for the case of partial response constant envelope CPM systems. The effective bandwidth is used as a measure of spectral efficiency.

1. Minimum distance and effective bandwidth.

In this work the problem of the determination of the pulse shape for optimum spectral behaviour with a prescribed bit error rate performance is afforded in the case of partial response constant envelope signaling with binary input data. The spectral performance is described by means of effective bandwidth, while the error probability is characterized by the minimum Euclidean distance, which is a meaningful figure of merit when the signal-to-noise ratio is sufficiently high. The transmitted signal is

$$(1) \qquad v(\mathbf{a},t) = V_0 \operatorname{Re}\{\exp j[\omega_0 t + \varphi(\mathbf{a},t)]\}$$

where V_0, ω_0 are the carrier amplitude and frequency and φ has the form

$$(2) \qquad \varphi(\mathbf{a},t) = 2\pi h \sum_{-\infty}^{\infty} a_n q(t - nT) + \varphi_0 \qquad a_n \in \{-1, 1\}$$

In (2) $\{a_n\}$ is a sequence of independent, equiprobable binary data transmitted at a rate $1/T$; $q(t)$ is a causal, continuous baseband pulse shape ($1/2$ for $t > rT$, r integer); h is the modulation index and φ_0 is an arbitrary phase.

The distance D_e between two different signals associated with two sequences $\mathbf{a}_1$ and $\mathbf{a}_2$, using the narrowband hypothesis, can be written as

$$(3) \qquad D_e^2 = \int_{-\infty}^{\infty} [v(\mathbf{a}_1,t) - v(\mathbf{a}_2,t)]^2 \, dt = 2V_0^2 \int_{-\infty}^{\infty} \sin^2 \frac{\varphi(\mathbf{a}_1,t) - \varphi(\mathbf{a}_2,t)}{2} \, dt$$

Introducing the difference sequence $\{\gamma_n\}$, we obtain

---

(*) - This work is supported by Ministero della Pubblica Istruzione of Italy.

(4) $$D_e^2 = 2V_0^2 \int_{-\infty}^{\infty} \sin^2[\pi h \sum_{-\infty}^{\infty} \gamma_n q(t-nT)] dt \qquad \gamma_n = a_{1n} - a_{2n}, \; \gamma_n \in \{-2,0,2\}$$

The minimum value D of D_e is obtained for a suitable difference sequence which must have only a finite number of nonzero terms with zero sum [1]. Denoting with k the subscript of the last nonzero term, we have

(5) $$D^2 = 2V_0^2 \int_{-\infty}^{\infty} \sin^2[\pi h \sum_{0}^{k} \gamma_n q(t-nT)] dt$$

Under the assumption that the minimizing sequence consists only of two terms, in (5) we will take $k = 1$, $\gamma_0 = 2$ and $\gamma_1 = -2$. Normalizing D^2 to the bit energy $V_0^2 T/2$ and introducing the functions

(6) $$q_p(t) = q(t + pT) \qquad (0 \leq t \leq T)$$

we have

(7) $$d^2 \triangleq \frac{2D^2}{V_0^2 T} = \frac{4}{T} \sum_{p=0}^{r} \int_0^T \sin^2 2\pi h[q_p(t) - q_{p-1}(t)] dt$$

We observe that for a spectrally efficient signaling, the baseband pulse must not change too much in a single signaling interval, so that an approximate expression of d^2 can be used

(8) $$d^2 \simeq \frac{16\pi^2 h^2}{T} \int_0^T \sum_{p=0}^{r} [q_p(t) - q_{p-1}(t)]^2 dt$$

The effective bandwidth is defined as

(9) $$B_e^2 = 4 \int_0^{\infty} (f-f_0)^2 W_v(f) df \Big/ \int_0^{\infty} W_v(f) df$$

where $W_v(f)$ is the spectral density of $v(a,t)$. It is easily shown that [2]

(10) $$B_e^2 = -\frac{1}{\pi^2} \frac{d^2 R_x}{d\tau^2}\Big|_{\tau=0} \Big/ R_x(0)$$

where $R_x(\tau)$ is the time-ensemble averaged autocorrelation function of the baseband complex signal $x(t) = V_0 \exp[j\varphi(a,t)]$. An explicit evaluation of the autocorrelation function $R_x(\tau)$ leads to

(11) $$B_e^2 = \frac{4h^2}{T} \int_0^{rT} \dot{q}^2(t) dt = \frac{4h^2}{T} \int_0^T \sum_{p=0}^{r-1} \dot{q}_p^2(t) dt$$

2. Determination of the optimum pulse shape.

To obtain the pulse shape for the optimum spectral performance, the functional (11) must be minimized with the constraint of a prescribed value of d^2 as given by (8); so that the first variation of the functional

$$(12) \quad F = B_e^2 T^2 - \lambda d^2 = 4h^2 \int_0^T \left\{ \sum_{p=0}^{r-1} T \ddot{q}_p^2(t) - \frac{4\lambda \pi^2}{T} \sum_{p=0}^{r} [q_p(t) - q_{p-1}(t)]^2 \right\} dt$$

must be imposed to be zero for arbitrary variations of $q_p(t)$ and h. The vanishing of the variation with respect to h leads to

$$(13) \quad F = B_e^2 T^2 - \lambda d^2 = 0$$

so that

$$(14) \quad \lambda = \frac{B_e^2 T^2}{d^2}$$

Defining

$$(15) \quad \vartheta = \frac{2\pi B_e}{d} t \quad \text{and} \quad \tau = \frac{2\pi B_e}{d} T$$

the functional F becomes proportional to

$$(16) \quad S = \frac{1}{2} \int_0^\tau \left\{ \sum_{p=0}^{r-1} \ddot{q}_p^2(\vartheta) - \sum_{p=0}^{r} [q_p(\vartheta) - q_{p-1}(\vartheta)]^2 \right\} d\vartheta$$

so that the extremum conditions for S must be found with the constraints:

(17,a) $q_{-1}(\vartheta) \equiv 0$ $q_r(\vartheta) \equiv 1/2$

(17,b) $q_0(0) = 0$ $q_{r-1}(\tau) = 1/2$

(17,c) $q_h(\tau) = q_{h+1}(0)$ (h = 0,1,...,r-2)

Eq. (17,a) follows from the definition of q(t), while (17,b) and (17,c) follow from the hypothesis of continuity of the pulse shape.

Taking into account the following position (18) in (16), we have

$$(18) \quad y_h = q_h - \frac{h+1}{2(r+1)}$$

$$(19) \quad S = \frac{1}{2} \int_0^\tau \left\{ \sum_{p=0}^{r-1} \ddot{y}_p^2 - \sum_{p=0}^{r} [y_p - y_{p-1}]^2 \right\} d\vartheta - \frac{\tau}{8(r+1)}$$

with the conditions

(20,a) $\quad y_{-1}(\vartheta) = y_r(\vartheta) = 0$

(20,b) $\quad y_0(0) = -\dfrac{1}{2(r+1)} \qquad y_{r-1}(\tau) = \dfrac{1}{2(r+1)}$

(20,c) $\quad y_h(\tau) = y_{h+1}(0) + \dfrac{1}{2(r+1)} \quad (h = 0,1,\ldots,r-2)$

Making the following transformation of variables (21), S takes the form (22)

(21) $\quad y_h(\vartheta) = \sum_{k=1}^{r} Q_k(\vartheta) \sin \dfrac{\pi k(h+1)}{r+1}$

(22) $\quad S = \dfrac{r+1}{4} \int_0^\tau \sum_{k=1}^{r} (\dot{Q}_k^2 - \omega_k^2 Q_k^2) \, d\vartheta - \dfrac{\tau}{8(r+1)} = \int_0^\tau L \, d\vartheta$

being

(23) $\quad L = \dfrac{r+1}{4} \sum_{k=1}^{r} [\dot{Q}_k^2(\vartheta) - \omega_k^2 Q_k^2(\vartheta)] - \dfrac{1}{8(r+1)} \quad$ with $\quad \omega_k^2 = 4\sin^2 \dfrac{\pi k}{2(r+1)}$

The extremum condition for (22) furnishes

(24) $\quad \ddot{Q}_k + \omega_k^2 Q_k = 0 \quad (k = 1,2,\ldots,r)$

The general solution of (24) involves 2r arbitrary integration constants. Therefore a further optimization must be performed with respect to these constants, taking into account the boundary conditions.

The variation of S due only to the integration constant variations is

(25) $\quad dS = \sum_{k=1}^{r} (P'_k \, dQ'_k - P_k \, dQ_k)$

where

(26) $\quad Q_k = Q_k(0) \quad Q'_k = Q_k(\tau) \quad P_k = \dfrac{\partial L}{\partial \dot{Q}_k}\bigg|_{\vartheta=0} \quad P'_k = \dfrac{\partial L}{\partial \dot{Q}_k}\bigg|_{\vartheta=\tau}$

Owing to (24), the quantities P_k and P'_k are linear functions of Q_k and Q'_k whose expressions can be obtained from the general solution of (24)

(27) $\quad Q_k(\vartheta) = A_k \cos\omega_k \vartheta + B_k \sin\omega_k \vartheta$

and from the definition (23) of L. More specifically, the constants A_k and B_k can be expressed, using (27), in terms of Q_k and Q'_k as follows

(28) $$A_k = Q_k \qquad B_k = \frac{Q'_k - Q_k \cos\omega_k \tau}{\sin\omega_k \tau}$$

and, taking into account (26), we obtain

(29) $$P_k = \frac{r+1}{2} \omega_k \frac{Q'_k - Q_k \cos\omega_k \tau}{\sin\omega_k \tau} \qquad P'_k = \frac{r+1}{2} \omega_k \frac{Q_k \cos\omega_k \tau - Q'_k}{\sin\omega_k \tau}$$

Substituting (29) in (25) and integrating the form dS, we have

(30) $$S = \frac{r+1}{4} \sum_{k=1}^{r} \omega_k \frac{(Q_k^2 + Q_k'^2)\cos\omega_k \tau - 2Q_k Q'_k}{\sin\omega_k \tau} - \frac{\tau}{8(r+1)}$$

In order to exploit the boundary conditions (20,b) and (20,c) we return to the old variables $y_k = y_k(0)$ and $y'_k = y_k(\tau)$ using the inverse transformation

(31) $$Q_k(\vartheta) = \frac{2}{r+1} \sum_{h=0}^{r-1} y_h(\vartheta) \sin\frac{\pi k(h+1)}{r+1}$$

The expression of S in terms of these variables is

(32) $$S = \frac{1}{r+1} \sum_{k=1}^{r} \sum_{m=0}^{r-1} \sum_{n=0}^{r-1} \omega_k \frac{(y_m y_n + y'_m y'_n)\cos\omega_k \tau - 2y_m y'_n}{\sin\omega_k \tau}$$

$$\cdot \sin\frac{\pi k(m+1)}{r+1} \sin\frac{\pi k(n+1)}{r+1} - \frac{\tau}{8(r+1)}$$

It is easily seen that S has the following properties

(33,a) $S(\mathbf{y},\mathbf{y'}) = S(\mathbf{y'},\mathbf{y})$

(33,b) $S(y'_0, y'_1, ..., y'_{r-1}, y_0, y_1, ..., y_{r-1}) = S(y'_{r-1}, y'_{r-2}, ..., y'_0, y_{r-1}, y_{r-2}, ..., y_0)$

(33,c) $S(\mathbf{y},\mathbf{y'}) = S(-\mathbf{y},-\mathbf{y'})$

As a consequence, the function S is invariant under the change of variables

(34) $$(y_0, y_1, ..., y_{r-1}, y'_0, y'_1, ..., y'_{r-1}) \to (-y'_{r-1}, -y'_{r-2}, ..., -y'_0, -y_{r-1}, -y_{r-2}, ..., -y_0)$$

We note also that the conditions (20,b,c) are invariant under the same change of variables, so that the unique solution must obey the conditions

(35) $$y_0 = -y'_{r-1}, \; y_1 = -y'_{r-2}, ..., y_{r-1} = -y'_0$$

which allow a substantial reduction of variables. In fact from (35) we can express the variables $\mathbf{y'}$ in terms of the variables $\mathbf{y}$. Therefore, using a matrix notation, the function S becomes

$$(36) \quad S(\mathbf{y},\mathbf{y}') \stackrel{\Delta}{=} \Phi(\mathbf{y}) = \frac{2}{r+1} \sum_{m=0}^{r-1} \sum_{n=0}^{r-1} y_m y_n \left\{ \sum_{k=1}^{r} \omega_k \frac{\sin\frac{\pi k(m+1)}{r+1} \sin\frac{\pi k(n+1)}{r+1}}{\sin\omega_k \tau} \right.$$
$$\left. \cdot \left(\cos\omega_k \tau + (-1)^{k+1}\right) \right\} - \frac{\tau}{8(r+1)} = \mathbf{y}^T \mathbf{A} \mathbf{y} - \frac{\tau}{8(r+1)}$$

The constraints (20,b) and (20,c) become

$$(37) \quad y_0 = -\frac{1}{2(r+1)} \qquad y_p + y_{r-p} = -\frac{1}{2(r+1)} \quad (p=1,2,\ldots,r-1)$$

At this point it is necessary to distinguish between the cases r odd and r even. When r is odd, putting

$$(38) \quad r = 2\nu + 1$$

$$(39) \quad \mathbf{y} = \mathbf{b} + \boldsymbol{\eta} \quad \text{with} \quad b_0 = -\frac{1}{2(r+1)} \quad b_h = -\frac{1}{4(r+1)} \quad (h \neq 0)$$

the conditions (37) become

$$(40) \quad \eta_0 = 0 \qquad \eta_p + \eta_{r-p} = 0 \qquad (p=1,2,\ldots,r-1)$$

From (36) we have

$$(41) \quad \Phi = \boldsymbol{\eta}^T \mathbf{A} \boldsymbol{\eta} + 2\mathbf{b}^T \mathbf{A} \boldsymbol{\eta} + \mathbf{b}^T \mathbf{A} \mathbf{b} - \frac{\tau}{8(r+1)}$$

The constraints (40) can be taken into account by the introduction of a vector **u** of ν free variables by writing

$$(42) \quad \boldsymbol{\eta} = \mathbf{T}^T \mathbf{u} \quad \text{with} \quad \mathbf{T} = \begin{bmatrix} \mathbf{0} & \mathbf{I}_\nu & -\mathbf{E}_\nu \end{bmatrix}$$

where **0** denotes the ν - component zero column vector, $\mathbf{I}_\nu$ is the identity matrix of order ν and $\mathbf{E}_\nu$ is the square matrix of order ν

$$(43) \quad \mathbf{E}_\nu = \begin{bmatrix} 0 & 0 & \ldots & 0 & 1 \\ 0 & 0 & \ldots & 1 & 0 \\ & & \ldots & & \\ 1 & 0 & \ldots & 0 & 0 \end{bmatrix}$$

In terms of the free variables **u** the function Φ becomes

$$(44) \quad \Phi = \mathbf{u}^T \mathbf{T} \mathbf{A} \mathbf{T}^T \mathbf{u} + 2\mathbf{b}^T \mathbf{A} \mathbf{T}^T \mathbf{u} + \mathbf{b}^T \mathbf{A} \mathbf{b} - \frac{\tau}{8(r+1)}$$

The extremum condition for (44) can be written as

$$(45) \quad \mathbf{T} \mathbf{A} \mathbf{T}^T \hat{\mathbf{u}} + \mathbf{T} \mathbf{A} \mathbf{b} = \mathbf{0}$$

which allows determining the optimum value $\hat{\mathbf{u}}$ of the unknown **u**.

A tedious but straightforward calculation furnishes the vector of constants

TAb and the matrix of coefficients **TAT**^{T}. We have, for $\alpha = 1,2,\ldots,\nu$

$$(46) \quad \mathbf{TAb}_\alpha = -\frac{1}{(r+1)^2} \left\{ \sum_{\substack{n=1 \\ (n \text{ odd})}}^{r} \cos\frac{\pi n(\alpha+1/2)}{r+1} \sin\frac{\pi n}{r+1} (2 - \cos\frac{\pi n}{r+1}) \cot\frac{\omega_n \tau}{2} \right.$$

$$\left. - \sum_{\substack{n=1 \\ (n \text{ even})}}^{r} \sin\frac{\pi n(\alpha+1/2)}{r+1} \sin^2\frac{\pi n}{r+1} \operatorname{tg}\frac{\omega_n \tau}{2} \right\}$$

and furthermore for $\alpha, \beta = 1,2,\ldots,\nu$

$$(47) \quad \mathbf{TAT}^T_{\alpha\beta} = -\frac{4}{r+1} \left\{ \sum_{\substack{n=1 \\ (n \text{ even})}}^{r} [\sin\frac{\pi n\alpha}{r+1} + \sin\frac{\pi n(\alpha+1)}{r+1}][\sin\frac{\pi n\beta}{r+1} + \sin\frac{\pi n(\beta+1)}{r+1}] \right.$$

$$\sin\frac{\pi n}{2(r+1)} \operatorname{tg}\frac{\omega_n \tau}{2} - \sum_{\substack{n=1 \\ (n \text{ odd})}}^{r} [\sin\frac{\pi n\alpha}{r+1} - \sin\frac{\pi n(\alpha+1)}{r+1}][\sin\frac{\pi n\beta}{r+1} - \sin\frac{\pi n(\beta+1)}{r+1}]$$

$$\left. \sin\frac{\pi n}{2(r+1)} \cot\frac{\omega_n \tau}{2} \right\}$$

The condition (13), taking into account (44) and (45), furnishes

$$(48) \quad \mathbf{b}^T \mathbf{AT}^T \hat{\mathbf{u}} + \mathbf{b}^T \mathbf{Ab} - \frac{\tau}{8(r+1)} = 0$$

which allows to determine τ, considering its smallest positive solution, as it follows from the meaning of τ as expressed in (15).

The explicit calculations for $\mathbf{b}^T \mathbf{Ab}$ can be performed. The result is

$$\mathbf{b}^T \mathbf{Ab} = \frac{1}{4(r+1)^2} \left\{ \sum_{\substack{n=1 \\ (n \text{ odd})}}^{r} \left(\frac{\cos\frac{\pi n}{2(r+1)}}{2\sin^2\frac{\pi n}{2(r+1)}} + \cos\frac{\pi n}{r+1} \right) \sin\frac{\pi n}{r+1} (2 - \cos\frac{\pi n}{r+1}) \cot\frac{\omega_n \tau}{2} \right.$$

(49)

$$\left. - \sum_{\substack{n=1 \\ (n \text{ even})}}^{r} \left(\frac{1}{2\sin\frac{\pi n}{2(r+1)}} + \sin\frac{\pi n}{2(r+1)} \right) \sin^2\frac{\pi n}{r+1} \operatorname{tg}\frac{\omega_n \tau}{2} \right\}$$

The previous relations allow determining the optimum signal. In fact, from (48) we calculate the parameter τ. From (45) the value of **u** is determined corresponding to the value of τ. Using (39) and (42) we obtain the vector **y**. The vector **y'** is therefore fixed by (35). The variables Q_k and Q'_k are evaluated using (31). Expressions (28) are then used to find A_k and B_k. Eq. (27) furnishes therefore the expressions of $Q_k(\vartheta)$. Eq. (21) gives the expression of $y_k(\vartheta)$ and successively, through (18), $q_k(\vartheta)$. The signal $q(t)$ is finally determined using (15).

When r is even, a similar calculation can be performed. The result is that, putting $r = 2\nu$, all the previous relations hold provided that the labels α and β run from 1 to $\nu - 1$. In the exceptional case $r = 2$ ($\nu = 1$), no unknown vector **u** of free variables is defined, so that eq. (48) for the determination of τ reduces to the last two terms. The vector η must be taken equal to zero. The remaining procedure for the determination of the optimum signal is exactly the same as the one previously discussed.

[1] T.Aulin, N.Rydbeck, C.-E.W.Sundberg, "Continuous phase modulation - Part II: Partial response signaling". *IEEE Trans. Commun.*, vol. COM-29, pp. 210-225. March 1981.

[2] M.Campanella, U.LoFaso, G.Mamola, "Bandwidth - distance optimization in full response constant envelope modulation systems". *Alta Frequenza* vol. LVI, n.4, pp. 205-213. June 1987.

LAND MOBILE SATELLITE CHANNEL - MODEL AND ERROR CONTROL [*]

Erich Lutz, Frank Dolainsky, Wolfgang Papke

German Aerospace Research Establishment, DFVLR
Digital Networks Division
D-8031 Wessling, Federal Republic of Germany

The fading behaviour of the land mobile satellite channel has been recorded for different satellite elevations, antennas, and types of environment. The statistical analysis of the recordings include the power spectral density of the received signal and bit error burst statistics. The important parameters of a channel model are discussed in dependence on satellite elevation. The application of automatic repeat request for reliable data transmission is addressed. It is shown that the throughput efficiency can be maximized by a proper selection of code rate and signalling speed.

1. INTRODUCTION

Satellite communication with landmobile terminals is limited by strong variations of the received power level which are caused by multipath fading and signal shadowing.

In order to investigate the time varying behaviour of the satellite channel several recording experiments have been performed. An unmodulated test carrier was transmitted from the ESA ground station in Villafranca (Spain) and relayed by the geostationary satellite MARECS in the L-band (1.54 GHz). The test carrier was received by a van equipped with different antennas: a conical spiral antenna C3 (3 dBi nominal gain) with hemispherical characteristic, a drooping crossed dipole antenna D5 (5 dBi gain) and a cylindrical slot antenna S6 (6 dBi gain), both having a toroidal characteristic with vertical selection. The received carrier was mixed into base band, and its inphase and quadrature components were recorded on magnetic tape.

The measurements were conducted in different environments (city, suburbs, rural roads, forest, and highway) and in areas with different satellite elevations: Stockholm ($13°$), Copenhagen ($18°$), Hamburg ($21°$), Munich ($24°$), Barcelona ($34°$), and Cadiz ($43°$).

The recorded time varying behaviour of the land mobile channel can be reproduced in amplitude and phase for various purposes (stored channel principle [1]). In section 2 error burst statistics are given as an example for the statistical evaluation of the channel recordings. Section 3 discusses the main parameters of a channel model and their dependence on satellite elevation.

---

[*] Part of this work was performed under contract to Racal-Decca Advanced Development Ltd., England, as part of a larger study carried out for the European Space Agency. The permission by INMARSAT to use the MARECS satellite is acknowledged. Views expressed in this paper are not necessarily those of INMARSAT.

In order to provide the mobile users with reliable (low rate) data transmission, automatic repeat request (ARQ) is necessary. The throughput performance of ARQ has been evaluated by software simulation based on the recorded land mobile channels. The influence of FEC code rate and signalling rate on the throughput efficiency of ARQ is discussed in section 4.

2. STATISTICAL EVALUATION OF CHANNEL RECORDINGS

The analog magnetic tape recordings were digitized and transferred to a mainframe computer which offers various possibilities for a statistical evaluation of the recorded data.

Fig. 1a shows a typical example of the received power level during a channel recording in the city of Munich. The figure shows a high-frequency fading process which is superimposed on a low-frequency shadowing process. For instance, in the time interval 928 sec to 932 sec a crossroad permits an unobstructed "view" of the satellite, while before and after this period the satellite is hidden by multi-storey flats. Relatively "good" and very "bad" channel periods can be distinguished, having a mean level difference of approx. 15 dB. Fig. 1b shows the received power level during a recording on a highway. Here, most of the time only small level variations predominate. Total shadowing is caused by a bridge at 684 sec.

During the recordings the vehicle velocity v was kept constant in order to allow a conversion of the time scale to other velocities or into meters: The distance of the fading events (in meters) are determined by the stationary electromagnetic field and therefore are independent from the mobile velocity. According to the relation:
velocity = distance/time, the time duration of the fading events is inversely proportional to the velocity of the mobile terminal.

Fig. 2 shows the power spectral density (normalized to unit power) of the received signal in the city of Munich. The slow frequency component below approx. 4 Hz corresponds to the shadowing process. The component with cutoff frequency 8 Hz represents the multipath fading process. The pronounced cutoff frequency agrees well with the

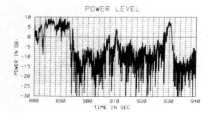 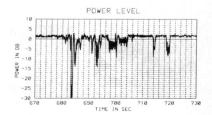

FIGURE 1
Received power level. 0 dB = mean received power.
a) City, antenna S6, v = 10 km/h
b) Highway, antenna S6, v = 60 km/h.

relation $f_c = v/\lambda$ (wavelength λ = 0.2 m) [2].

Detailed statistics of power level, fade duration and connection duration are given in [3].

A hardware channel simulator with an error analyzer has been used to investigate the bit error performance of the recorded channel. Error bursts start and end with errors and are defined to be separated by at least 50 error free bits.

Fig. 3 shows a measured histogram of the error burst statistics of a recording in the Munich city, using antenna S6. White Gaussian noise has been added to produce a mean E_s/N_o = 32 dB. The upper histogram gives the relative occurrence of bursts with a certain length. The burst length is shown on a semilogarithmic scale with intervals 1,2,...,9,10,11-20,21-30,...,101-200,... The lower histogram shows the error density within the bursts.

It can be seen that error bursts with length up to 301-400 bits occur. 1 per cent of all bursts are longer than 100 bits. The error rate within long bursts is around 10 percent. The high error rate in short bursts indicates synchronization losses.

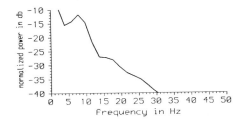

FIGURE 2
Power spectral density of received signal. Munich city, elevation 24°, antenna D5, v = 5 km/h.

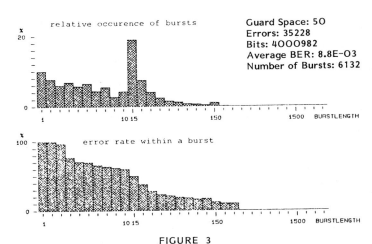

FIGURE 3
Error burst statistics. Munich city, antenna S6, E_s/N_o = 32 dB, DECPSK modulation with 1200 bit/sec.

3. CHANNEL MODEL

Starting from the statistical evaluation of the channel recordings a model of the land mobile channel can be derived. In order to describe the probability density of the received power it is useful to distinguish areas with unobstructed "view" of the satellite (not shadowed areas) from areas in which the direct satellite signal is shadowed by an obstacle.

If no shadowing is present the direct signal is superpositioned by reflected components. The momentary received power S obeys a Rician density:

$$p_{Rice}(S) = c\, e^{-c(S+1)}\, I_o(2c\sqrt{S}). \tag{1}$$

Here c is the direct-to-multipath signal power ratio (Rice factor) and I_o is the modified Bessel function of order zero. The power of the unfaded satellite link is normalized to unity resulting in a mean received total power $E\{S \mid \text{not shadowed}\} = 1 + 1/c$.

If shadowing is present it is assumed that no direct signal path exists and that the multipath fading has Rayleigh characteristic with a lognormally distributed time-varying short-term mean received power S_o.

Such a Rayleigh-lognormal distribution is often used for the terrestrial land mobile channel [4].

In order to get the resulting probability density function of the received power the Rice and Rayleigh-lognormal densities must be properly combined. To this end the time-share of shadowing, A is defined, and the resulting probability density function becomes

$$p(S) = (1-A) \cdot p_{Rice}(S) + A \cdot p_{R-LN}(S). \tag{2}$$

p(S) is independent of vehicle velocity v (v is assumed constant).

The full channel model including parameters for the severeness of shadowing, as well as mean durations of shadowed and not shadowed intervals has been described in [3]. It can be argued, however, that a transmission during shadowed intervals may not be successful with a realistic link budget (cf. Fig. 1). Therefore, the most important parameters of the land mobile satellite channel are Rice factor c, Eq.(1) and time-share of shadowing, A. High values of c indicate few multipath fading ($c \to \infty$ for the Gaussian channel) resulting in good transmission behaviour, whereas low values of c correspond to severe fading (c = o for the Rayleigh channel). 1-A is the time-share during which a transmission is possible and therefore represents a rough estimate of the achievable gross throughput for an ARQ system.

Figs. 4 and 5 show the parameters 1-A and c for different satellite elevations, respectively. It should be noted that the results for 34° and 43° elevation are preliminary.

The lines drawn in Fig. 4 show that the time share of not shadowed intervals, 1-A, tends to increase with elevation (the shadowing decreases). The values of 1-A indicate that on highways substantial data throughput can be achieved for all elevations tested, whereas in cities the throughput is rather small for low elevation.

The lines drawn in Fig. 5 show the dependence of the Rice factor c in not shadowed areas for the different antennas tested. For antenna C3 the Rice factor on highways increases rather continuously with satellite elevation whereas it seems to remain constant in cities. Antennas D5 and S6 show good behaviour for $20°...30°$ elevation corresponding to the shape of their vertical pattern.

On highways the Rice factor is more than 10 dB for all elevations tested, which corresponds to a rather good-natured transmission channel. In cities there is a lot of multipath fading, especially for antenna C3.

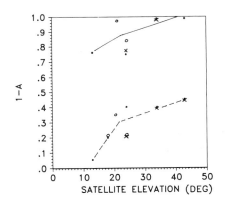

FIGURE 4
Time share of not shadowed intervals versus satellite elevation. Lines represent all antennas.

- antenna C3
o antenna D5
x antenna S6
— highway
-- city

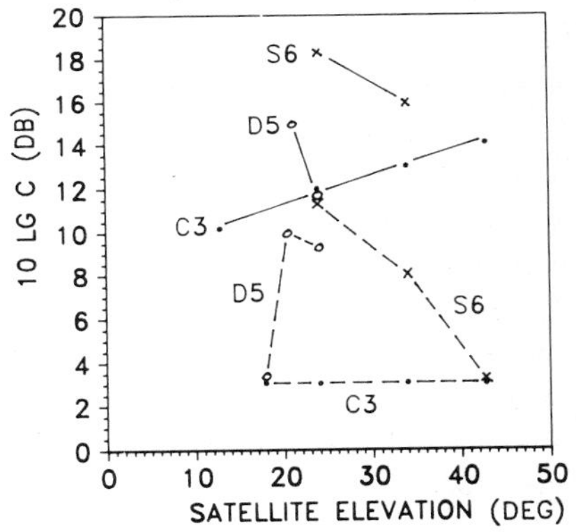

FIGURE 5
Rice factor for not shadowed intervals versus satellite elevation.

- antenna C3
o antenna D5
x antenna S6
— highway
-- city

4. DATA TRANSMISSION WITH AUTOMATIC REPEAT REQUEST

If a return channel is available, reliable data communication with mobile users can be achieved by using Automatic Repeat Request (ARQ). The receiver acknowledges each reliably received data block through the return channel. An incorrectly received block is negatively acknowledged, and the transmission is repeated. In this way ARQ exploits time intervals with good channel characteristics for transmission. Because a data block should be contained in a connection interval, no interleaving should be used. For ARQ with selective-repeat (SR) strategy, only those blocks for which no positive acknowledgement (ACK) comes back to the transmitter, are retransmitted.

Transmission errors are detected at the receiver by using block codes. In a Type-I hybrid ARQ system the redundancy of the code is used partially for error correction and partially for error detection. If only a few errors have occured they are corrected. If too many errors have occured the receiver asks for a retransmission.

The reliability of an ARQ scheme is limited by the probability that transmission errors are not detected by the receiver, and a false message is committed to the user. In order to keep the probability of this event sufficiently low, a specified number of parity bits should be used for error detection in every single codeblock. In addition, a redundancy check sum should be added to the total message [5]. Some parity bits should also be used for error detection in the ACK/NAK blocks.

The throughput efficiency η of an ARQ system is defined as the ratio of the average number of information bits successfully accepted by the receiver to the total number of bits that could be transmitted during the same time period [6].

The mean information rate of the ARQ system is

$$R_i = \eta R \tag{3}$$

with R being the signalling rate of the forward data link. R_i tells how fast the system can transmit information reliably over a given link.

In order to investigate the throughput of ARQ a software simulation program has been written which retrieves stored samples of the recorded time-varying satellite channel. Random bit errors are produced according to the time-varying signal-to-noise ratio. Waiting times caused by free space propagation and data processing are included in the simulation. The fading channel is simulated for the forward link, as well as for the return link. With this simulation method the performance of data transmission schemes can be evaluated in a realistic and reproducible way.

For the simulation ideal carrier recovery (if necessary), ideal bit timing and correct block synchronization are assumed. However, the simulation includes all the memory of the land mobile channel.

In the simulation recordings at $24°$ elevation with the S6 antenna were used. The simulated velocities are 30 km/h in city and 120 km/h on highway.

The data blocks are considered as extended Reed-Solomon (RS) code words with blocklength n symbols and k information symbols. The overhead is assumed as h = 32 bits (e.g. unique word). The code words contain n-k parity symbols. 4 parity symbols (4 m bits) are assumed to be reserved for error detection. The remaining parity symbols can be used to correct $t = \lfloor (n-k-4)/2 \rfloor$ symbol errors. If a data block contains more errors, a retransmission is necessary. In ARQ with pure error detection the data blocks have n-k = 4 parity symbols.

Fig. 6 shows the influence of code rate k/n on the throughput of the data link. For the considered RS-code with n = 64 the code rate k/n = 60/64 = 0.9375 corresponds to ARQ with pure error detection. With a hybrid ARQ scheme (resulting in a lower code rate) the throughput of the data link can be increased, especially for low signal-to-noise ratios. There is an optimum code rate which maximizes the throughput. If the code rate is too low, the throughput decreases because of the remaining overhead. For the examples shown in Fig. 6 the optimum code rate is in the range 0.65 to 0.85. It is low for poor signal-to-noise ratios and depends only slightly on the environment of the mobile terminal.

Fig. 7 considers the mean information rate R_i of the data link for varying signalling rates R. For a fixed code and constant link power C there is an optimum signalling rate which maximizes R_i. Faster signalling suffers from poor signal-to-noise ratio E_s/N_o.

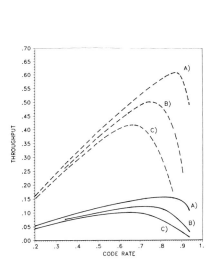

FIGURE 6
Throughput efficiency η_o of the data link as a function of code rate k/n.
RS-codes with n = 64, signalling rate: 2.4 kb/s.

Signal-to-noise ratio E_s/N_o of the unfaded satellite link:
A) 7 dB
B) 5 dB
C) 4 dB

——— city
------ highway

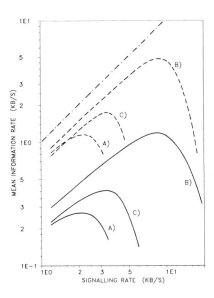

FIGURE 7
Mean information rate R_i of the data link versus signalling rate R.
t error correcting RS codes with n = 64 and k information symbols.

curve	C/N_o	k	t
A	40.8 dBHz	60	0
B	46.8 dBHz	60	0
C	40.8 dBHz	54	3

——— city
------ highway
–·–·–·– $R_i = R$

corresponding to the relation $E_s = C/R$. For pure error detecting ARQ the optimum signalling rate corresponds to $E_s/N_o \simeq 7$ dB. The curves A and B indicate that increasing the link power by x dB increases the optimum signalling rate and the achieveable information rate approximately by the factor $10^{x/10}$. Curve C shows that the use of hybrid ARQ has a similar effect as increasing the link power.

REFERENCES

[1] Hagenauer, J., Papke, W., Data Transmission for Maritime and Land Mobiles Using Stored Channel Simulation, Proc. IEEE 32nd Vehicular Technology Conference 1982, San Diego, pp. 379-383.
[2] Clarke, R.H., A Statistical Theory of Mobile-Radio Reception. Bell Systems Techn. Journ. 47 (1968), pp. 957-1000.
[3] Lutz, E., Papke, W., Plöchinger, E., Land Mobile Satellite Communications - Channel Model, Modulation and Error Control, Proc. 7th Intern. Conf. on Digital Satellite Communic., 1986, Munich, pp. 537-543.
[4] Hansen, F., Meno, F., Mobile Fading - Rayleigh and Lognormal Superimposed, IEEE Trans. on Vehicular Technology, VT-26 (1977), pp. 332-335.
[5] Comroe, R.A., Costello, D.J., ARQ Schemes for Data Transmission in Mobile Radio Systems, IEEE J. on Sel. Areas Comm. SAC-2 (1984), pp. 472-481.
[6] Lin, S. et al, Automatic-Repeat-Request Error Control Schemes, IEEE Commun. Magazin 22, December 1984, pp. 5-16.

A Tool For Evaluating On-board Processing Satellite Architectures

S.V. Vaddiparty, W. Doong, T.V. Nguyen, T.Q. Nguyen

Ford Aerospace and Communications Corp.
3939 Fabian Way
Palo Alto, CA. 94303.

ABSTRACT

A generic software model that evaluates the end-to-end bit error rate (BER) and certain other parameters of an on-board processing satellite was developed. The model can accommodate a variety of RF link parameters, uplink/downlink access schemes and system configurations such as on-board modulation/demodulation, bit regeneration, forward error correction (FEC) encoding/decoding and baseband switching. A unique feature of the model is the capability provided to the user to interactively analyze various configurations and perform system optimization. This software was developed as part of an in-house effort to evaluate various on-board processing satellite architectures.

1. INTRODUCTION

Satellites that can have the capability to no longer just re-transmit an analog signal, but rather, perform digital signal processing functions on board are presently being implemented. These satellites could incorporate regeneration, buffering, switching and FEC encoding and decoding functions. To easily ascertain system level characteristics of such a communication satellite, a software model that can, with a minimal amount of user interaction, emulate the end-to-end communication link under various propagating conditions and on-board system configurations was developed. This menu driven model, written in LOTUS 1-2-3 (a spreadsheet software) for IBM PC's or compatibles, can simulate the complete path of a QPSK modulated signal through a satellite. A host of RF link parameters may be modified and the following uplink/downlink access schemes selected: FDMA/FDMA, TDMA/TDMA, and FDMA/TDM-FM. These were designed to simulate INTELSAT's IBS (International Business Service), TDMA, and IDR (Intermediate Data Rate) services, respectively. It is completely versatile in that it can simulate a purely transparent satellite system (bent-pipe) to a regenerative satellite with on-board encoding and decoding, and all possible combinations thereof. Furthermore, encoding and decoding of the data at the earth stations can also be accounted for in the model. This variability in the system configuration may be best described by the simplified system block diagram of Figure 1 where a desired configuration is set up by closing the appropriate switches. In what follows, we shall provide a brief analytical background for a signal propagating through either a transparent or regenerative satellite system and then expand on the functional description of each of the subsystems. The algorithms used to evaluate the end-to-end BER between the transmit and receive earth stations for each of the various configurations will then be elucidated upon.

2. ON-BOARD PROCESSING SATELLITE SYSTEMS

2.1 Background

The effect of demodulating and remodulating (regenerating) the bit stream on board the satellite is that the uplink and downlink are completely decoupled, and as consequence, each link may be independently optimized. Furthermore, the downlink is free of any uplink noise, unlike in a transparent system. The only uplink errors that could appear at the destination earth station are those that are made in the bit detection process on board the satellite. The overall performance advantage of a regenerative satellite system is proportional to the relative balance between the uplink and downlink performance. For a highly unbalanced regenerative or transparent satellite system, the dominant link will be the one with the lowest E_b/N_0 (worst BER). A completely balanced regenerative system will yield a maximum E_b/N_0 improvement of 3dB over a transparent system. Assuming an on-board error free detection process, the end-to-end bit error rates for a regenerative satellite can be derived as follows.

In a regenerative system, the probability of correctly transmitting a bit from end-to-end, P_{bc}, is product of the individual uplink and downlink probabilities:

$$P_{bc} = (1 - BER_{up})(1 - BER_{down}) \tag{1a}$$

where BER_{up} and BER_{down} are the probability of bit error on the uplink and downlink, respectively. The end-to-end probability of error, $BER_{End-to-End}$, is evaluated as:

$$BER_{End-to-End} = 1 - P_{bc}$$
$$= BER_{up} + BER_{down} - BER_{up} BER_{down} \tag{1b}$$

For small values of uplink and downlink bit error rates, the product term is negligible and the resultant end-to-end BER is seen to be just the sum of the uplink and downlink bit error rates:

$$BER_{End-to-End} = BER_{up} + BER_{down}$$
$$= f(U_{up}) + f(U_{down}) \tag{1c}$$

where $f(U)$ is the probability of error transformation, Lindsey and Simon [2], and U is the energy per noise power density, (E_b/N_0). For a transparent satellite system, the end-to-end BER is

$$BER_{End-to-End} = f(U_{End-to-End}) \tag{2a}$$

where $U_{End-to-End}$, the end-to-end bit energy per noise power density, $(E_b/N_0)_{End-to-End}$ is:

$$\left[\frac{E_b}{N_0}\right]^{-1}_{End-to-End} = \left[\frac{E_b}{N_0}\right]^{-1}_{up} + \left[\frac{E_b}{N_0}\right]^{-1}_{down} \tag{2b}$$

2.2 The Subsystems

Before discussing the algorithms used in generating the end-to-end bit error rates under the various system configurations, a brief description of each of the individual modules as well as their effects in system performance degradation is presented. The overall system modeled by the software package is shown below. As an aid in the following discussions, the bit error rates are indicated by the notation B_i.

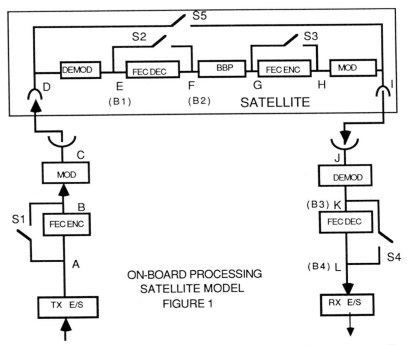

ON-BOARD PROCESSING
SATELLITE MODEL
FIGURE 1

Modems: The transmit earth station establishes a bit stream of information rate IR which is coded by an FEC encoder of code rate r and forms a bit stream of transmission rate IR/r bits/sec. The modulator is modeled as a unit that takes in this encoded bit stream and generates an N-ary modulated signal with a symbol rate SR of $(2/\log_2 N)*(IR/r)$ symbols/sec and modulates it to the transmit frequency. The demodulator performs the exact inverse operation in that it generates an encoded bit stream of rate (IR/r) bits/sec given the symbol rate generated by the modulator. The bit error rates B_1 and B_3 at the outputs of the demodulators in Figure 1 represent the numerical evaluation of the probability of error transformation, f(U), given U equals the E_b/N_0 value at the demodulator inputs, nodes D and J, respectively. The modulator and demodulator are assigned a certain performance (modem) degradation due to the various TWTA/HPA non-linearities in the modulators, non-ideal receive filter and carrier and phase recovery circuitry in the demodulator. This degradation is subtracted from the E_b/N_0 value available at the demodulator input. Note that the modem degradation is a function of the demodulator's input E_b/N_0 value and may be set by the user in the model. The adjacent channel and co-channel interference effects can also be explicitly accounted for in the model.

Encoders and Decoders: As stated above, the encoder is modeled as a device that takes in a digital bit stream of information rate IR and generates a coded bit stream at rate IR/r, r being the FEC code rate. This is then fed to the modulator for transmission over the channel. The decoders have as their inputs the error rates B_1 and B_3, and generate an improved bit error rate B_2 and B_4, respectively, as indicated in Figure 1. This may be envisioned as a decoding transformation that represents the performance advantage of using a particular decoding scheme. Three basic types of encoding/decoding schemes are simulated by the model: BCH, soft decision Viterbi decoding, and hard decision Viterbi decoding. The code rates of the Viterbi decoders may be either 1/2 or 3/4, and the block codes implemented are BCH(128,112,5) and BCH(255,239,5). The transformation representing these three decoding techniques will now be expanded upon.

To derive the probability of error for a linear block code with parameters (n,k,d), where n is the length of the block, k is the number of information bits, and d is the minimum distance, let us begin by assuming that the decoder can correct at most t errors, $t = [(d-1)/2]$. This is an example of a "bounded distance" decoder, a class to which BCH decoders belong to. Then, P_{bt}, the probability that more than t errors occurred in a block is given by

$$P_{bt} = \frac{1}{n} \sum_{i=t+1}^{n} \binom{n}{i} p^i (1-p)^{n-i} \qquad (3)$$

where p is the bit error rate at the decoder input. The probability of a bit error, P_{be}, assuming a systematic code, is then found to be Odenwalder [4]:

$$P_{be} = \frac{1}{n} \sum_{i=t+1}^{n} \beta_i \binom{n}{i} p^i (1-p)^{n-i} \qquad (4)$$

where β_i is the average number of symbol errors that remain to be corrected in the received sequence given that i errors have occurred. Note that a decoder can correct at most t errors and add at most t errors for i>t. Thus, β_i is found to be bounded:

$$i - t \leq \beta_i \leq i + t \qquad (5)$$

When β_i is not determined for a particular code, a good approximation is to set $\beta_i = i$. Equation (4) is then simplified to:

$$P_{be} = \frac{1}{n} \sum_{i=t+1}^{n} i \binom{n}{i} p^i (1-p)^{n-i} \qquad (6)$$

Hence, P_{be} is the decoder transformation that represents the improved BER that is achieved by using a BCH(n,k,d) decoder given p, the input BER.

The decoder transformation for the various Viterbi codes was generated from data representing the output BER as a function of input E_b/N_0 supplied by a vendor of these decoders. Thus, this data entered into the model represents the actual hardware performance as distributed by this outside source.

BBP: The Baseband Processor (BBP) represents the buffering and switching functions of the regenerative satellite. Interconnectivity between the various beams may be achieved with an appropriate amount of internal buffering and a switching configuration. Assuming that these components are space qualified, the BBP will affect only the timing and memory requirements. Therefore, the link degrading effects of the BBP are negligible and are not accounted for in this software package.

2.3 The System

There exist seven possible system configurations that can be simulated by the model by selectively closing the appropriate switches in Figure 1. In what follows, we shall describe the algorithms used in modeling each of these configurations. Specifically, we are interested in evaluating the end-to-end BER provided to the user at the receive earth station--node L in Figure 1-- for a given set of link parameters and system configuration. The term "end-to-end" coding is used below to represent encoding at the source earth station and decoding at the destination earth station only.

Config. A (S1,S4,S5 closed): In this transparent case, the end-to-end E_b/N_0 value is found by performing a power combination of the appropriate uplink and downlink values as stated in (2b). The end-to-end BER is found by mapping this end-to-end E_b/N_0 through the probability of error transformation f(U).

Config. B (S5 closed): This case is similar to configuration A except that end-to-end coding is now in place. The end-to-end BER provided to the user is at the output of the earth station's decoder and is derived by applying the decoding transformation to the BER of the transparent case.

Config. C (S1,S2,S3,S4 closed): For this regenerative case without on-board encoding or decoding, the end-to-end BER is found by adding the uplink and downlink bit error rates as given in (1c).

Config. D (S2,S3 closed): In this regenerative with end-to-end coded configuration the end-to-end BER is generated by feeding the BER of config. C through the decoding transformation and deriving an improved BER at the output of the earth station's decoder.

Config. E (S3,S4 closed): The end-to-end BER in this regenerative with uplink coding configuration is found by summing B_2, the improved BER generated by the on-board decoder, with an RF downlink (uncoded) BER, B_3.

Config. F (S1,S2 closed): This configuration is similar to configuration E except that, with the uplink uncoded and the downlink coded, the end-to-end BER is found by summing B_4, the improved BER generated by the decoder given the RF downlink BER B_3, with an uplink (uncoded) BER B_1.

Config. G (all open): This configuration is the full service case where on-board regeneration and both uplink and downlink coding exist. Note that each of the links may be coded differently with unique code rates assuming compatibility of received and transmitted transponder capacities. Here, the bit error rates B_2 and B_4 are the improvements incurred by driving the bit error rates B_1 and B_3 through the decoding transformation, respectively. The end-to-end BER is formed by summing B_2 with B_4.

We shall now provide a brief description of the software and its general nature which allows for its great versatility.

3. THE SOFTWARE MODEL

3.1 Software Description

The software model designed to emulate the communication link of an on-board processing satellite system features a variety of functions including different uplink and downlink access, forward error correction, and system configuration schemes. Heavy utilization is made of the MACRO feature of LOTUS 1-2-3 in program flow control and a menu driven execution of the model as well as output display control. The set up of the variety of system configurations, coding, decoding and access schemes are just some examples of the use of this MACRO feature in program control. The model in its present form is capable of emulating a QPSK modulation scheme but can be easily expanded to model any N-ary scheme. Execution of the model consists of initiating the software and performing the following steps:

a. Select the uplink and downlink access schemes from the home window. The options here include FDMA-QPSK/FDMA-QPSK, TDMA-QPSK/TDMA-QPSK, and FDMA-QPSK/TDM-FM.
b. Choose the system configuration to be modeled and the coding and decoding schemes to incorporated, if any, from the configuration window.
c. Vary the appropriate transmission rate and RF link parameters by moving the cursor through the user modifiable fields in both the uplink and downlink budget windows.
d. Execute the program and move through the uplink, downlink, or summary windows to view the results.

3.2 The Generality of the Model

The model is of a sufficiently general nature that it can model the signal path through transponders of varying size and spectral densities. This is handled by giving the user the capability to define transponder occupancy and guard band factors. To see this, let us visualize an FDMA/FDMA situation where the transponder bandwidth is equally divided into N regions. The bandwidth of each region, defined as a carrier spacing parameter, consists of the spectrum of the carrier adjusted by some scaling factors:

$$\text{Carrier Spacing} = \left[\frac{IR}{r}\right] * [a \, b] \qquad (7)$$

where IR is the information rate, a is the guard band factor, and b is a spectral occupancy factor. From bit conservation considerations, we have

$$IR_{up} * N_{up} = IR_{down} * N_{down} \qquad (8)$$

that is, #bits transmitted on the uplink equals #bits re-transmitted on the downlink. The model is capable of simulating a signal path through one transponder associated with the uplink and back to a destination earth station via a separate transponder, possibly of different bandwidth. The scale factors a and b are transponder utilization specifications and are generally dictated for a given service. Hence, within the constraints of (7), (8) and RF link parameters, the user can adjust the information rate, coding rate and number of uplink and downlink carriers.

3.3 An Example

To demonstrate the model, a relatively balanced system is set up with the excess margin parameter iterated upon such that the end-to-end BER is the same (1E-6) for each of the configurations. This excess (pseudo) margin is only used for comparing the different configurations and performing system optimization. The contrast between each of the configurations is then observed via the different excess link margin values given that all other link parameters are held constant. An increase in this excess margin can be used to decrease the required transmitter power, antenna size, or a multitude of other system parameters. The decoder modeled here incorporates the output BER verses input E_b/N_o data acquired from a vendor for a Viterbi rate 1/2, soft decision (8-level) device.

We begin by setting the uplink and downlink margins on the full service (config. G) to achieve an end-to-end BER of 1E-6. With only downlink coding (config. F), the uplink excess margin is reduced by 5.3dB to achieve this same BER of 1E-6. Similarly, with only uplink coding (config. E), the downlink excess margin is reduced by 5.25dB. Comparing the full service configuration to a regenerative system with end-to-end coding (config. D), we see that the latter has approximately 1dB less excess margin. This indicates that on-board encoding and decoding does not yield a significant improvement in this particular situation over encoding and decoding at the earth stations. The full service system does however show an 5.4dB improvement in the uplink and an 5.2dB improvement in the downlink margins as compared to the regenerative system (config. C).

The regenerative system exhibits a 3.4dB and 1.55dB improvement in the uplink and downlink margins over the transparent case, respectively. Due to the decoupled nature of a regenerative system, a given link can be designed to exhibit a greater than 3dB improvement but the end-to-end advantage will be less than 3dB.

The transparent system with end-to-end coding (config. B) has a 6.00dB and 4.25dB advantage in the uplink and downlinks, respectively, over an uncoded transparent system (config. A). Similarly, the regenerative system with end-to-end coding (config. D) is seen to have approximately 4.35dB advantage in both links over one without end-to-end coding (config. C). This expected advantage is attributed to the earth station encoding and decoding. The regenerative system with end-to-end coding has, as expected, a slight advantage over a transparent one with end-to-end coding.

Note that the regenerative system (config. C) is worse than a transparent one with end-to-end coding (config. B). Such a result questions the need for on-board regeneration--why not simply encode and decode at the earth stations? In response to this, we note again the inherent advantages of on-board regeneration, namely, the simplification in the system design based on the decoupling of the uplink and downlinks, and the capability to perform switching in space for maximal interconnectivity among various beams. The relative cost of performing encoding and decoding at a multitude of earth stations versus incorporating regeneration on a single satellite must also be traded off.

The prevailing result of these simulations is that better performance can be achieved at the expense of system complexity and of course, cost. The model can be used to identify this improvement and iterated upon to perform system trade-off analysis. Particular parameters of the communication link may be adjusted within the constraints of overall link performance criterion.

4. CONCLUSIONS

The results of a study of on-board processing satellite were presented. A brief background followed by a description of the various modules and overall system was then presented. The software itself was then described followed by a comparative analysis of the various system configurations. The advantage of these various configurations was demonstrated via example in terms of increased excess link margins that may be used to perform system trade-offs.

	Uplink			Downlink			End-To-End		
	Ex. Mar	C/No+Io	BER	Ex. Mar	C/No+Io	BER	C/No+Io	BER(b)	BER(a)
A	8.00	76.26	N/A	3.00	74.14	N/A	72.04	1E-6	N/A
B	14.00	70.20	N/A	7.25	69.89	N/A	67.03	3E-2	1E-6
C	11.4	72.80	5E-7	4.55	72.57	5E-7	N/A	N/A	1E-6
D	15.80	68.40	2E-2	8.80	68.34	2E-2	N/A	4E-2	1E-6
E	16.80	67.40	5E-7	4.50	72.64	5E-7	N/A	N/A	1E-6
F	11.50	72.70	5E-7	9.75	67.39	5E-7	N/A	N/A	1E-6
G	16.80	67.40	5E-7	9.75	67.39	5E-7	N/A	N/A	1E-6

Notes:
1. Bit Error Rates rounded off
2. BER (a) is at node L, Fig. 1
3. BER (b) is at node K, Fig. 1
4. Carrier to noise power density in dB-Hz
5. The Uplink and Downlink BER in config. E, F, and G are at the output of the decoders, where applicable.

FIGURE 2-- COMPARISON OF VARIOUS CONFIGURATIONS

ACKNOWLEDGEMENTS

The authors wish to express their gratitude for the whole team involved in the in-house study. This paper is but a by-product of this effort, the main purpose of which was to further educate ourself in this area of great future importance. Specifically, we acknowledge the contributions of Dr. Chuck Windett, Dr. Vijaya Gallagher, Mr. Steve Ames, Mr. Robert Nicholas, and Mr. Paul Monte who developed the decoder and demodulator models. We thank the management at Ford Aerospace and Communications Corporation for funding the project, particularly Mr. Ed Hirshfield, our department manager.

REFERENCES

[1] Gagliardi, R.M. **Satellite Communications,** Lifetime Learning Publishers, Belmont, California 1984.
[2] Lindsey, W.C., and M.L. Simon, **Telecommunication Systems Engineering,** Prentice Hall, Englewood Cliffs, N.J., 1973.
[3] Nuspl, P.P., R. Peters, and T. Abdel-Nabi, "On-board Processing For Communications Satellite Systems, ICDSC-7, Munich, W. Germany, May, 1986.
[4] Odenwalder, J.P., "Error Control Coding Handbook", Linkabit Corp, San Diego, California, Prepared for the U.S. Airforce under contract number #F44620-76-C-0056, 1976.

DRS TRANSMISSION SYSTEM: ARCHITECTURAL TRADE-OFFS AND CODING

A. Arcidiacono, R. Giubilei

Selenia Spazio S.p.A. - Roma Italy

1. INTRODUCTION

The NASA Tracking and Data Relay Satellite (TDRSS) for Low-Earth-Orbiting (LEO) satellites and the management of scientific platforms have proven to be operationally advantageous. Based on this market trend ESA initiated a preliminary study program for European Data Relay Satellite (DRS).
The basic objective of the DRS system is to support the future European space programs which will provide users with:
- transfer of data from LEO terminals and launchers to ground
- communications between LEO users and their ground control stations
- telemetry and telecommand
- ranging operations for orbit and position determination for spacecraft in orbit and for launchers during ascent.

The subjects referred to this paper were specifically studied to give a wide scenario of possible system architectures in order to optimize high data rate (up to 500 Mb/sec) data communications on K band from LEO terminal, through DRS, to the ground station (Return link) /1/.
A comparative study was carried out on Bit Error Rate (B.E.R.) evaluation for a conventional repeater satellite and a regenerative repeater (with on-board demodulation and high speed soft-decoding).
A through comparitive analysis was also performed between coding architectures in order to minimize required EIRP both on the LEO terminal and the DRS /2/.
The use of convolutional codes with Viterbi high speed soft decoding was analyzed /8/.
The problems of hardware implementation in a VLSI paralleled structure were also considered in order to reach the maximum allowable coding gain compatible with economic and technological constraints.
Finally the use of punctured codes to solve problems of bandwidth occupancy and concatenated Reed Solomon (RS) + Convolutional codes to meet the requirements of higher coding gains were considered.

This work was developed under ESA contract N° 6498

2. CHANNEL ARCHITECTURES AND MODELING

A data Relay satellite (DRS) communication system includes three basic elements:
- the relaying payload in geostationary earth orbit (GEO);
- the terminal of the user satellite flying on a low earth orbit (LEO);
- the ground stations which receive the data and support the control of the system.

In this paper we evaluate the performance of the return link in K-band between the LEO terminal and the ground station via DRS.

The maximum transmission capacity on the return link is up to 500 Mbps and it may be achieved by means of parallel transmission at lower rate on several RF carriers, (2 or 4 channels in parallel) thus alleviating the problems concerning the hardware implementation of co-decoders at high data rate.

For the satellite repeater the classical transparent solution has been considered as baseline and comparisons have been carried out with the regenerative solution which envisages complete demodulation on board and data remodulation toward the ground station. The former solution is robust and flexible, the latter is very complex in terms of hardware implementation and less flexible, but more effective in terms of power saving.

2.1 TRANSPARENT CHANNEL

In table I a set of possible architectures is reported.

Table I: Possible architectures for transparent link

- **INFORMATION RATE:** 524 Mb/sec
- **MODULATION FORMAT:** QPSK

INFORMATION RATE	OF PARALLEL CHANNELS	CODING	CHANNEL RATE PER CHANNEL	TOTAL CHANNEL BANDWIDTH
2 x 262 Mb/sec	2	CONV. K=7 R=1/2	262 Msym/sec	850 MHz
2 x 262 Mb/sec	2	CONV. K=7 R=7/8 PUNCTURED	150 Msym/sec	500 MHz
2 x 262 Mb/sec	2	REED SOLOMON (255, 243)	137 Msym/sec	460 MHz
4 x 131 Mb/sec	4	CONV. K=7 R=7/8	75 Msymb/sec	527 MHz
2 x 262 Mb/sec	2	REED SOLOMON (255, 243) CONCATENATED CONV. R = 1/2	275 Msymb/sec	890 MHz

In the following points we summarize the detailed characteristics (for a single channel) of LEO transmitter, DRS repeater and ground receiver:

LEO Transmitter

Number of channel	:	2
Information Rate	:	524 (2 x 262) Mb/sec
Channel Rate	:	2 x 262 Msymb/sec
Pulse shaping	:	Raised Cosine $\alpha = 0.5$
Coding	:	Convolutional $K = 7$ $R = 1/2$
TWT characteristics	:	Backoff = 0 dB AM/PM = 5 deg/dB
Spacing between carriers	:	$1.7/T_s = 450$ MHz

DRS Repeater

Input Filter	:	Butterworth 5 order BTs = 1.6, B = 420 MHz
TWT characteristics	:	Backoff: 0 dB AM/PM : 5 deg/dB
Output Filter	:	Butterworth 4 order BTs = 1.5, B = 400 MHz

Ground Receiver

Front End Filter	:	Butterworth 4 order BTs = 1.8 B = 470 MHz
Matched Filter	:	Raised Cosine $\alpha = 0.5$
Decoder	:	Viterbi soft decoder (3 bit quantization)

For the other architectures we have mantained a similar structure in terms of pulse shaping (Nyquist-type, roll-off = 0.5), TWT characteristics (Backoff: 0 dB, AM/PM : 5 deg/dB) spacing between carriers (1.7/Ts, Ts: symbol duration) and filter characteristics (type, order and BTs product).

Fig. 1a shows the performance of proposed architectures: the performance is evaluated in terms of Eb/No-down vs. Eb/No-up for a fixed Bit Error Rate (in this case BER = 10-5).

The concatenated scheme gives the best performance in terms of power saving: considering a balanced channel it needs about 7.2 dB in up and down link, instead of 8.7 dB for the architecture with only convolutional coding (R = 1/2).

This improvement is at the expense of hardware complexity and bandwidth increasing (890 MHz vs. 850 MHz).

2.2 REGENERATIVE CHANNEL

The significant system feature in on board regeneration is the separation of uplink and down link. This feature prevents the accumulation of the uplink thermal noise over the downlink, and allows an independent optimization of each link using different access schemes, modulation formats and data rates /3/.

In this paragraph we discuss the performance of four possible architectures for a regenerative link. These architectures are summarized in table II.

For the performance evaluations the following parameters are used:

DRS Receiver
Front End filter   :   Butterworth 4 order BTs = 1.9
Detection filter   :   Butterworth 2 order BTs = 1.0
DRS Transmitter
Pulse shaping   :   Butterworth 4 order BTs = 1.5
TWT   :   Backoff: 0 dB   AM/PM : 5 deg/dB
Output filter   :   Butterworth 4 order BTs = 1.9

Ground Receiver
Front End filter   :   Butterworth 5 order   BTs = 1.9
Detector filter   :   Butterworth 2 order   BTs = 1.0

LEO Transmitter
Pulse shaping   :   Butterworth 4 order BTs = 1.5
TWT   :   Backoff 0 dB   AM/PM : 5 deg/dB
Output filter   :   Butterworth 4 order BTs = 1.9
Channel spacing   :   1.8/Ts

Fig. 1b shows the obtained performance.

TABLE II : Possible architectures for a Regenerative link.

- INFORMATION RATE : 524 Mb/sec
- MODULATION: QPSK

OF CHANNELS		CODING		BANDWIDTH MHz	
UP	DOWN	UP	DOWN	UP	DOWN
2	1	CONVOLUTIONAL R = 1/2	NONE	850	400
2	2	CONVOLUTIONAL R = 1/2	CONVOLUTIONAL R = 1/2	850	850
2	2	CONVOLUTIONAL R = 1/2	CONVOLUTIONAL R = 7/8	850	500
2	1	CONVOLUTIONAL R = 1/2 CONCATENATED REED SOLOMON (255, 243) VITERBI DECODING ON BOARD DRS		890	420
2	2	CONVOLUTIONAL R = 1/2	CODED 8-PSK	850	430

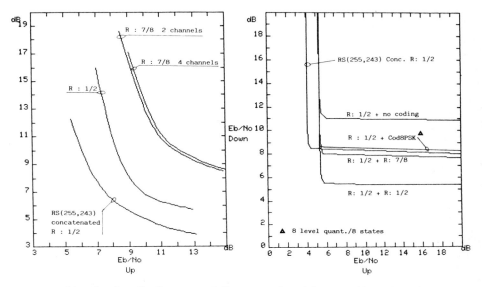

Fig. 1a, b - Performance of Transparent and Regenerative channel

3.   CODING TECHNIQUES

The set of considered coding system was restricted to codec structures with a coding gain of 4 to 5 dB, or more.
The use of concatenated block codes can be considered when the transmitting protocols always have a block format. In the case of using the concatenated block schemes proposed by Lin, a coding gain up to 4 dB is obtainable with an acceptable hardware complexity /11/. In the near future an higher coding gain using concatenated block codes could also be reached using RS + Soft decoded Golay code as proposed in /4/. When the transmitting protocols are not block formatted the use of convolutional codes is more effective than the use of block codes. In this case the three levels of synchronization required from the concatenated block schemes are not required because of the self synchronizing structure of the convolutional codes decoded with a Viterbi decoder. A coding gain of 5 dB can be achieved using a K = 7, R = 1/2 convolutional code. VLSI implementation of Viterbi decoder in a parallel structure in order to reach 500 Mbit/sec operating speed is under advanced development /8/.
If we need a switchable data rate, punctured codes are available with a range of coding gain from 3 to 5 dB corresponding to a coding range from 15/16 to 1/2. In this case synchronization of the bit selector is necessary at the decoder level.
The use of trellis codes /5/ was also considered: a coding gain of 4.5 db is achievable without bandwidth expansion if we use a 4-8 PSK coding scheme with R = 2/3 and 64 states. Modems operating at 120 to 140 Mbit/sec are presently under advanced development by Mitsubishi /6/ and COMSAT /7/.
If we want to obtain a coding gain up to 7 dB the use of a concatenated coding

scheme is mandatory with the use of a convolutional (K = 7, R = 1/2) inner code with Viterbi soft decoding. In this case two levels of synchronization are necessary for the RS encoder and a further level of synchronization is required for the interleaver because of the burst error output characteristic of the inner code.
This problem does not arise if concatenated block codes are used.
The use of a RS code as an outer code is also necessary when we want to establish an intercomputer link with design B.E.R. from 10-9 to 10-10.
A summarizing comparison among block codes, convolutional and concatenated codes performance is given in fig. 2.

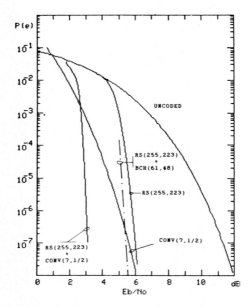

Fig. 2 - Performance of various coding schemes

3.1 ON BOARD ENCODING AND DECODING

The impact of co/decoder hardware has a great importance in the trade-offs between transparent and regenerative solution.
If we intend to use a transparent repeater solution the efforts will be limited to the implementation of on board encoders that do not represent a technological problem because space qualified technology is available at the required operating speed. Considering Regenerative links, several problems arise for the implementation of an high speed on board Viterbi decoder.
The high speed requirement leads directly to a VLSI implementation able to perform at high speed (10 to 30 Mbit/sec) in a single chip configuration and at very high speed in a parallel implementation.
This problem has been recently addressed by Qualcomm /8/ and Mitsubishi /6/ who have developed a CMOS-VLSI chip operating at maximum speed of 17 Mbit/sec

and 30 Mbit/sec respectively, both with approximately 2 watts of power consumption. These VLSI implementations have been possible at this high speed using some further simplifications of the classical Viterbi Decoding algorithm without affecting system performance.
There are three the critical areas in the design of a Viterbi decoder looking at the minimization of the required number of gates:
- ACS (add, compare and select) processing /6/ /8/
- Accumulated metrics handling /6/ /10/
- Surviving path sorting /10/

The first can be solved using a look-up table architecture where all possible input bit configuration are stored.
The second using a continuous rescaling method /8/, a periodic subtraction of a fixed value /6/ or an autoadaptive rescaling /4/.
The third problem can be solved using a three structure with a couple comparison of accumulated metrics /10/ obtaining a regular and easily implementable VLSI structure.
A power consumption of approximately 40 W will be required for a complete decoder operating at 140 Mbit/sec. This value takes into account the power for the multiplexing circuitry (using GaAs technology) self test and has built in redundancy.
Finally we have to consider problems due to radiation hardening for on board implementation. The CMOS technology intended to be used provides good protection against low-energy radiation (up to 50-100 Krad total).
If additional protection is required special packaging concepts (RAD-PAK) could be employed to increase the radiation hardness of the circuit.
Another possibility in the use of an board soft decoding could be represented by the use of the Golay (24, 12) shown in /4/. VLSI hardware complexity of this decoder at the operating speed is comparable with hardware complexity of the Viterbi soft-decoder. Nevertheless the use of convolutional codes instead of Golay codes seems to be more efficient because of the higher coding gain obtained and self synchronizing characteristic of the Viterbi decoder.
The use of the Golay code could be of interest in the case of a transparent channel with concatenated encoding because in this case at the output of the Golay decoder we have no burst errors as at the output of the Viterbi decoder. The introduction of a further level of synchronization due to the Golay soft-decoder is balanced by eliminating the interleaving section.

4. CONCLUSION

In Fig. 3 a summary of the final results has been given.
To carry out some realistic evaluations not only the degradations due to ISI (Intersymbol Interference) and the distortion effects of the bandpass nonlinearities (TWTA) have been taken into account, but also the transponder and on ground receiver imperfections such as:
- Phase ripple and Amplitude ripple in filtering sections
- Incidental AM produced by voltage fluctuations in the TWTA power supply
- Oscillator Phase Noise

Such evaluations have been carried out by a simplified analysis using the method proposed by Weinberg /9/ .

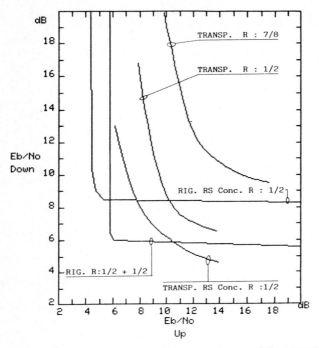

Fig. 3 - Realistic performance of Regenerative and Transparent channel

Looking at fig. 3 , the best performance in terms of power saving is obviously of the Regenerative channel with Convolutional Coding (R = 1/2, K = 7) in up and down link (Viterbi Soft decoding).
When the Eb/No increases the Transparent channel with Concatenated Coding has better performance because the Regenerative channel is like a single hop channel with Convolutional Coding (R = 1/2), while the Transparent one saves the Concatenated (RS + Conv. R = 1/2) structure.
Anyway, assuming that the up-link is more critical than the down-link, the advantage in terms of power of the Regenerative structure is evident.
The disadvantages of Regenerative structure are:

o   less flexibility because it is not possible to change (in a simple way) modulation format and/or channel coding. On the contrary the Transparent channel has only a bandwidth constraint and it is quite insensitive to the modulation and/or coding scheme used.

o   More complexity in terms of hardware

o   Less robustness in terms of reliability due to the increasing of the hardware devices.

o   More weight/volume and DC consumption

Taking into account this remarks the Regenerative Solution seems a favorable solution only when is mandatory a big power saving in the up-link.
Finally some guidelines can be given about the coding techniques. The use of convolutional code with Viterbi soft decision concatenated with an outer RS code seems to be the best solution for a transparent (or on board Viterbi decoded) repeater. With this solution we match at the same time requirements of: good performance at high input B.E.R., for the inner channel, high level of coding gain (up to 7 dB), good performance at very low B.E.R. (10-10), good bandwidth efficiency on down link, in the case of on board Viterbi decoding, and immunity to eventual burst errors. The same results could be obtained with the use of the concatenated block codes, with Golay soft-decision at the inner level, at the expense of about 1 dB in overall coding gain.
An alternative new solution to this approach may be represented by the use of convolutional code plus Trellis code in a regenerative system architecture.

REFERENCES

/1/ R. Giubilei "DRS Transmission System - K-band Channel: final results" Technical Report Selenia Spazio S.p.A., SES/ENG/TN-638/87.

/2/ A. Arcidiacono "DRS Transmission System: Coding Techniques" Technical Report Selenia Spazio S.p.A., SES/ENG/TN-622/86.

/3/ T. Izumisawa, S. Kato and T. Kohri", Regenerative SCPC satellite communication systems," presented at AIAA '84, 1984.

/4/ Y. Be'ery, J. Synders "A new soft-decision decoding method for the Golay (24, 12) code" 15th IEEE Conference on Electronic Eng. in Israel April 1987.

/5/ G. Ungerboeck et alii "Coded 8-PSK modem for the INTELSAT SCPC system" Proc. ICDSC-7 Munchen May 1986.

/6/ T. Fujino et alii "A 120 Mb/sec 8-PSK modem with soft-decision Viterbi decoder" Proc. ICDSC-7 Munchen May 1986.

/7/ R. Fang: "A coded 8-PSK system for 140 Mbit/sec informaion rate transmission over 80 MHz nonlinear transponder" Proc. ICDSC-7 Munchen May 1986.

/8/ Qualcomm Inc. "High speed error control techniques: final report WP 2400" under Selenia Spazio contract.

/9/ A Weinberg "The effects of Transponder Imperfections on the Error Probability Performance of a Satellite Communication System" IEEE Trans. on Commun. VOL. COM-28 pagg. 858-872- June 1980.

/10/ A. Arcidiacóno "Technical key points to reduce VLSI Viterbi implementation" Internal Technical Note Selenia Spazio.

/11/ S. Lin "Coding considerations on DRS" NASA Grant NAG 5-407 October 1985.

ROUND TABLE

COMPARISON OF CONVENTIONAL SYSTEMS
(presently available or under development)
AND FUTURE INTEGRATED SYSTEMS

Chairman: B.G. Evans *(University of Surrey, UK)*

ROUND TABLE DISCUSSION

PANELISTS

Professor B. G. Evans (University of Surrey, U.K.) Chairman, P. Bartholomè (ESA, The Netherlands), Dr. S. J. Campanella (COMSAT Labs, U.S.A.), D. P. Dharmadasa (EUTELSAT, France), S. De Padova (S.I.P., Italy), Dr. S. Kato (NTT, Japan), Dr. F.M. Naderi (Nasa, U.S.A.), Dr. P. Nuspl (INTELSAT, U.S.A.), S. Tirrò (Telespazio, Italy).The round table addressed the following questions which had arisen from the Workshop sessions:

* What are the future markets and services and how will they shape future networks?
* VSAT's or not V'SAT's?
* OBP - when and why? Effects on the network.
* ISL's - how will they change the network?
* Role of mobiles:- separate networks or integrated?
* What role do satellites have in a future integrated network?

The following is a summary of the major discussion items and conclusions on the above. It has been impossible to represent the views of all individuals who contributed to the discussion and we attempt here to give a flavour of the meeting with major conclusions.

1. FUTURE MARKETS AND SERVICES AND HOW THEY MAY EFFECT NETWORKS

It was generally recognised that to date satellite communications had been dominated by P and T's and high/medium capacity international traffic. Services had mainly been multiplexed multi-channel telephony and T.V. with data communications increasing in recent times. Satellites and the way that they interfaced with networks had reflected this and been shaped by the demands. Success in using existing systems occurred where the service matched the current network architecture e.g. leased T.V.. However, there appeared to be demands from new areas; DBS, small fixed businesss users and mobiles (land and aeronautical). It was felt that satellites would always win out where there was no economic alternative, and the mobile areas were obvious examples of this, as was the quick start up facilities that could be offered to ISDN. It was clear that the cost of services was all important and that satellites need to undercut terrestrial if they were to succeed. Evidence of this was the early success of VSAT's in the U.S.A.. It was agreed that new services such as land

and aeronautical mobile, point-to-multipoint data base dissemination, multi-point-to-point data collection, global positioning etc. had demonstrated market demand. Whether such services could be met cost-effectively by existing satellites was doubted, and this was seen as a major role for OBP satellites in the future. Institutional constraints and policy were thought to be a key feature in future systems. This was particularly true in Europe where the choise scemed to be between going to a super-PTT situation with centralised control or to liberalise completely and adopt the U.S. free market pattern. It was obvious that the slow take up of VSAT's in Europe was due to the difficulty in achieving cross-national communications agreements with a large number of PTT's. However, it was stated that V-SAT systems in Europe would be up and running within the next year.

2. VSAT's or not VSAT's

It was considered that VSAT's were perhaps the trail blazers as far as the new era of satellite communications were concerned. They were opening up the new markets and creating further demands. However, the situation in the U.S.A. was different to that in Europe. In the U.S.A. the main use had been as 'by-pass' links, whereas in Europe true point-to-multipoint etc. was foreseen as the major market. It was true that European PTT's saw them as a threat and that some agreement through EUTELSAT seemed necessary if they were to go ahead. The point was made that many early systems had been rapidly constructed and that the protocols and software had given problems, particularly when used in the broadcast mode. These problems need to be overcome prior to the European launch. The tariffing problem was considered difficult as, at the moment, the European PTT's had control of all up-links and so could price services out of the market. Despite all of the problems, the consensus was that VSAT's would form a major new satellite business service and that their use in future rural distribution systems was also potentially attractive.

3. OBP - WHEN AND WHY? EFFECTS ON THE NETWORK

This topic produced the dilemma of whether to stick with good old flexible transparent transponders, of which there were plenty available, or to move to the potentially attractive, but higher risk, on-board processing variety. It was considered that it was a risk that we just could not afford not to take.

The benefits were enormous and it would be the only way that satellites would effectively be part of the future ISDN. However, it was perhaps an act of faith to push into the area now (the fact that the Japanese had not taken the leap was noted) but the consequences for satellite communications were dire if it was not taken. Looking back (perhaps at a future Tirrenia conference in 2010) on the development of OBP it was considered that a scenario might be as follows:

SS-TDMA: MSM	88
SS-FDMA	
REGENERATION	) mid 90's
BSM - BASEBAND SW. MATRIX	)) mid 90's
MCD - MULTI- CARRIER DEMOD.	))
DECODING	) late 90's
BBP (TST)	)
ISL - OPTICAL	late 90's

Many of the above innovations were already with us, SS-TDMA with OLYMPUS, ITALSAT, INTELSAT VI, ACTS and Japanese satellites. It was difficult to forecast ahead and many of the experimental satellites had got themselves locked into fixed bit rates, which were probably too high. Future operational systems would have to go down to VSAT bit rates. This meant flexible MCD's and SCPC/FDMA (TDMA) up and TDM down systems. The other techniques were in the R and D stage and were all the subject of papers in this Workshop. Thus the technology was in fact seen to be coming into place. The major exception perhaps was the fault-tolerant software problems which would possibly prove to be the most difficult to solve if OBP goes ahead.

It was considered that effects on the network would be evolutionary and occur in a phased format. They had the advantage of being compatible with the ISDN, but signalling and delay problems needed to be addressed and enhancements could be readily added. The disadvantages were in the complexity and the resistance to change by the network operators. A satellite acting on a multi-rate switching node posed many new problems to the Network operators and these must be studied prior to its acceptance within the overall network infrastructure.

Some doubters still posed the question of whether OBP was necessary. The answer to this lay in the fact that most, if not all, of the potentially new and attractive services required it if they were to be economic. So it wasn't if it was when!

4. ISL's CHANGE IN THE NETWORK

The picture was painted of a truly global network with full interconnection via ISL's and the dropping off of circuits to earth when needed. This was perhaps a 2000's scenario and was related to non-speech traffic due to the delays involved. It was considered that ISL's had an immediate and important part to play in data relay satellites (LEO -- GEO) and in future communications to space stations and platforms. GEOS - GEOS links would undoubtedly improve the connectivity in International systems and open up the possibility of interesting interconnections of business traffic. However, the latter is intimately tied up with the existence of OBP.

5. ROLE OF MOBILES: "SEPARATE OR INTEGRATED" NETWORKS?

There appeared two interpretations of this question. Integration between mobiles (land, maritime and aeronautical) or integration of mobile and terrestrial networks.

It was pointed out that although maritime and aeronautical services had current frequency allocations in L-band, the LMS did not. The current WARC-MOB (Sept 87) was addressing this problem and a division between those who wanted a complete free-for-all and those who wanted a planned allocation, would need to be resolved. The U.S.A. and Canada appeared particularly adamant that if things went against a free-for-all they would still go ahead.

It was considered that there was some advantage in intergrating maritime and transoceanic aeronautical services. However, the land mobile service required completely different coverage and satellite types. Such satellites may best cover inland waterways and domestic aeronautical services. However, commonality of equipment was highly desirable and if true integration was to be the aim, then this must be sought now.

Looking ahead to third generation mobiles, the synergy between satellites and terrestrial cellular to provide an integrated personal communication system made a great deal of sense. Satellites would fill in the out of urban gaps as well as acting as personal communicator locators and providers of paging services. Even in the shorter term, it was considered that business communication systems must extend the office to the mobile and thus mobile and business systems must come together.

Institutional and political problems, particularly in the area of mobile 'by-pass' systems, were difficult to resolve. However, it was considered that the PTT's and satellite operators must come much closer together in order to work out the future integrated network.

6. ROLE OF SATELLITES IN FUTURE INTEGRATED NETWORKS?

It was stressed that satellites must be taken seriously by future network planners. In the planning of future networks, satellites were merely one technique. However, the planners should make use of their unique advantages;

(i)   Quick digital start-up
(ii)   Flexible routing to allow study of traffic patterns

(iii)   Demande assignment and multiple access flexibility for traffic
(iv)   OBP and ISL's including multi-rate switching to simplify terrestrial infrastructure.

It was true that satellites provided ISDN services now! The integration of mobiles into future integrated networks would necessitate the use of satellites. The use of OBP and on-board switching opened up the possibility of considerable simplification of terrestrial infrastructure. It was true that there were problems associated with the delay on satellite networks, especially for speech services. However, there was no technical reason that satellites should not provide as good (if not better) quality than other transmission means. It was considered essential that when planning for future packet-switched networks, the protocol definitions were made so as not to exclude satellites. The latter's advantages to the overall integrated network would far outweigh small modifications to terrestrially oriented protocols. Satellites were also considered to have a major role in future IBCN's (Integrated broad band communication networks) such as being researched in RACE. There appeared to be a dangerous move to exclude them at present.
In conclusion the vision of a three dimensional network including satellites in a truly integrated services network involving mobile systems should be the real aim.

Professor B. G. Evans

BGE/LSJC/16.10.87/3D